Quantum Chromodynamics
(La Jolla Institute, 1978)

AIP Conference Proceedings
Series Editor: Hugh C. Wolfe
Number 55
Particles and Fields Subseries, No. 18

Quantum Chromodynamics

(La Jolla Institute, 1978)

Editors
William Frazer and Frank Henyey
University of California, San Diego

American Institute of Physics
New York **1979**

L.C. Catalog Card No. 79-54969
ISBN 0-88318-154-1
DOE CONF- 7808111

TABLE OF CONTENTS

QUANTUM CHROMODYNAMICS AND DYNAMICS OF HADRONS 1
Stanley J. Brodsky, SLAC and Stanford University

APPLICATION OF QUANTUM CHROMODYNAMICS 97
R.D. Field, California Institute of Technology

THE PHASES OF A GAUGE THEORY 243
L. Susskind, Stanford University

SEMICLASSICAL METHODS IN QUANTUM CHROMODYNAMICS:
TOWARD A THEORY OF HADRON STRUCTURE 269
Curtis G. Callan, Jr., Princeton University
Roger F. Dashen, The Institute for Advanced Study
David J. Gross, Princeton University

PREFACE

The notes in this volume are based on lectures given at the La Jolla Institute's Summer Workshop on Quantum Chromodynamics August 1978. The lecture topics cover three very active areas of research: QCD perturbative phenomenology, quark binding, and lattice field theories. In the interest of relatively intensive coverage and detailed presentations, breadth of coverage had to be sacrificed. We hope that the reader will find the omission of his favorite approach to jet calculations or confinement theory more than compensated by the more detailed treatment of the topics covered.

The Workshop received support from the La Jolla Institute and partial support from the National Science Foundation and the Department of Energy. Without the generous support of the La Jolla Institute and the organizing efforts of its director, Dr. Adolf R. Hochstim, this Workshop could not have taken place.

William Frazer
Frank Henyey
Physics Department
University of California,
San Diego
La Jolla, California

QUANTUM CHROMODYNAMICS AND TGE DYNAMICS OF HADRONS

Stanley J. Brodsky
SLAC and Stanford University

I. INTRODUCTION 2

II. EXCLUSIVE PROCESS AND THE EXCLUSIVE-INCLUSIVE CONNECTION IN QUANTUM CHROMODYNAMICS (IN COLLABORATION WITH G. PETER LEPAGE) 12

III. TWO-PHOTON COLLISION AND SHORT-DISTANCE TESTS OF QUANTUM CHROMODYNAMICS 48

IV. HADRON AND PHOTON PRODUCTION AT LARGE TRANSVERSE MOMENTUM AND THE DYNAMICS OF QCD JETS 65

ISSN: 0094-243X/79/550001-95$1.50

QUANTUM CHROMODYNAMICS AND THE DYNAMICS OF HADRONS*

Stanley J. Brodsky
Stanford Linear Accelerator Center
Stanford University, Stanford, California 94305

Abstract

The application of perturbative quantum chromodynamics to the dynamics of hadrons at short distance is reviewed, with particular emphasis on the role of the hadronic bound state. A number of new applications are discussed, including

(a) the modification to QCD scaling violations in structure functions due to hadronic binding;
(b) a discussion of coherence and binding corrections to the gluon and sea-quark distributions;
(c) QCD radiative corrections to dimensional counting rules for exclusive processes and hadronic form factors at large momentum transfer;
(d) generalized counting rules for inclusive processes;
(e) the special role of photon-induced reactions in QCD, especially applications to jet production in photon-photon collisions, and photon production at large transverse momentum.

We also present a short review of the central problems in large p_T hadronic reactions and the distinguishing characteristics of gluon and quark jets.

I. INTRODUCTION

In quantum chromodynamics the fundamental degrees of freedom of hadrons and their interactions are the quanta of quark and gluon fields which obey an exact internal SU(3) symmetry. It is possible (but by no means certain!) that quantum chromodynamics is <u>the</u> theory of the strong interactions in the same sense that quantum electrodynmaics accounts for the electromagnetic interactions. In many ways the present period in theoretical physics parallels the 1930's. Although the structure of quantum electrodynamics was known at that time, the lack of a consistent computational scheme allowed only the simplest (Born approximation) aspects of the theory to be understood. Eventually, with the advent of the covariant renormalization program, the full quantum theory could be developed and tested. For example, the QED prediction for the electron's gyromagnetic ratio including sixth order corrections has been confirmed by experiment to 10 significant figures! The fact that we can understand a

---

* Work supported by the Department of Energy under contract number EY-76-C-03-0515.

fundamental parameter of nature to such precision of course encourages our optimism that there is an analogous local gauge field theoretic basis for hadrons. It is well known that the general structure of QCD meshes remarkably well with the facts of the hadronic world especially quark-based (especially charm) spectroscopy, current algebra, the approximate parton-model structure of large momentum transfer reactions, logarithmic scale violations, the scaling and magnitude of $\sigma(e^+e^- \to \text{hadrons})$, jet-production as well as the narrowness of the Ψ as Υ. However, because of the difficulties of computation, it is difficult to obtain rigorous, quantitative predictions beyond leading order, asymptotic limits.

It is clearly crucial to find critical, unassailable tests of QCD. If there is even one bonafide failure in any area of hadronic phenomena, the theory is wrong.

In these lectures, I will concentrate on the application of QCD to hadron dynamics at short distances, where asymptotic freedom allows a systematic perturbative approach. A main theme of this work will be to incorporate systematically the effects of the hadronic wavefunction in deep inelastic reactions. Although it is conventional to treat the hadron as a classical source of on-shell quarks, there are important dynamical effects due to color coherence and constituent off-shell behavior which modify the usual predictions, and lead to a broader testing ground for QCD. We will also discuss QCD predictions for exculsive processes and form factors at large momentum transfer in which the short distance behavior and the finite compositeness of the hadronic wavefunction play crucial roles.

There are a number of excellent introductory and review articles on quantum chromodynamics that I used in preparing these lectures, especially

"Inelastic Processes in QCD"
Y. L. Dokshitser, D. D'yakanov, S. Troyan[1]

"Jets and QCD"
C. H. Llewellyn Smith[2]

"Applications of QCD"
J. Ellis[3]

"Parton-Model Ideas and QCD"
C. T. Sachrajda,[4]

and the Physics Reports by H. Politzer[5] and W. Marciano and H. Pagels.[6] Some of the new topics discussed here are based on work done in collaboration with others, particularly G. P. Lepage, R. Blankenbecler, C. Carlson, Y. Frishman, T. DeGrand, J. Gunion, H. Lipkin, and C. Sachrajda, and I am grateful for their help.

We begin by reviewing the fundamental principles and assumptions of quantum chromodynamics.[7]

A. Quarks are the fundamental representations of

$$SU(n_f) \otimes SU(3)_C \quad ,$$

i.e.: there are n_f flavors ⊗ three colors of quarks.[8] Although the flavor symmetry is broken by the weak and electromagnetic interactions, color symmetry is exact; there is no way to distinguish color - all directions in color space are equivalent.

B. Hadrons are color-less states

$$|M\rangle \sim \sum_{i=1}^{3} |\bar{q}_i q_i\rangle \quad , \quad |B\rangle = \sum \epsilon_{ijk} |q_i q_j q_k\rangle \quad (1.1)$$

C. $SU(3)_C$ is an exact local symmetry: rotations in color space can be made independently at any point. The mathematical realization of this is the (Yang-Mills) gauge field theory.

D. The Lagrangian density of QCD is

$$\mathscr{L}_{QCD}(x) = \bar{q}_i(x)\,\gamma^\mu\left(i\frac{\partial}{\partial x_\mu}\delta_{ij} + \frac{g}{2}A^a_\mu(x)\,\lambda^a_{ij}\right)q_j(x)$$

$$-\frac{1}{4}\left(\frac{\partial}{\partial x_\mu}A^a_\nu(x) - \frac{\partial}{\partial x_\nu}A^a_\mu(x) + g\,f_{abc}A^b_\mu A^c_\nu\right)^2$$

$$i,j = 1,2,3 \quad ; \quad a = 1,2,\ldots,8 \quad (1.2)$$

(A quark mass term and sum over flavors is understood.) Here the λ^a are the eight Gell-Mann SU(3) matrices with $Tr[\lambda^a,\lambda^b] = 2\delta^{ab}$ (conventional normalization). We can contrast this with

$$\mathscr{L}_{QED} = \bar{q}(x)\,\gamma^\mu\left(i\frac{\partial}{\partial x_\mu} + eA_\mu(x)\right)q(x)$$

$$-\frac{1}{4}\left(\frac{\partial}{\partial x_\mu}A_\nu(x) - \frac{\partial}{\partial x_\nu}A_\mu(x)\right)^2 \quad (1.3)$$

We can also use the more compact notation

$$\mathscr{L}_{QCD} = \bar{q}(x)\,\not{D}\,q(x) - \frac{1}{4}\,Tr\,F^2_{\mu\nu} \quad (1.4)$$

where

$$D_\mu = i\partial_\mu - gA_\mu$$

$$F_{\mu\nu} = \partial_\mu A_\nu - \partial_\nu A_\mu - ig[A_\mu , A_\nu] \quad (1.5)$$

where $A_\mu \equiv \sum_a \frac{\lambda_a}{2} A_\mu^a$, D_μ and $F_{\mu\nu}$ are 3×3 SU(3) color matrices.
Local gauge invariance and color symmetry follows from the invariance of $\mathscr{L}_{QCD}$ under the general gauge transformation

$$A_\mu(x) \rightarrow U(x)\, A_\mu(x)\, U^{-1} + \frac{i}{g}\, U(x)\, \partial_\mu\, U^{-1}(x)$$

$$q(x) \rightarrow U(x)\, q(x) \qquad (1.6)$$

where U is any unitary matrix $U = \exp i \sum_a \theta_a(x) \frac{\lambda_a}{2}$. Note that F is in general not invariant: $F_{\mu\nu}(x) \rightarrow U(x)\, F_{\mu\nu}(x)\, U^{-1}(x)$ since the field strength, like the gluon field is in the adjoint representation of $SU(3)_{color}$.

The Feynman rules of QCD are similar to QED with the $q\bar{q}g$ coupling

$$\gamma_\mu\, \lambda^a_{ij}\, \frac{g}{2} \; .$$

The tri-gluon and quartic gluon coupling color factors are (see Ref. 9)

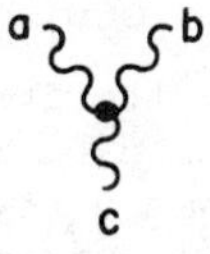

$g\, f_{abc}$

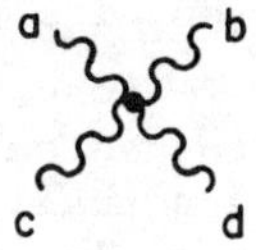

$g^2\, f_{aeb}\, f_{ced} + \cdots$

where $[\lambda^a, \lambda^b] = 2if^{abc}\, \lambda^c$.

The dimensionless coupling constant is $\alpha_s = g^2/4\pi$. A convenient graphical method for evaluating the color algebra has been given by Cvitanovic.[9] The main rules are

(1) a closed quark loop  gives $\mathrm{Tr}\,[I] = n_c = 3$.

(2) A gluon propagator  is equivalent to  minus $1/n_c$ times the identity (to remove the U_3 singlet). Thus

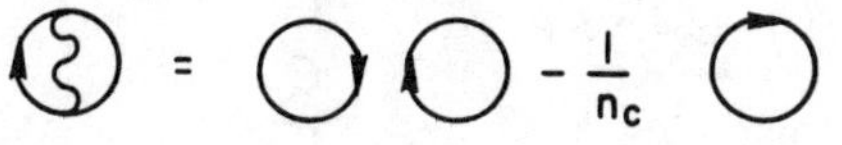

$$= n_c^2 - 1 = 8$$

(times the coupling constant $\left(\frac{g}{2}\right)^2 \mathrm{Tr}\,[\lambda^8 \lambda^8] = g^2/2$.)

Additional rules allow the graphical reduction of the tri-gluon vertex. In typical perturbative calculations (e.g., soft radiation) we have the simple replacement

$$\alpha_{QED} \rightarrow \begin{cases} C_F\,\alpha_s = \frac{4}{3}\,\alpha_s & \text{quark current} \\ C_A\,\alpha_s = 3\alpha_s & \text{gluon current} \end{cases} \qquad (1.7)$$

Effectively "e_g^2" = 9/4 "e_q^2".

Despite the parallels with QED perturbation theory, the postulated absence of asymptotic colored states implies that a perturbative expansion in terms of free - or even dressed - quark and gluon states does not exist in QCD. However, we shall assume that amplitudes with off-shell quark and gluon external legs - corresponding to processes which occur within the hadronic boundaries - do have a perturbative expansion. For such amplitudes, our experience with QED is directly applicable. It is interesting to note that practically all of the predictions recently made for QCD at short distances have a direct analogue in QED, with positronium atoms replacing mesons, etc. In fact, many QCD results for radiative corrections (e.g., structure function moments) and large p_T exclusive or inclusive processes involving bound states actually provide new elegant treatments of QED problems. Conversely, almost every phenomenon known in QED and atomic physics has its parallel in QCD.

In my own research in QCD, I always use the criterion of whether a given prediction or approach to hadron dynamics can be carried over to QED. In some cases, one actually finds that a model-dependent assumption used in QCD leads to incorrect results in electrodynamics. Particularly problematic is the often-used device of replacing an incoming hadron by a probabilistic classical distribution of on-shell constituents. This leads to incorrect QED predictions, and, as we shall see, misses interesting hadronic physics.

We normalize the 3-point vertices at a common off-shell (space-like) mass: $p_i^2 = -\mu^2$, $i = 1,2,3$

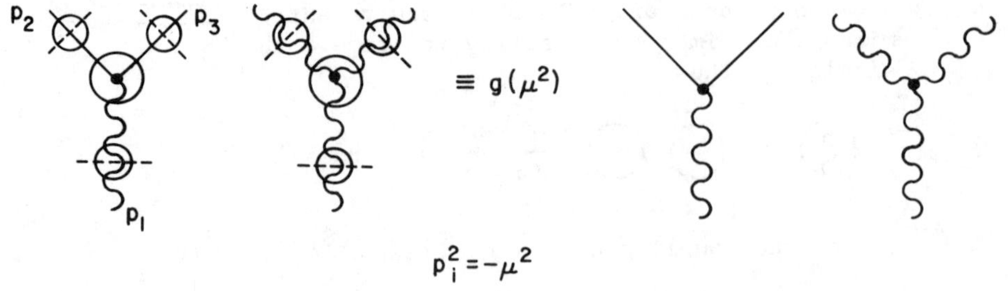

The circles in the figure indicate vertex and self-energy insertions to all orders. The dividing dotted lines indicate that a square root of propagator renormalization constant is to be associated with "wavefunction" renormalization. Notice that we use the same value of $-\mu^2$ at all legs to keep gauge-invariance for the total Compton amplitude.

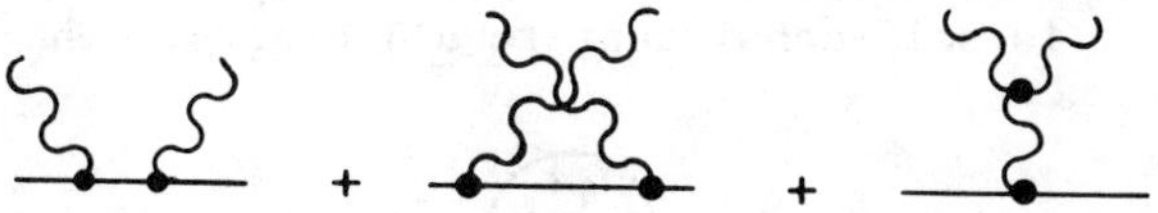

The renormalized amplitude with all vertex and self-energy insertions at $p_i^2 = -\mu^2$ reduces to the Born amplitude with $g^2 = g^2(\mu)$. The choice of μ^2 is arbitrary. Once $\alpha_s(\mu^2)$ is given at any point μ^2, the theory determines α_s at all other values through a renormalization group equation. In terms of diagrams (using axial gauge)

$$\alpha_s(q^2) = \frac{\alpha_s(\mu^2)}{[1 - \pi(q^2,\mu^2,\alpha_s(\mu^2))]} \tag{1.8}$$

where π is the irreducible gluon self-energy insertion. The lowest order diagrams give ($|q^2|$, $|\mu^2| \gg m_q^2$)

$$\pi^{(2)}(q^2) = \frac{\alpha_s(\mu^2)}{4\pi} \log\left(\frac{-q^2}{\mu^2}\right)\left[\frac{2}{3}n_f + 5 - 16\right] \tag{1.9}$$

where (in the Coulomb gauge) the three terms correspond to the indicated intermediate states.

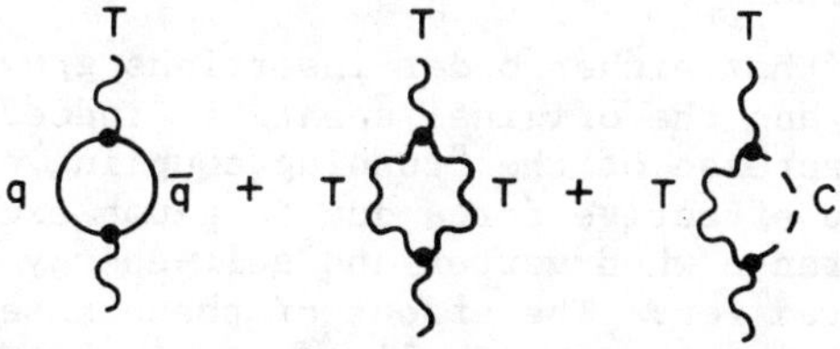

Although the $q\bar{q}$ term must be positive (it is related by unitarity to $e^+e^- \to q\bar{q}$) the crucial Coulomb plus transverse gluon term does not correspond to the production of physical quanta and can indeed be negative.[14]

Thus to lowest order, $\alpha_s(q^2)$ decreases logarithmically (if $n_f \le 16$)

$$\alpha_s(q^2) = \frac{\alpha_s(\mu^2)}{\left[1 + \frac{\alpha_s(\mu^2)}{4\pi} \log \frac{-q^2}{\mu^2} (11-2/3\; n_f)\right]} \tag{1.10}$$

We shall assume that this is the correct asymptotic limit and verify that the result is self-consistent to all orders. The next order diagrams

gives, as in QED, $\pi^{(4)} \sim O[\alpha_s^2(\mu^2) \log q^2/\mu^2]$. However, we can include the effects of the self-energy insertions associated with the exchanged gluon by utilizing $\alpha_s(k^2)$: we have the effective replacement:

$$\alpha_s^2(\mu^2) \log \frac{q^2}{\mu^2} \Longrightarrow \alpha_s(\mu^2) \int_{\mu^2}^{q^2} \frac{dk^2}{k^2} \alpha_s(k^2)$$

$$\sim \alpha_s(\mu^2) \log \log q^2 \tag{1.11}$$

assuming $\alpha_s(k^2) \sim 1/\log q^2$ asymptotically. Thus we have

$$\frac{4\pi}{\alpha_s(q^2)} = \frac{4\pi}{\alpha_s(\mu^2)} + (11-2/3\; n_f) \log \frac{-q^2}{\mu^2} + O(\log \log q^2) \tag{1.12}$$

It is easy to see that higher order insertions grow even less strongly with q^2, and the original ansatz is indeed self-consistent. The logarithmic decrease of the "running coupling constant" $\alpha_s(q^2)$ indicates that the effective force due to gluon exchange becomes weak at short distance when vertex and self-energy insertions of all orders are accounted for. The effect of these insertions is also to weaken the ultraviolet growth of all loop calculations compared to lowest order perturbation theory.

Unlike QED where α can be fixed directly by Coulomb scattering, the empirical determination of α_s at any renormalization point is non-trivial. It is conventional to use the form

$$\tilde{\alpha}_s(q^2) = \frac{4\pi}{(11-2/3\; n_f) \log \frac{-q^2}{\Lambda^2}} \tag{1.13}$$

and attempt to determine Λ^2 phenomenologically. However this form can only be used for $q^2 >> \Lambda^2$ and $\log q^2/\Lambda^2 >> \log \log q^2$; in particular, the pole at $-q^2 = \Lambda^2$ is incorrect. Many analyses unfortunately tend to determine Λ^2 by fitting to the rapid rise of α_s at $q^2 = -\Lambda^2$.

The actual form of α_s can only have singularities at q^2 timelike where the cuts corresponding to gluon and quark production begin. A convenient simple form which moves the pole to $q^2 = 0$ is[15]

$$\tilde{\alpha}_s(q^2) = \frac{4\pi}{(11-2/3\, n_f)\log\left(1-\frac{q^2}{\Lambda^2}\right)} \tag{1.14}$$

or perhaps

$$\frac{4\pi}{\tilde{\alpha}_s(q^2)} = 11\log\left(1-\frac{q^2}{\Lambda^2}\right) - \frac{2}{3}\sum_f \log\left(1-\frac{q^2-4m_f^2}{\Lambda^2}\right) \tag{1.15}$$

which also takes into account heavy quark thresholds.

An amusing but heuristic feature of the form (1.14) is that it automatically produces a confining linear potential at large distances ($V_{eff} \to C/\vec{q}^{\,4}$, $V_{eff}(r) \to \tilde{C}r$) as well as any asymptotically free form ($V_{eff} \to C'/q^2 \log q^2$, $V_{eff}(r) \sim \tilde{C}'/r \log r$) at short distance. Richardson[15] has shown that using this result as a Schroedinger potential gives an excellent representation of the charm and upsilon spectra. The linear potential agrees with the string model Regge slope $\alpha_R'(0) = 0.90\ \text{GeV}^{-2}$ with $\Lambda = 0.436$ GeV and $n_f = 3$ in Eq. (1.14).

The above speculations on the form of $\alpha_s(q^2)$ are of course only meant to be suggestive. Any non-perturbative effects are expected to be important in the long distance domain. The form of the effective potential between quarks with a hadron is also affected by gluon exchange and retardation effects not included in a naive potential. Further, the gluon and quark pair self-energy insertions in the gluon propagator are themselves affected by higher order corrections, probably giving an effective mass to the gluon intermediate states and weakening the singularity of $\alpha_s(q^2)$ at $q^2 = 0$.

Despite these complexities, there is evidently a unique form for $\alpha_s(q^2)$ determined by the theory.

Another aspect of the non-perturbative nature of QCD is its novel, non-trivial structure of the vacuum state - often described as a dilute gas of instantons (classical solutions of the gauge field sector of the theory). We shall assume that for processes which occur at short distances, i.e.: probe 4-momentum squared Q^2 greater than typical hadronic masses, the non-perturbative effects can be numerically neglected. Estimates of instanton effects which have appeared in the literature support this view.[10] In addition, one can imagine further non-perturbative effects due to initial or final state interactions; e.g., in the Drell-Yan process $pp \to \ell\bar{\ell}X$ the nucleons could influence each other even at large $Q^{\mu} = (\ell + \bar{\ell})^2$. On the other hand, Witten[11] has argued (on the basis of results from soluable gauge field theories) that instantons do not play an important role in physical processes once quantum corrections are taken into account. In any event it is clearly of interest to develop and test the predictions based on short-distance perturbation theory as far as possible.

As is well-known, it is the asymptotic freedom,[12,13,14] nature of QCD which allows a perturbative approach to short distance hadronic physics. It is paradoxical that at this time the most important detailed tests of QCD have come from its predictions for scale-breaking corrections to Bjorken scaling for deep inelastic lepton scattering. This is analogous to trying to first verify QED from the radiative correction to a given scattering process, rather than the cross sections itself. However, the most direct test of QCD, to check the form of quark quark or gluon quark scattering at high momentum transfer, at present suffers from a number of experimental and theoretical complications (see Chapter IV). As we shall argue in Chapter II, the most conclusive evidence that the basic Born structure of the theory is correct comes at present from high momentum transfer exclusive processes, particular form factors.

A striking feature of the rigorous QCD operator product analysis of scale breaking effects in deep inelastic processes is the fact that the asymptotic predictions for the q^2 variation of moments, etc. are independent of the nature of target, whether it is a quark, gluon, meson, proton, or nucleus. Although these results are very powerful, they are strictly true only for $q^2 \to \infty$, and the question of non-asymptotic corrections, as well as the nature of the hadronic distribution functions themselves is left unanswered.

In these lectures we shall consider the "synthesis" problem - matching on the QCD scale-breaking form to the hadronic wavefunctions. The analysis given here is based on a collaboration with G. Peter Lepage. Among the questions we shall consider are

(1) What can be predicted for QCD for the form of the structure functions; i.e., what controls the "initial" distributions?
(2) What is the origin of the sea and gluon distribution in QCD?
(3) What are the corrections to the naive probabilistic treatment of the hadron as a classical distribution of the on-shell quarks?
(4) What is the physics and role of higher "twist" operators?

References

1. Yu. L. Dokshitser, D. I. D'yakanov and S. I. Troyan, SLAC-TRANS-183, translations for Proceedings of the 13th Leningrad Winter School on Elementary Particle Physics (1978).

2. C. H. Llewellyn Smith, Oxford preprint 47/48, Lectures presented at the XVII Internationale Universitatswochen für Kernphysik, Schladming, Austria (1978), published in Acta Physica Austriaca, Suppl. XIX, 331-397 (1978).

3. J. Ellis, SLAC-PUB-2121 (1978); and Proceedings of the SLAC Summer Institute on Particle Physics (1978).

4. C. T. Sachrajda, CERN-TH-2492 (1978); presented at the XIII Rencontre de Moriond.

5. H. Politzer, Phys. Reports 14C (1974).

6. W. Marciano and H. Pagels, Phys. Reports 36C (1978).

7. H. Fritzsch, M. Gell-Mann and H. Leutwyler, Phys. Lett. B47, 365 (1973). For other references, see Refs. 5 and 6.

8. O. W. Greenberg, Phys. Rev. Lett. 13, 598 (1967).

9. P. Cvitanovic, Phys. Rev. D14, 1536 (1976), and references therein.

10. N. Andrei and D. J. Gross, Phys. Rev. D18, 468 (1978); R. D. Carlitz and C. Lee, Phys. Rev. D17, 3238 (1978); L. Baulieu, J. Ellis, M. K. Gaillard and W. J. Zakrzewski, Phys. Lett. 81B, 41 (1979) and 77B, 290 (1978); and T. Appelquist and R. Shankar, Phys. Rev. D18, 2952 (1978).

11. E. Witten, Harvard preprint HUTP-78/A058 (1978).

12. H. Politzer, Phys. Rev. Lett. 30, 1346 (1973).

13. D. J. Gross and F. Wilczek, Phys. Rev. Lett. 30, 1343 (1973).

14. A physical discussion of asymptotic freedom is given by V. F. Weisskopf, Proceedings of the Erice Summer School, Erice Subnucl. Phys. 1974; and R. Field, this volume. An early calculation was given by I. B. Khriplovich, Yad. Fiz. 10, 409 (1969).

15. J. L. Richardson, SLAC-PUB-2229 (1978). See also W. Celmester and F. S. Henyey, Phys. Rev. D17, 3268 (1978).

II. EXCLUSIVE PROCESSES AND THE EXCLUSIVE-INCLUSIVE CONNECTION IN QUANTUM CHROMODYNAMICS

This chapter was written in collaboration with
G. Peter Lepage

1. Introduction

Although the structure of large momentum transfer inclusive reactions is now well understood in perturbative quantum chromodynamics, exclusive processes involving large transfer of momentum have received relatively little attention since they involve detailed, explicit features of hadronic wavefunctions. On the other hand, the dimensional counting[1,2] ansatz, which is based on the premise that the high momentum tail of wavefunctions can be computed from the first iteration of a scale-invariant Bethe-Salpeter kernel, leads to a number of successful predictions for form factors, hadron scattering and photoproduction at large angles, as well as the $x \to 1$ dependence of the structure functions.

In brief, the dimensional counting rules are:

(a) (Spin-averaged) form factors at $|t| \gg M^2$ (Fig. 1a)

$$F_H(t) \sim \frac{1}{t^{n-1}} \qquad (1.1)$$

where n = number of constituent fields in H.

(b) Large angle scattering at $s \gg M^2$, t/s fixed (Fig. 1b)

$$\frac{d\sigma}{dt}(AB \to CD) \sim \frac{1}{s^{n-2}} f(t/s) \qquad (1.2)$$

where n = total number of constituent fields in A,B,C and D.

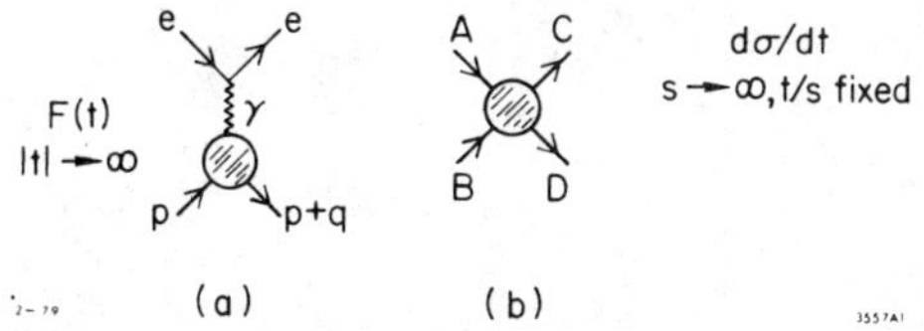

Fig. 1. Hadronic form factors and fixed angle exclusive scattering.

(c) Structure functions at $Q^2 >> M^2$, $\mathcal{M}^2 = \frac{1-x}{x} Q^2$ fixed:

$$F_{2H}(x,Q^2) \sim (1-x)^{2n_s - 1}$$

where n_s = number of spectator fields in H.[3]

In each case the minimum number of fields dominates the asymptotic behavior. The rules follow simply from tree graphs in any renormalizable field theory if the four-momentum of each hadron is partitioned among its constituents.[1]

A comparison of the counting rule prediction $t^{n-1}F_H(t) \to$ const. for F_π, G_{M_p}, G_{M_n}, and F_D is shown in Figs. 2 and 3.

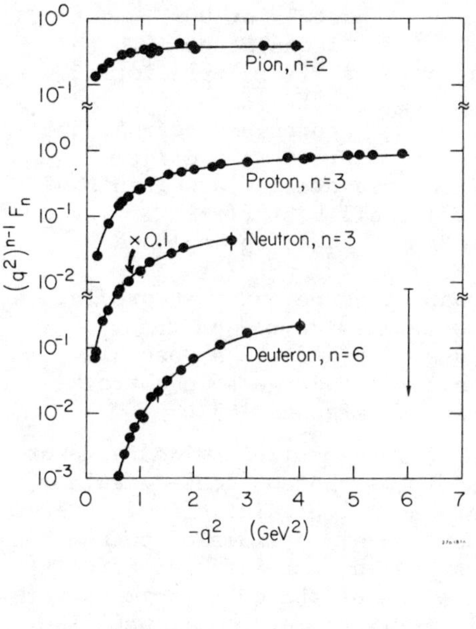

Fig. 2. Hadronic form factors multiplied by $(q^2)^{n-1}$. (From Ref. 5.)

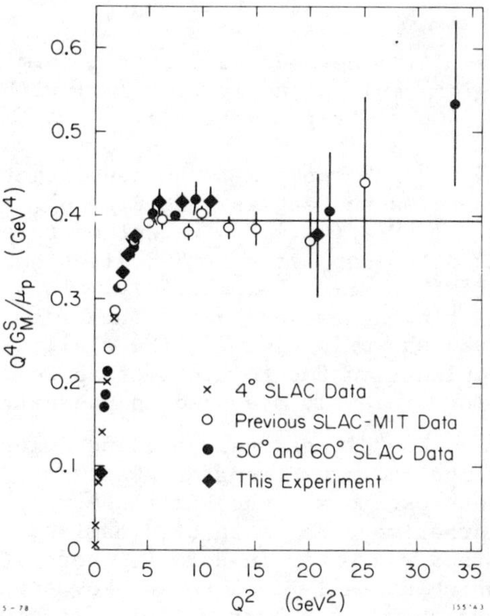

Fig. 3. The proton form factor G_M multiplied by $(Q^2)^2$. (From Ref. 6.)

In the case of the dueteron, the neutron and proton components evidentally each receive a momentum transfer ~q/2. It is thus convenient to define the "reduced" form factor[4,5]

$$f_D(Q^2) = \frac{F_D(Q^2)}{F_N^2(Q^2/4)} \tag{1.4}$$

The prediction $Q^2 f_D(Q^2) \to$ const. is compared with the data[7] in Fig. 4.

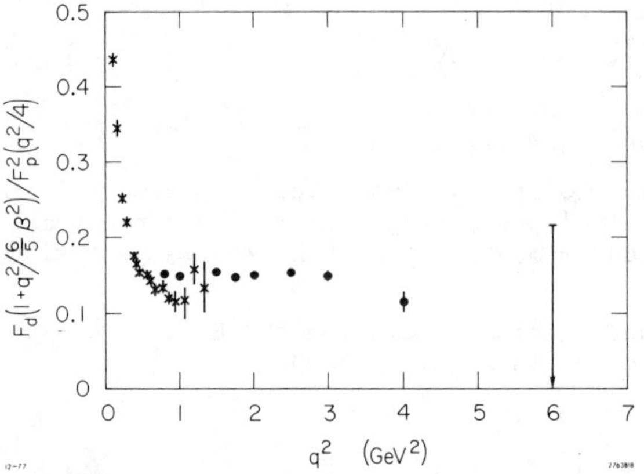

Fig. 4. The reduced deuteron form factor $f_D(q^2)$ multiplied by $(1-q^2/m_0^2)$ with $m_0^2 =$ 0.5 GeV^2. (From Ref. 5.)

In the case of photoproduction at large angles, one predicts $d\sigma/dt(\gamma p \to \pi^+ n) \sim s^{-7} f(\theta_{cm})$, since the photon is an elementary field. The best fit to the 90° data[8] (Fig. 5a) is $s^{-7.3 \pm 0.4}$. The expectation that $d\sigma/dt(\gamma p \to \gamma p)/d\sigma/dt(\gamma p \to \pi^0 p)$ increases with energy at fixed angle is borne out by the data[9] shown in Fig. 5b. Meson-baryon scattering data[10] at 90° is not inconsistent with the $s^{-8} f(\theta_{cm})$ prediction. In the case of proton-proton scattering, the prediction $d\sigma/dt \sim s^{-10} f(\theta_{cm})$ appears to be consistent with data over a large range of energies and angles (see Fig. 6). The overall best fit[11] is $s^{-9.7 \pm 0.5} f(\theta_{cm})$. A recent measurement[12] of $d\sigma/dt(pp \to pp)$ and $d\sigma/dt(np \to np)$ at $\theta_{cm} = 90°$ gives best fits $s^{-9.81 \pm 0.05}$ and $s^{-10.40 \pm 0.34}$, respectively. This apparent agreement with prediction is flawed however by the factor of two oscillations[13] in the data shown in Fig. 7. (We shall discuss in Section 7 a possible explanation for this effect as well as for the large spin correlations recently measured in pp-scattering at Argonne.[14])

The dimensional counting rules neatly summarize a wide range of large momentum transfer phenomena, and it is thus of considerable interest to see whether these rules are actual predictions of perturbative quantum chromodynamics. The general outlines of such a proof was given in Ref. 1, where it was noted that infrared effects which normally lead to Sudakov suppressions of the quark form factor are absent for hadron (color singlet) matrix elements. It was also conjectured that QCD would lead to logarithmic corrections to the scaling predictions, but the nature of these corrections was not understood.

In this talk we will present an outline of a new analysis of exclusive processes in QCD. (A detailed report will appear elsewhere.[15,16]) The main elements of this work involve a consistent Fock space decomposition of the hadronic wavefunction, plus evolution

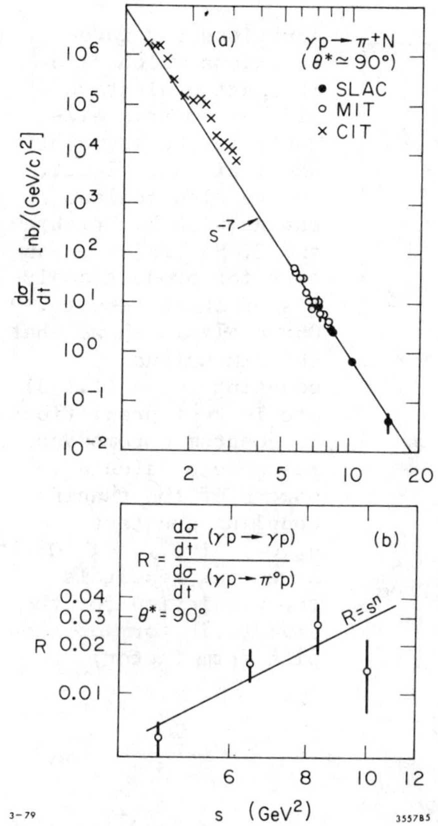

Fig. 5. (a) Pion photoproduction data at 90°. (From Ref. 8.) (b) Comparison of 90° Compton scattering and π^0 photoproduction. The best fit to the ratio is s^n with $n = 2.1 \pm 0.6$. (From Ref. 9.)

Fig. 6. Proton-proton elastic scattering at fixed θ_{cm}. (From Ref. 11.)

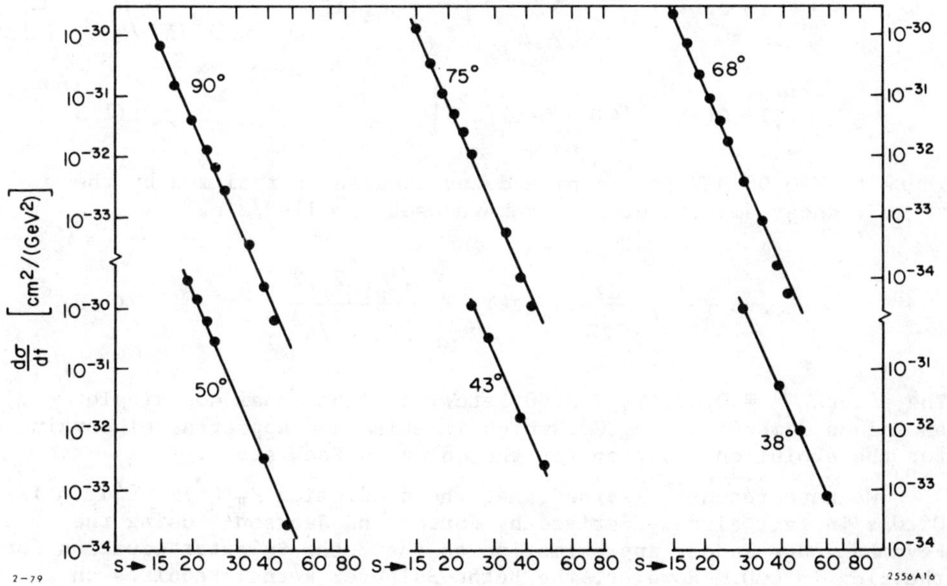

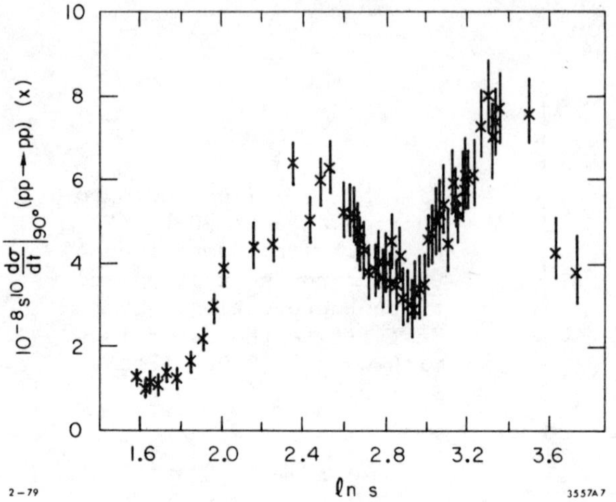

Fig. 7. Proton-proton scattering cross section at 90° multiplied by s^{10}. (From Ref. 13.)

equations for wavefunctions which allow an exact evaluation of hadronic matrix elements in the asymptotic short distance limit. We are also collaborating with Y. Frishman and C. Sachrajda on an operator product analysis of these results.[16] Our analysis shows that the dimensional counting rules (1.1-3) are in fact predictions of quantum chromodynamics modulo calculable powers of the running coupling constant $\alpha_s(Q^2)$, i.e.: $\log^{-1} Q^2/\Lambda^2$. A central result is the prediction for the asymptotic form of the pion form factor:

$$F_\pi(Q^2) = 16\pi \, f_\pi^2 \, \frac{\alpha_s(Q^2/4)}{Q^2}$$

$$\cdot \left[1 - c_2(\lambda^2)\left(\frac{\log Q^2/4\Lambda^2}{\log \lambda^2/\Lambda^2}\right)^{-\gamma_2} + c_4(\lambda^2)\left(\frac{\log Q^2/4\Lambda^2}{\log \lambda^2/\Lambda^2}\right)^{-\gamma_4} + \ldots\right]^2$$

$$\cdot \left[1 + O(\alpha_s(Q^2/4)) + O(m^2/Q^2)\right] \tag{1.5}$$

where $f_\pi \cong 0.95$ MeV is the pion decay constant normalized by the $\pi^+ \to \mu^+\nu$ decay amplitude, and we have used ($\beta = 11 - 2/3\ n_f$)

$$\exp \frac{\beta}{4\pi} \int_{-\lambda^2}^{-Q^2} \frac{d\ell^2}{\ell^2}\, \alpha_s(\ell^2) \cong \frac{\log Q^2/\Lambda^2}{\log \lambda^2/\Lambda^2} \tag{1.6}$$

The powers $\gamma_2 \cong 0.62$, $\gamma_4 \cong 0.90$, etc., are the usual non-singlet anomalous dimensions in QCD, which in this case appear as eigenvalues for the evolution equation for the $q\bar{q}$ meson Fock state.

We have recently learned that the prediction $F_\pi(Q^2) \to c f_\pi^2 \alpha_s(Q^2)/Q^2$ was in fact already derived by Farrar and Jackson[17] using the Feynman gauge ladder approximation to the Bethe-Salpeter equation for the pion in QCD. However, the Bethe-Salpeter kernel requires an infinity of crossed ladder diagrams in covariant gauge even when working in the leading log approximation. This result also appears in a recent report by Efremov and Radyushkin[18] who assume the validity of a particular short distance operator product expansion of the hadronic form factor. The possibility of utilizing products of local fields as hadronic interpolating fields to derive the dimensional counting rules has also been discussed by Polyakov.[19] A conformal group analysis has been given by Menotti.[20]

The graphical analysis utilized here builds on earlier work by Appelquist and Poggio[21] for ϕ^3 in six dimensions. Other early work includes analyses of Yukawa theories by Goldberger, Guth and Soper[22] and QCD in 2 dimensions by Brower, Einhorn, Ellis, Weis[23] and others.

2. Exclusive Processes in QCD

A convenient framework for the analysis of hadronic matrix elements in QCD is time-ordered perturbation theory in the infinite momentum frame; i.e.: quantization on the light-cone. The meson Fock state can be represented as a column vector Ψ with an infinite number of components corresponding to $\langle q\bar{q}|\Psi\rangle$, $\langle q\bar{q}g|\Psi\rangle$, etc. The bound state equation[24]

$$\Psi = S K \Psi \qquad (2.1)$$

is thus an infinite set of coupled equations where the matrix K is the irreducible kernel (see Fig. 8a). The meson form factor receives contributions from each Fock state component as indicated in Fig. 8b. The two particle $q\bar{q}$ Fock state component of the pion corresponds to the usual Bethe-Salpeter amplitude evaluated at equal relative "time" $x^+ = x^0 + x^3$. As we shall see, this amplitude is normalized by the local decay amplitude $\pi^+ \to W^+ \to \mu^+\nu$. The existence of components corresponding to a finite number of constituents is only possible for a color singlet state.[25]

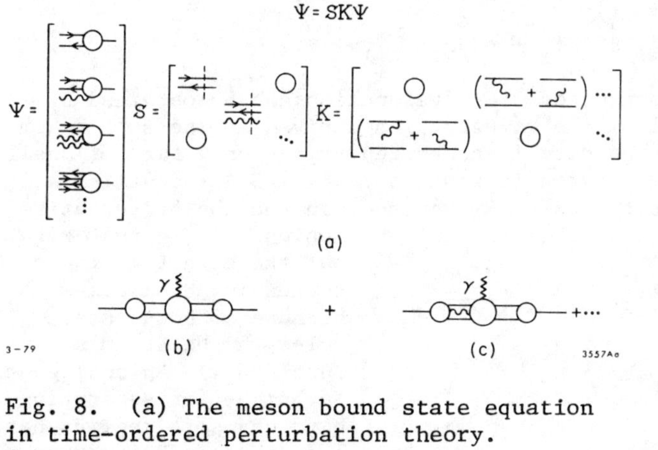

Fig. 8. (a) The meson bound state equation in time-ordered perturbation theory.
(b) The meson form factor in TOPTh.

The n-particle propagator in momentum space is

$$S_{(n)}^{-1} = M^2 - \sum_i^n \frac{\vec{k}_{\perp i}^2 + m_i^2}{x_i} + i\varepsilon \tag{2.2}$$

where $x_i = x_i^+ = \frac{(k_0 + k_3)_i}{(p_0 + p_3)}$ is the fractional momentum variable:

$$\sum_{i=1}^n x_i = 1$$

We can separate hard and soft components in the wave function by defining the propagator

$$S_\lambda = \begin{cases} 0 & \text{if} \quad \left| M^2 - \sum \frac{\vec{k}_\perp^2 + m^2}{x} \right| > \lambda^2 \\ S & \end{cases} \tag{2.3}$$

Thus S_λ vanishes for virtual states far-off the energy shell; in particular $S_\lambda = 0$ if any constituent has large transverse momentum $k_\perp > \lambda$ relative to the direction of the bound state. We then can write

$$\begin{aligned} \Psi &= S_\lambda K \Psi + (S - S_\lambda) K \Psi \\ &= \Psi_\lambda + \Delta S K \Psi \\ &= \Psi_\lambda + \Delta S K \Psi_\lambda + \Delta S K \Delta S K \Psi_\lambda + \ldots \end{aligned} \tag{2.4}$$

i.e.: the "hard" component of the wavefunction can be obtained by perturbation theory from the wavefunction derived in the soft regime (see Fig. 9). This expansion is convergent since only far off shell propagation occurs in intermediate states: $\Psi_\lambda = S_\lambda K \Psi$ contains all non-perturbative effects; $\Delta S = S - S_\lambda$ vanishes in the non-perturbative region. (Note that without the cutoff λ^2 the series in ΔS diverges because of bound state poles in the Green's functions.) An analogous technique for separating hard and soft regimes has been used in Ref. 26 to systematically treat high order radiative corrections to the positronium spectrum. The renormalization program can be

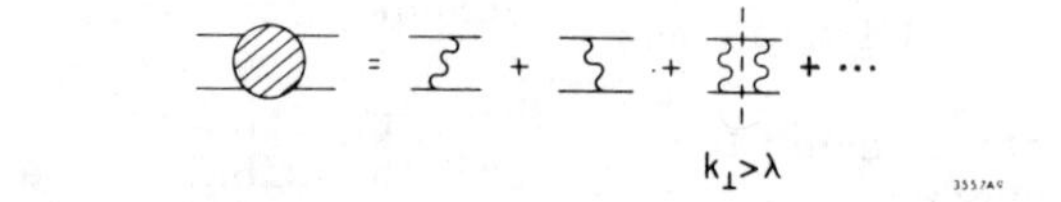

Fig. 9. The $q\bar{q}$ Fock state wavefunction of the meson using Eq. (2.4). The two particle reducible amplitudes in the kernel only contribute for $k_\perp > \lambda$.

implemented as discussed in Refs. 26 and 27.

The pion form factor can now be represented as a sum of matrix elements between Ψ_λ states (see Fig. 10). Notice that all connected diagrams (reducible and irreducible) contribute to the hard scattering amplitude. As we shall show, there are no singularities from the phase-space factors at $x_i \to 0$. By definition the transverse momenta in the amplitudes are regulated by λ. All nonperturbative effects in the form factor amplitude are contained in Ψ_λ.

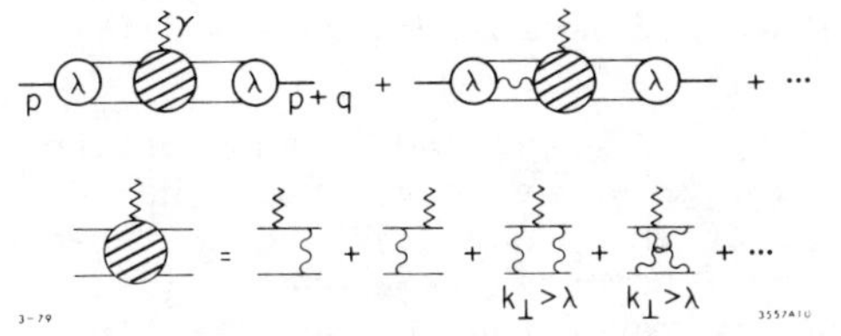

Fig. 10. The meson form factor in QCD using the S division. All Fock states contribute.

It is easy to check by dimensional analysis/power counting that Fock components which contain more than the minimum number of quarks in Ψ_λ give contributions to the form factor which are power law suppressed: the nominal power is (see Fig. 11)

$$F(Q^2) \sim \left(\frac{1}{Q^2}\right)^{n_q - 1} \tag{2.5}$$

An analogous statement also holds for extra gluons in the Fock state, but the interpretation depends on the gauge choice. First we note that diagrams where a gluon connects the initial and final-state soft constituents are suppressed by a power of Q^2 because the hadron is a singlet and true infrared divergences cancel. (In the case of colored states such contributions lead to the analogue of the Sudakov form factor.) Diagrams where a soft gluon from Ψ_λ interacts with a hard quark as in Fig. 11b are also suppressed by a power of q^2 unless the gluon has polarization ε_+. In fact, contributions from any number of such gluons can be resummed (via the Ward identity) into an effective two particle wavefunction. Alternatively, one can simply choose the light-cone gauge $\eta \cdot A = 0$ for which $\varepsilon_+ = 0$, and the anomalous contributions are suppressed by higher power of Q^2.

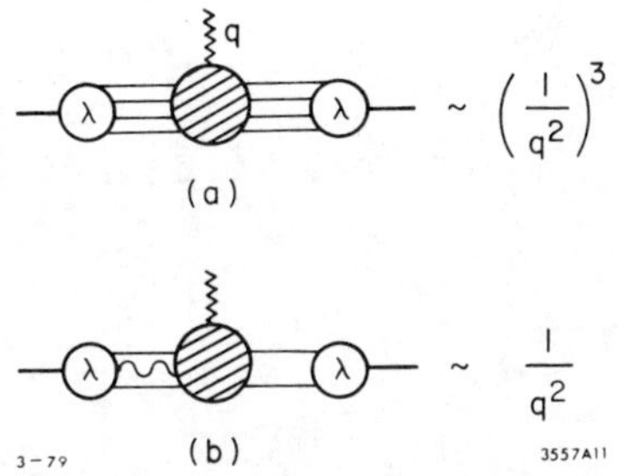

Fig. 11. Fock state contributions to the meson form factor. The $|qq\bar{q}\bar{q}\rangle$ amplitude (a) obeys Eq. (2.5). The $|q\bar{q}g\rangle$ amplitude (b) gives an anomalous contribution $O(Q^{-2})$ if $\varepsilon = \varepsilon_+$; in light cone gauge its contribution is $O(Q^{-4})$.

The calculation of the form factor have been performed in general covariant gauges as well as in the light-cone gauge, although for simplicity we will only present the light-cone gauge calculation here. All of the final results are gauge invariant. Note that Eq. (2.5) refers to Fock state constituents that are forced to change directions. Higher Fock state components in which the constituents annihilate before the exchange of the hard momentum Q only modify the form of the soft wavefunction Ψ_λ.

Thus contributions to the leading $1/Q^2$ power dependence of the meson form factor only arise (in the light-cone gauge) from the quark-antiquark Fock state component of the meson wavefunction (see Fig. 12). It is convenient to choose a Lorentz frame where q_μ is transverse to the direction of the incident meson ($-q^2 = Q^2 = 4\vec{q}_\perp^{\,2}$). It is then straightforward to identify the leading logarithmic corrections to the $1/Q^2$ power law in each order of perturbation theory. In fact, by using the light-cone gauge only the ladder diagrams are required. The neglected terms will be of relative order $\alpha_s(Q^2)$ and m^2/Q^2.

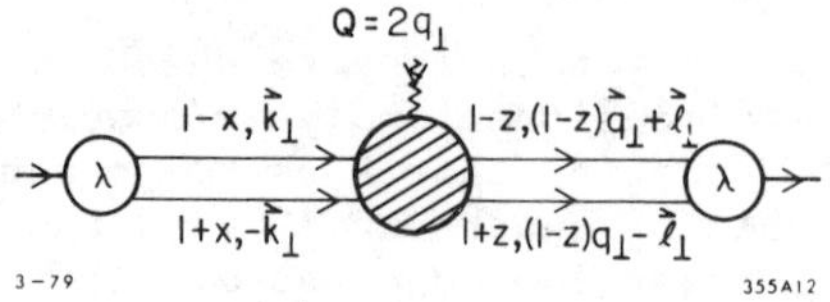

Fig. 12. Two-particle Fock state contribution to the meson form factor; the kinematics are chosen so that $x, z, \vec{k}_\perp$ and $\vec{\ell}_\perp$ vanish in the zero binding limit. Here $Q^2 = (2\vec{q}_\perp^{\,2})$.

To this order, the pion form factor in QCD takes the form (see Fig. 13) ($q^2 = -Q^2$)

$$F_\pi(Q^2) = \int_{-1}^{1} dx \int_{-1}^{1} dz \; \phi^+(z, Q^2/4) \; T_B(z,x,Q^2) \; \phi(x, Q^2/4) \qquad (2.6)$$

where ($C_F = 4/3$)

$$T_B = \frac{16\pi\alpha_s(Q^2/4)C_F}{Q^2} \frac{1}{(1+x)} \frac{1}{(1+z)} \qquad (2.7)$$

is the result expected from the simplest impulse approximation.

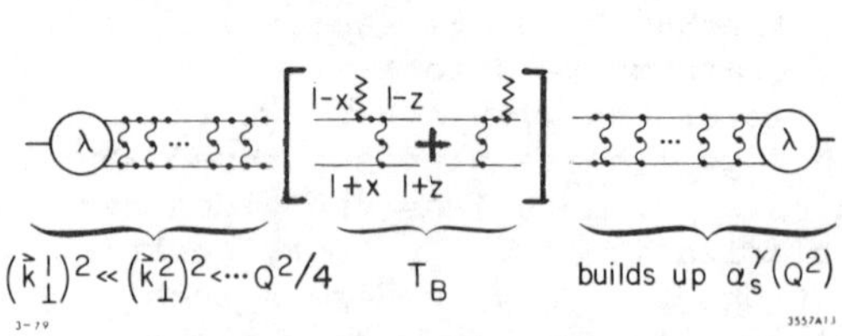

Fig. 13. Leading logarithm contributions to the meson form factor. The dominant momentum flow is through T_B. The ladder and self-energy insertions yield the evolution Eq. (2.14) and the anomalous dimensions in Eq. (2.15).

As we shall see, the wavefunction $\phi(x,Q^2)$ in Eq. (2.6), satisfies an evolution equation in QCD, and leads to non-trivial logarithmic modification of dimensional counting. To leading order in $\alpha_s(Q^2)$, it is sufficient to compute the two-body wavefunction integrated over transverse momentum $\vec{k}_\perp^2 < Q^2$

$$\tilde{\phi}(x,Q^2) = \int_0^{Q^2} \frac{d\vec{k}_\perp^2}{16\pi^2} \psi(x, \vec{k}_\perp^2) \qquad (2.8)$$

since the dominant momentum flow occurs through T_B. The leading contribution to order $\alpha_s(Q^2)$ can now be computed simply from the ladder graph structure with strong ordering of the loop momentum and the inclusion of all vertex and self-energy insertions as indicated in Fig. 13.

The wavefunction ϕ appearing in Eq. (2.6) is given by

$$\phi(x,Q^2) = \tilde{\phi}(x,Q^2) \left[\frac{\log Q^2/\Lambda^2}{\log \lambda^2/\Lambda^2}\right]^{-C_F/\beta} \qquad (2.9)$$

where the last factor with $C_F = 4/3$ and $\beta = 11-2/3\, n_f$ is due to vertex and fermion self-energy corrections in T_B. By the Ward identity, such factors are process independent.

Because of the strong ordering of the $\vec{k}_\perp^2$ integrations, the integrated wavefunction obeys the inhomogeneous equation,

$$\tilde{\phi}(x,Q^2) = \tilde{\phi}(x,\lambda^2) + \int_{\lambda^2}^{Q^2} \frac{d\vec{k}_\perp^2}{\vec{k}_\perp^2} \frac{C_F}{4\pi} \alpha_s(\vec{k}_\perp^2) \int_{-1}^{1} dy\, V(x,y)\, \tilde{\phi}(y,\vec{k}_\perp^2) \qquad (2.10)$$

where the kernel V is

$$V(x,y) = \frac{1-x}{1-y}\left(1 + \frac{2}{(x-y)_+}\right)\theta(x>y) + \frac{1+x}{1+y}\left(1 - \frac{2}{(x-y)_+}\right)\theta(y>x) \qquad (2.11)$$

(This is the kernel for the quark and antiquark with opposite helicities, as required for pseudoscalar and $\lambda = 0$ vector meson states. In the case of parallel helicities, only the $2/(x-y)_+$ terms contribute.) The distributions in V are defined for integrals over x as

$$\frac{1-x}{1-y}\frac{2}{(x-y)_+}\theta(x>y) = \frac{1-x}{1-y}\frac{2}{x-y}\theta(x-y) - \delta(x-y)\int_y^1 dz\, \frac{1-z}{1-y}\frac{2}{z-y} \qquad (2.12)$$

Thus there are no actual singularities as $x = y$. This infrared cancellation is due to the self-energy corrections to the q and $\bar{q}$ legs in Fig. 13; it would not occur if the constituents formed a non-singlet.

If we define

$$\xi = \frac{\beta}{4\pi} \int_{\lambda^2}^{Q^2} \frac{d\vec{k}_\perp^2}{\vec{k}_\perp^2} \alpha_s(\vec{k}_\perp^2) \cong \log \frac{\alpha_s(\lambda^2)}{\alpha_s(Q^2)} , \tag{2.13}$$

then ϕ satisfies an evolution equation:

$$\frac{\partial}{\partial \xi} \phi(x,\xi) = \frac{C_F}{\beta} \int_{-1}^{1} dy\ V(x,y)\ \phi(y,\xi) - \frac{C_F}{\beta} \phi(x,\xi) \tag{2.14}$$

These results are analogous to the evolution equations[28] derived for the ξ-dependence of non-singlet structure functions. In each case, infrared divergences cancel and procedures are available to evaluate the higher order corrections in $\alpha_s(Q^2)$. An essential difference is that the short-distance operator T_B is an integral over x rather than a delta-function.

The physics of the evolution equation for $\phi(x,\xi)$ is contained in the eigenvalues and eigenfunctions of $V(x,y)$. The general solution is

$$\phi(x,\xi) = (1-x^2) \cdot \sum_{n=0,2,4} a_n C_n^{3/2}(x) \left(\frac{\log Q^2/\Lambda^2}{\log \lambda^2/\Lambda^2} \right)^{-\gamma_n} \tag{2.15}$$

Only terms even under $x \to -x$ occur in the pion wave function when SU(2) symmetry is assumed.

Here $C_n^{3/2}(x)$ is a Gegenbauer polynomial, e.g.,

$$C_n^{3/2}(x) = \begin{cases} 1 & n = 0 \\ -3/2(1-5x^2) & n = 2 \\ 15/8(1-14x^2+21x^4) & n = 4 \end{cases} \tag{2.16}$$

The weights a_n are determined from the initial wavefunction

$$a_n = \frac{2n+3}{2(2+n)(1+n)} \int_{-1}^{1} dx\ \phi(x,\lambda^2)\ C_n^{3/2} \tag{2.17}$$

and the γ_n are the standard non-singlet anomalous dimensions (taking $\beta = 9$)

$$\gamma_n = \frac{C_F}{\beta}\left\{1 + 4\sum_{2}^{n+1}\frac{1}{k} - \frac{2}{(n+1)(n+2)}\right\} \cong \begin{cases} 0 & n=0 \\ .62 & n=2 \\ .90 & n=4 \end{cases} \quad (2.18)$$

(The only assumption needed here is that the initial wavefunction $\phi(x,\lambda^2)$ falls as $(1-x^2)^\varepsilon$ with $\varepsilon > 0$ at $x = \pm 1$, which is in fact required by self-adjointness of the kinetic energy for the bound state wavefunction. This boundary condition can in fact be regarded as the condition of compositeness of the meson; if there were an elementary field with the quantum numbers of the meson then $\phi(x,\lambda^2)$ would be constant as $x^2 \to 1$ due to the possible $\bar{\psi}\gamma_5\psi$ coupling. Of course in this case $F_\pi(Q^2) \to$ const. In fact this type of analysis is required in the case of the photon structure function and transition form factors in QCD.)

In the asymptotic limit $Q^2 \to \infty$, we have from Eq. (2.15)

$$\phi(x,Q^2) \to \frac{3}{4} k_\pi (1 - x^2) \quad (2.19)$$

where

$$k_\pi = \lim_{Q^2 \to \infty} \left[\frac{\log Q^2/\Lambda^2}{\log \lambda^2/\Lambda^2}\right]^{-C_F/\beta} \int_{-1}^{1} dx \int^{Q^2} \frac{d^2k_\perp}{16\pi^2} \psi(x,\vec{k}_\perp) \quad (2.20)$$

Notice that the singularity of T_B at $x = -1$ or $z = -1$ is damped by the wavefunction. This should be contrasted with the situation for theories such as QCD in two-dimensions or ϕ^3 field theory where $T_B \sim (1+x)^{-2}(1+z)^{-2}$ and power-law modifications to the form factor can and do occur from this infrared, long distance region.[23]

We thus see from Eqs. (2.6) and (2.7) that in the asymptotic limit

$$F_\pi(Q^2) \underset{Q^2 \to \infty}{\to} \frac{16\pi\alpha_s(Q^2/4)\, C_F}{Q^2}\left(\frac{3}{2} k_\pi\right)^2 \quad (2.21)$$

In fact, we can determine the normalization from the weak decay amplitude for $\pi \to \mu\nu$:

$$f_\pi = \sqrt{n_c}\, k_\pi \sim .94 \text{ MeV} \quad (2.22)$$

i.e.:

$$F_\pi(Q^2) \underset{Q^2\to\infty}{\rightarrow} \frac{36\pi\alpha_s(Q^2/4)}{Q^2}\,\frac{f_\pi^2 C_F}{n_c} \qquad (2.23)$$

where n_c is the number of colors.

Although Eq. (2.23) gives the asymptotic form factor, the anomalous dimension terms result in sizeable corrections until very large momentum transfer, and in fact tend to compensate for the fall-off due to $\alpha_s(Q^2)$ over a wide range of Q^2. If we assume that $\phi(x,\lambda^2)$ is peaked at $x \sim 0$ (equal momentum partition) as is characteristic of non-relativistic bound states, then the evolution equation causes $\phi(x,Q^2)$ to broaden asymptotically out to the $(1-x^2)$ envelope where T_B is maximum (see Fig. 14). Thus one obtains Eq. (1.5) where $c_2 = -a_2/a_0$ and $c_4 = a_4/a_0$ are positive, and are computed from Gegenbauer moments of the soft wavefunction. The form factor is of course independent of the choice of λ^2. Figure 15 shows the prediction for $Q^2F_\pi(Q^2)$ assuming that $\phi(x,\lambda^2)$ is either strongly peaked at $x = 0$ or has a smooth $(1-x^2)$ dependence.

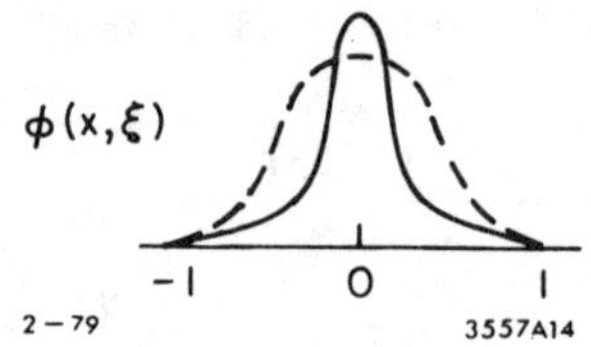

Fig. 14. Schematic representation of wavefunction evolution. As $Q^2 \to \infty$, $\phi(x,Q^2) \to (1-x^2)$.

It is of course significant that the same anomalous dimensions enter in the pion form factor and the moments of the non-singlet structure function. This can be attributed to the fact that the leading local operators which couple to the pion wavefunction at short distances $x_\mu \to 0$ are the usual twist 2 $\bar\psi\gamma_\mu\gamma_5(D_\mu)^n\psi$ operators for $O_F^{(n)}$. The absence of an anomaly in the axial vector operator implies that the asymptotic form factor is of the form of Eq. (2.23).

In the case of vector mesons with total helicity ±1, T_B vanishes as a power

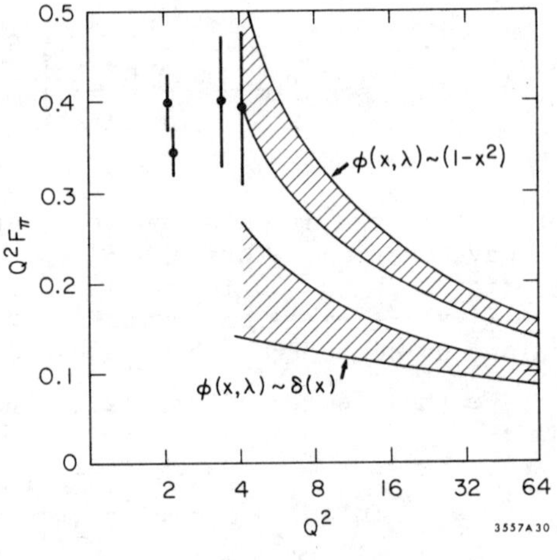

Fig. 15. QCD prediction for the meson form factor for two extreme cases: (a) $\phi(x,\lambda^2) \sim \delta(x)$, or (b) $\phi(x,\lambda^2) \sim (1-x^2)$. In the latter case the wavefunction is unchanged under evolution. The asymptotic predictions are absolutely normalized using Eq. (2.23). The bands correspond to $\pm\alpha_s/\pi$ with $\Lambda^2 = 0.25$ GeV^2.

of Q^2 faster than (2.7), and the potential only contains the $1/(x-y)_+$ term. Thus we obtain asymptotic forms

$$F_\rho(Q^2) \propto \begin{cases} F_\pi(Q^2) & \lambda_\rho = 0 \\ \dfrac{m^2(Q^2)}{Q^2}\ \dfrac{\alpha_s(Q^2)}{Q^2}\ [\alpha_s(Q^2)]^{2C_F/\beta} & \lambda_\rho = \pm 1 \end{cases} \qquad (2.24)$$

where $m(Q^2)$ is the "running mass" in QCD. Another contribution of order $1/Q^4$ can also evidently be obtained from the $n=3|q\bar{q}g\rangle$ Fock state component of the vector meson wavefunction.

The transition form factor of the photon, e.g., $\gamma(Q^2)+\gamma(k^2\sim 0) \to \pi^0$ can be analyzed in a similar fashion. We find

$$F_{\pi\gamma}(Q^2) = \frac{\left(e_u^2+e_d^2\right)2\sqrt{n_c}}{Q^2}\int_{-1}^{1} dx\, \frac{\phi(x,Q^2)}{1+x} \ . \qquad (2.25)$$

(The $\gamma^*\gamma\pi^0$ vertex is defined as $ie^2 F_{\pi\gamma}(Q^2)\, \epsilon_{\mu\nu\rho\sigma}\, p_\pi^\nu\, q^\rho\, \varepsilon^\sigma$.) This can be absolutely normalized in terms of the pion form factor

$$F_{\pi\gamma}(Q^2) = \frac{\left(e_u^2+e_d^2\right)2\sqrt{n_c}}{Q^2}\left[\frac{Q^2 F_\pi(4Q^2)}{4\pi C_F \alpha_s(Q^2)}\right]^{1/2} \ . \qquad (2.26)$$

The corrections are of order $O[\alpha_s(Q^2)]$.

The form factor of nucleons may be analyzed in a similar manner. The leading power law terms arise from the minimal 3 quark Fock component and gives $T_B \sim [\alpha_s(Q^2)/Q^2]^2$ (see Fig. 16). A three body kernel and evolution equation can then be obtained in parallel with the meson case. The asymptotic result is

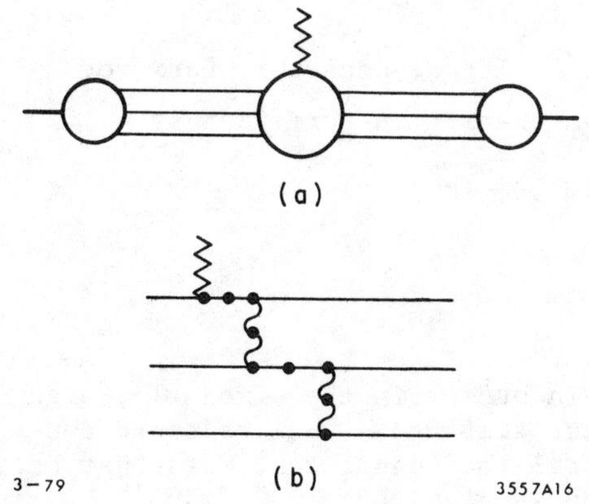

Fig. 16. Schematic representation of the nucleon form factor for the three quark Fock state.

$$F_1(Q^2) \sim \frac{\alpha_s^2(Q^2)}{(Q^2)^2}\left(\sum_j a_j(\log Q^2/\Lambda^2)^{-\gamma_j(N)}\right)^2 \qquad (2.27)$$

$$F_2(Q^2) \sim \frac{m^2(Q^2)}{Q^2} F_1(Q^2) \qquad (2.28)$$

The calculation of the nucleon anomalous dimensions $\gamma_j(N)$ is in progress.[15] These results agree with the dimensional counting predictions and verify that the empirical $G_M \sim G_E \sim 1/Q^4$ scaling laws are consistent with quantum chromodynamics, modulo over-all $\log Q^2$ corrections.

3. The Inclusive-Exclusive Connection

The above predictions for the form of the asymptotic form factor may seem somewhat paradoxical since QCD asymptotic freedom corrections to Bjorken scaling of hadronic structure functions appear to be relatively much stronger. In particular, as shown in Ref. 29, QCD predicts structure functions at large x of the form

$$F_2(x,Q^2) \sim (1-x)^{V+\tilde{\xi}(Q^2,k^2)}\, P(\tilde{\xi}) \qquad (3.1)$$

where $(1-x)^V$ is the effective power-behavior at $Q^2 \sim O(k^2)$, and

$$\tilde{\xi}(Q^2,k^2) = \frac{C_F}{\pi}\int_{k^2}^{Q^2} \frac{d\ell^2}{\ell^2}\,\alpha_s(\ell^2) \sim O(\log\log Q^2)\ . \qquad (3.2)$$

and $P(\tilde{\xi})$ is a normalization factor.[29] If one uses this form for fixed $\mathcal{M}^2 = \frac{1-x}{x}Q^2$; then one obtains transition form factors

$$F^2(Q^2) \sim \left(\frac{\mathcal{M}^2}{Q^2}\right)^{V+1+\tilde{\xi}(Q^2,k^2)} P(\tilde{\xi}) \qquad (3.3)$$

which fall faster than any power![30]

In fact, this "derivation" is incorrect in the fixed $\mathcal{M}^2$, high Q^2 domain because it ignores the fact that the struck hadronic constituent is far off-shell. In general the constituent mass that sets the lower limit in the $\tilde{\xi}$ integration is given by

$$k^2 = z\left[M_H^2 - \frac{k_\perp^2 + \tilde{m}^2}{1-z} - \frac{k_\perp^2}{z}\right] \qquad (3.4)$$

where $\tilde{m}^2$ is the square of the invariant mass of the remaining spectators and $\vec{k}_\perp$ and z are the struck constituent's light-cone coordinates in the hadronic wavefunction. Since $z > x$, and x is near 1,

$$-k^2 \sim \frac{k_\perp^2 + \tilde{m}^2}{1-x} \sim \frac{k_\perp^2 + \tilde{m}^2}{\mathcal{M}^2} Q^2 \qquad (3.5)$$

i.e.: $k^2 \sim O(Q^2)$ at fixed $\mathcal{M}^2$. Thus

$$\tilde{\xi} = \frac{4C_F}{11-2/3\, n_f} \log \frac{\alpha_s(k^2)}{\alpha_s(Q^2)} \qquad (3.6)$$

$$\Longrightarrow \frac{C_F}{\pi} \alpha_s(Q^2) \log \left(\frac{\mathcal{M}^2}{\vec{k}_\perp^2 + \tilde{m}^2} \right) \text{ at fixed } \mathcal{M}^2,\ Q^2 \to \infty.$$

i.e.: $\tilde{\xi}$ actually vanishes as $1/\log Q^2$ in the fixed $\mathcal{M}^2$ domain.

The behavior of structure functions in the large x region can be computed in leading order in α_s from the infinite set of diagrams indicated in Fig. 17. The infinite set of horizontal gluon ladder graphs above the quark leg labeled k^2 in the figure builds up the standard QCD corrections to Bjorken scaling and q^2 dependence of the structure function moments. The main power law dependence at $x \sim 1$ is given by the minimal number of (vertical) gluon exchanges required to stop the hadronic spectators. For the case of the nucleon the leading Fock state component is the $|qqq\rangle$ state, and two gluon exchanges with off-shell masses of order $k_x^2 \sim O((\vec{k}_\perp^2 + \tilde{m}^2)/(1-x))$ are required. These minimal hard gluon exchange diagrams give the analogue of T_B in the form factor calculation.

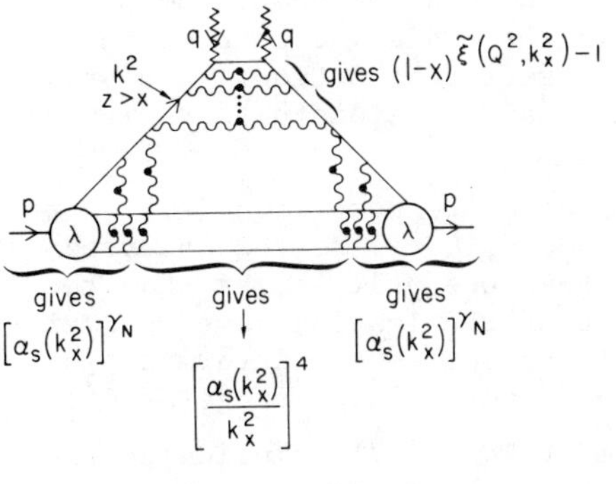

Fig. 17. Analysis of deep inelastic scattering (virtual Compton amplitude) to leading logarithmic order in perturbation theory. See Eq. (3.7).

In addition, the remaining infinite set of vertical gluon exchange diagrams (ordered, as usual, in momenta) leads to the evolution of the hadronic wavefunction from the soft region λ^2 to the off-shell value k_x^2. As in the form factor calculation, this leads to a series of anomalous logarithms $[\alpha_s(k_x^2)]^{\gamma_j^{(N)}}$ determined by the eigenvalues of the kernel for the Fock state. Combining factors, the leading behavior is given by

$$F_{2N}(x,Q^2) \sim C(\tilde{\xi})(1-x)^{3+\tilde{\xi}(Q^2,k_x^2)} \alpha_s^4(k_x^2)$$

$$\cdot \left[\sum_j a_j (\log k_x^2/\Lambda^2)^{-\gamma_j^{(N)}} \right]^2 \tag{3.7}$$

where the $\gamma_j^{(N)}$ are the leading anomalous dimension for the 3-quark nucleon wavefunction. Correction terms of higher power in (1-x) and $\alpha_s(Q^2)$ or $\alpha_s(k_x^2)$ are neglected. Notice that at fixed $\mathcal{M}^2$, $\tilde{\xi} \to 0$ and we obtain a perfect exclusive-inclusive connection with the corresponding form factor calculation Eq. (2.7), in agreement with the Drell-Yan-West relation.

Equation (3.7) is consistent with the standard evolution equations for QCD structure functions and moments and other results derived from the operator product expansion.[31,32] In the large x domain, however, the "initial" or "starting" structure function is no longer unknown but is directly determined from QCD perturbation theory and the wavefunction evolution equations at short distances. In a sense the most critical prediction from QCD is the nominal power law $(1-x)^3$ since the integer 3 reflects the existence of a 3 quark Fock state as well as nearly scale-invariant QCD quark-quark interactions within the nucleon. The logarithmic dependence from $\tilde{\xi}(Q^2,k_x^2)$ and $\alpha_s(k_x^2)$ in Eq. (3.7) yield the radiative corrections to the main dynamical dependence of the structure function. The predicted form for the structure function may prove useful for fits to data, at least for $x > 0.5$.

In practice, the expected values of $\tilde{\xi}$ are not large (e.g., for $Q^2 = 100$ GeV^2, $k_x^2 = 1$ GeV^2, and $\Lambda^2 = 1/3$ GeV^2, $\tilde{\xi} \cong 1$) so the observed (1-x) power should be typically less than one unit larger than the valence power. The direct measurement of the leading power behavior of the valence state structure function requires the determination of the structure function at fixed $\mathcal{M}^2$ over a large range of Q^2.

It should be noted that the form of Eq. (3.7) complicates the empirical analysis of moments at large N. In general

$$\mathcal{M}_N(Q^2) = \int_0^{1-\Delta M^2/Q^2} dx\, x^{N-1} F_2(x,Q^2)$$

$$= \left[\frac{\alpha_s(Q^2)}{\alpha_s(\bar{k}_N^2)} \right]^{\gamma_N} \tag{3.8}$$

(The elastic contribution gives a power-law correction to Bjorken scaling.) At large N, only large x is important and

$$\bar{k}_N^2 \sim O\left\langle \frac{\tilde{m}^2 + \vec{k}_\perp^2}{1-x} \right\rangle \sim O\left(\frac{\tilde{m}^2 + \vec{k}_\perp^2}{\Delta M^2}\right) Q^2 \qquad (3.9)$$

Thus $\mathcal{M}_N(Q^2)$ tends to scale in the large N limit, again because of the strong off-shell behavior of the struck constituent. The standard QCD prediction $\mathcal{M}_N(Q^2) \sim [\alpha_s(Q^2)]^{\gamma_N}$ only holds for Q^2 sufficiently large such that $\Delta M^2/Q^2 << N^{-1}$.

4. Quark Sea Phenomenology

The decomposition of the nucleon wavefunction in terms of its Fock state components $|qqq\rangle$, $|qqqg\rangle$, $|qqqq\bar{q}\rangle$, etc. leads to a new perspective on the origin of the phenomenological sea quark distributions (see Fig. 18). The "intrinsic" sea from the $|qqqq\bar{q}\rangle$ component has a nominal power dependence at large x

$$F_2^{\bar{q}} \propto (1-x)^{7+\tilde{\xi}} \qquad (4.1)$$

since there are 4 spectators.[34] The sea quarks which evolve in Q^2 due to lowest order pair production from the intrinsic gluons in the $|qqqg\rangle$ components lead to a contribution of the form

$$F_2^{\bar{q}} \propto \tilde{\xi}(1-x)^{5+\tilde{\xi}} \qquad (4.2)$$

Similarly, the sea quarks evolved from the $|qqq\rangle$ valence component via gluon bremsstrahlung and subsequent pair production lead to a contribution

$$F_2^{\bar{q}} \propto \tilde{\xi}^2(1-x)^{5+\tilde{\xi}} \qquad (4.3)$$

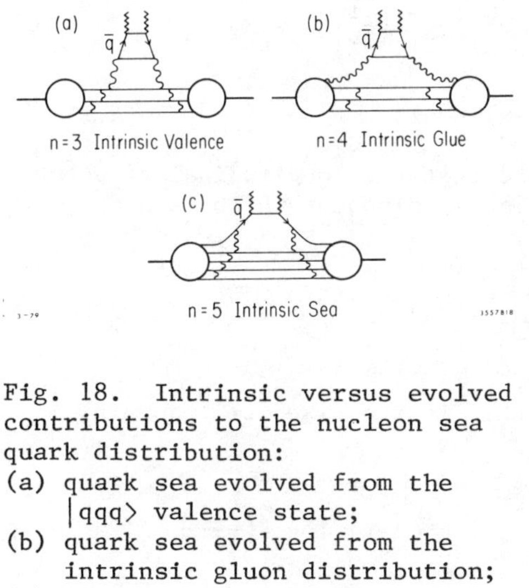

Fig. 18. Intrinsic versus evolved contributions to the nucleon sea quark distribution:
(a) quark sea evolved from the $|qqq\rangle$ valence state;
(b) quark sea evolved from the intrinsic gluon distribution;
(c) intrinsic sea contribution.

In general, the measured distribution should be a Q^2-dependent sum of such contributions.

At fixed $\mathcal{M}^2$, $\tilde{\xi} = \tilde{\xi}(Q^2, k_x^2) \to 0$, and thus despite the extra powers of (1-x) the intrinsic sea components could be particularly important in this domain. Further, as has been shown in Ref. 33, there are strong cancellations between various contributions to the gluon distribution in hadrons at low momentum due to the singlet nature of the source.

The intrinsic sea quark component may in fact be numerically dominant until very large Q^2. In that case we expect the nominal power of $F_2^{\bar{q}}(x,Q^2)$ at large x to decrease from $(1-x)^7$ to $(1-x)^5$ as the evolved component become relatively stronger.

5. Pion Structure Functions

The structure functions of mesons can also be analyzed at large Q^2 and small (1-x) in a manner similar to the analysis of nucleon structure functions. In this case, as first noted by Ezawa,[35] there is a kinematic suppression of the transverse cross section $\sigma_T(Q^2)$ by a power of (1-x) which can be attributed to the mismatch between the spin of the quark and spin of the meson. In our analysis this shows up in the corresponding suppression of the tree diagrams for T_B at x near 1. Since the meson wavefunction has a leading zero anomalous dimension,

$$F_2^{\pi}(x,Q^2) \sim (1-x)^{2+\tilde{\xi}} \log^{-2}\left(\frac{1}{1-x}\right)$$

$$Q^2 \to \infty \quad , \quad (1-x) \text{ small} \tag{5.1}$$

In addition, it is important to note that the longitudinal structure function has an anomalous (non-scaling) component which is nearly flat at large x:[36,37]

$$F_L^{\pi}(x,Q^2) \sim \frac{(1-x)^{0+\tilde{\xi}}}{Q^2} \log^{-2}\left(\frac{1}{1-x}\right) \tag{5.2}$$

where $\tilde{\xi} \sim \tilde{\xi}(Q^2,k_x^2)$. If we analyze this in the fixed $\mathcal{M}^2$ domain, then $\tilde{\xi} \to 0$ and

$$F_2^{\pi}(x,Q^2) \to F_L^{\pi}(x,Q^2) \sim (1-x) \log^{-2}\left(\frac{1}{1-x}\right)$$

$$Q^2 \to \infty \quad , \quad \mathcal{M}^2 \text{ fixed} \tag{5.3}$$

Thus, as in the nucleon case, there is a perfect exclusive/inclusive connection. The dominance of the longitudinal structure function in the fixed $\mathcal{M}^2$ limit for mesons is an essential prediction of perturbative QCD. It can be tested directly[36] in the $e^+e^- \to \pi X$ angular distribution near the kinematic boundary $(x \to 1)$. Perhaps the most dramatic consequence is in the Drell-Yan process $\pi p \to \mu^+\mu^- X$ (Fig. 19). In this case one predicts[38] that for fixed Q^2 pairs, the angular distribution of the μ^+ (in the pair rest frame) will change from the conventional $(1+\cos^2\theta_+)$ distribution to $\sin^2\theta_+$ for pairs produced at large longitudinal momentum, $x_L(\mu^+\mu^-) \to 1$. This is due to the

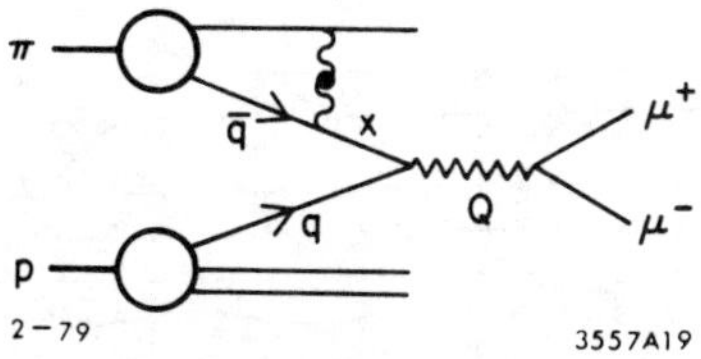

Fig. 19. Representative contribution to Drell-Yan $\pi p \to \mu^+\mu^- X$ cross section. The gluon exchange in the pion wavefunction is responsible for the power law fall off at $x \to 1$ [Eq. (5.1-3)] and the power law tail at large $k_\perp^2$ [Eq. (8.1)].

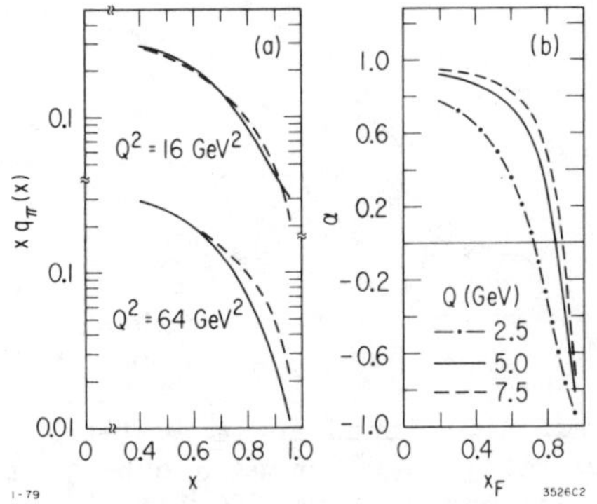

Fig. 20. (a) The pion structure function at large x. The solid line is the prediction $F_2^\pi \sim (1-x)^2 + C/Q^2$. (b) Prediction for the μ^+ angular distribution $1 + \alpha\cos^2\theta_+$ where θ_+ is measured relative to the incident pion in the $\mu^+\mu^-$ rest frame. (From Ref. 38.)

dominance of the meson's longitudinal structure function at large x and fixed Q^2 (see Fig. 20b). Figure 20a also shows that the predicted form of the structure function $F_2^\pi(x,Q^2) \sim (1-x)^2 + C/Q^2$ is not inconsistent with recent fits to the data. The dashed line is the experimental form $(1-x)^{1.01}$ given in Ref. 39.

6. Fixed Angle Scattering

The techniques which we have discussed for obtaining asymptotic results for form factors can be extended to the computations of any exclusive process involving large momentum transfer between color singlets. Here we shall focus on fixed angle hadronic scattering $d\sigma/dt(A+B \to C+D)$ as $s \to \infty$ at fixed t/s or θ_{cm}. In general, each hadron is represented by its Fock state decomposition; the leading power law dependence as $s \to \infty$ is obtained from the Fock state with the minimum number of interacting components. The analysis of fixed angle scattering is complicated by pinch singularities, so we must consider two different scattering mechanisms.

A. Hard Subprocesses

In this case the momentum transfer between constituents occurs through a single hard scattering amplitude T_B with all internal legs off-shell and proportional to $p_T^2 = tu/s$. The fixed angle amplitude is then to leading order in $\alpha_s(p_T^2)$ (see Fig. 21),

$$\mathcal{M}_{AB \to CD} = \int \prod_i dx_i \; \phi_C^+(x_c, p_T^2) \; \phi_D^+(x_d, p_T^2) \; T_B(x_i, p_T^2) \; \phi_A(x_a, p_T^2) \; \phi_B(x_b, p_T^2) \qquad (6.1)$$

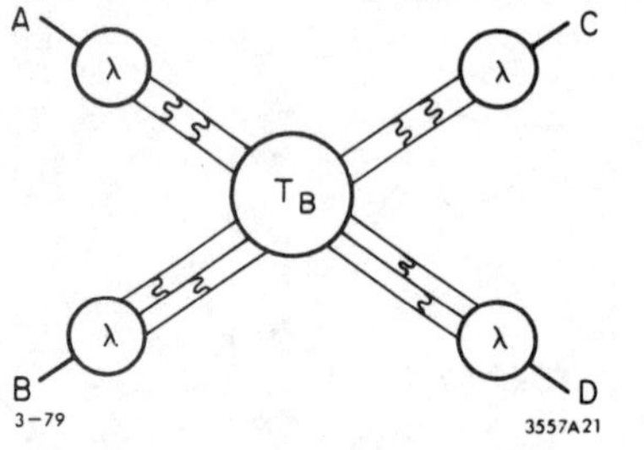

Fig. 21. Fixed angle scattering in QCD for hard subprocesses (see Eq. (6.1)).

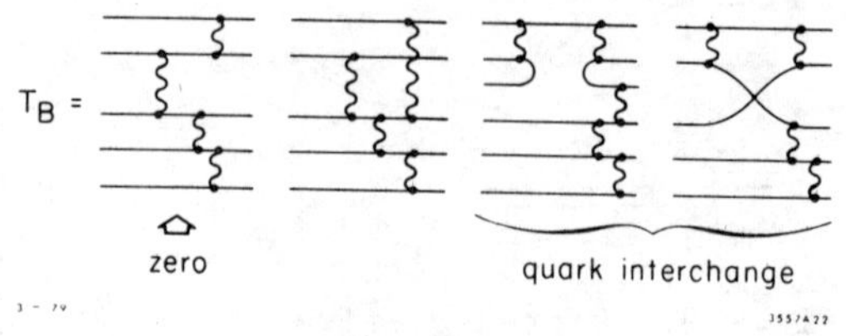

Fig. 22. Examples of hard scattering processes for πp elastic scattering.

The amplitude T_B yields the power-law fall-off $p_T^{-(n-4)}$ where n is the total number of constituents, in agreement with the dimensional counting rule.[1,2] Examples of the leading contributions to lowest order in $\alpha_s(p_T^2)$ for meson-baryon scattering are shown in Fig. 22. Single gluon exchange between color singlet hadrons is of course zero. The constituent interchange graphs,[40] Fig. 22c,d, are among the dominant contributions and lead to large flavor-exchanging amplitudes.

As in the form factor calculation, the evolution of the wavefunctions $\phi(x,\lambda^2)$ to $\phi(x,p_T^2)$ yields a series of terms with anomalous powers of $\alpha_s(p_T^2)$. The asymptotic cross section at $p_T^2 \to \infty$ has the form

$$\frac{d\sigma}{dt}(A+B \to C+D) \Longrightarrow \frac{[\alpha_s(p_T^2)]^{n-2+\sum_I \gamma_I}}{[p_T^2]^{n-2}} f(\theta_{cm}) \tag{6.2}$$

$$= \frac{\alpha_s^2(p_T^2)}{p_T^2} F_A(p_T^2)\, F_B(p_T^2)\, F_C(p_T^2)\, F_D(p_T^2)\, f(\theta_{cm}) \tag{6.3}$$

where γ_I is the leading anomalous dimension and $F_I(t)$ is the asymptotic form factor of hadron I. (The non-leading anomalous dimensions are not expected to be given correctly by Eq. (2.3) because of the different weighting of the x_i integrations.) The anomalous logarithms from each wavefunction is specific to each hadron, and thus gives a factorization theorem Eq. (2.1) for the logarithmic corrections to dimensional counting analogous to the factorization theorems for scale-violations to high p_T inclusive reactions.

B. Soft Subprocesses[16]

As first emphasized by Landshoff,[41] amplitudes with "pinch" singularities, analogous to Glauber scattering amplitudes may provide an alternative and possibly phenomenologically important source of hadrons at large p_T. The classic example is pp scattering, which can proceed via three successive (nearly on-shell) $qq \to qq$ elastic collisions each at $\theta_i \sim \theta_{cm}$ (see Fig. 23). The hadronic amplitude has the form ($\hat{s} \sim 1/9 s$, $\hat{t} \sim 1/9 t$)

$$\mathcal{M}_{pp \to pp} \sim \left[\frac{i}{\sqrt{stu\lambda^2}}\right]^2 \left[\mathcal{M}_{qq \to qq}(\hat{s},\hat{t})\right]^3 \qquad (6.4)$$

which gives nominally for spin 1 exchange

$$\frac{d\sigma}{dt}(pp \to pp) \sim \frac{1}{\lambda^4 t^8} f(s/t) \sim \frac{1}{t^8} \text{ for } s >> t \qquad (6.5)$$

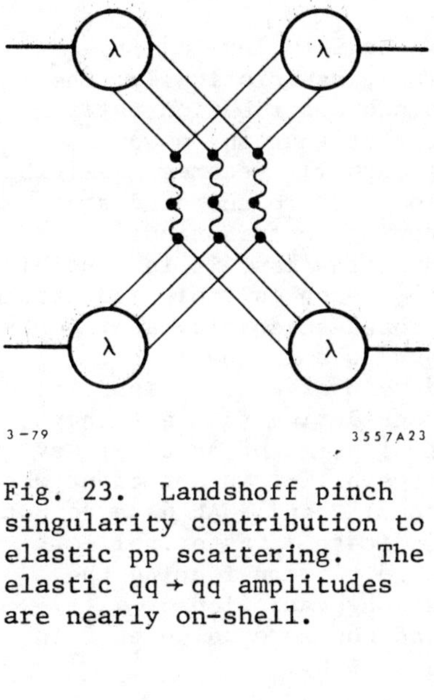

Fig. 23. Landshoff pinch singularity contribution to elastic pp scattering. The elastic $qq \to qq$ amplitudes are nearly on-shell.

The $i/\sqrt{stu\lambda^2}$ factor in Eq. (6.4) represents the probability amplitude for the scattered quarks to overlap with the final state hadrons. As noted by Donnachie and Landshoff,[42] data from the ISR and FNAL at $s > 800$ GeV^2, $4 < |t| < 10$ GeV^2 is compatible with Eq. (6.5). (See Fig. 24.)

The anomalous power law of the pinch singularities arises from exclusive $qq \to qq$ amplitudes where each intermediate state is nearly on-shell. Thus Fock state amplitudes Ψ_λ in the soft domain are required. Since gluon radiation is excluded in this exclusive reaction, $\mathcal{M}_{qq \to qq}$ is suppressed by the Sudakov quark form factor:

$$\mathcal{M}_{qq \to qq} \sim \mathcal{M}_{Born} \cdot \alpha_s(\hat{t})\, F_q^2(\hat{t}) \qquad (6.6)$$

The amplitude $F_q(t)$ can be computed from virtual gluon corrections,[43] or more simply from the exclusive-inclusive connection with the $G_{q/q}(x,q^2)$ distribution $P(\tilde{\xi})(1-x)^{\tilde{\xi}-1}$

$$F_q^2(\hat{t}) \sim [\lambda^2/\hat{t}]^{\tilde{\xi}}\, P(\tilde{\xi}) \qquad (6.7)$$

where

$$\tilde{\xi} = \tilde{\xi}(\hat{t},\lambda^2) = \frac{C_F}{\pi}\int_{\lambda^2}^{\hat{t}} \frac{d\ell^2}{\ell^2}\,\alpha_s(\ell^2) \quad . \qquad (6.8)$$

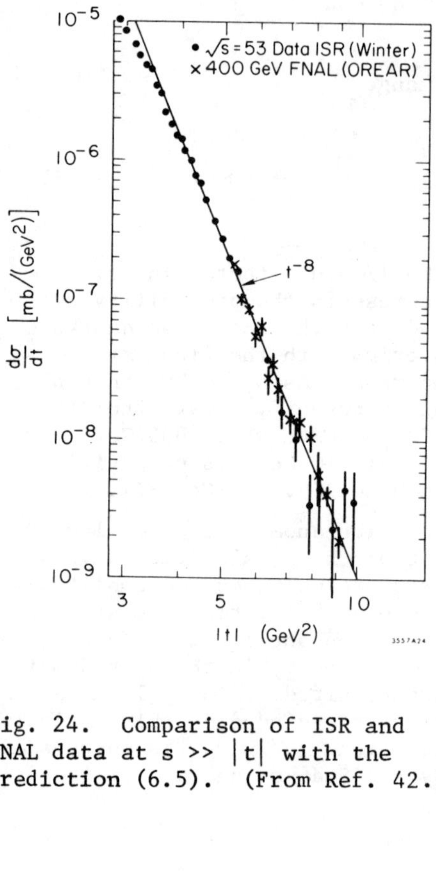

Fig. 24. Comparison of ISR and FNAL data at $s \gg |t|$ with the prediction (6.5). (From Ref. 42.)

Thus the leading pinch singularity contribution to pp-scattering falls as

$$\frac{d\sigma}{dt}(pp \to pp) \sim \frac{\alpha_s^6(\hat{t})}{\hat{t}^{8+6\tilde{\xi}}}\, P^6(\tilde{\xi})\, f(s/t) \qquad (6.9)$$

where $\tilde{\xi} \sim O(\log\log|\hat{t}|)$. Thus, asymptotically, the pinch contribution falls faster than any power and eventually becomes negligible compared to the hard scattering contributions. Nevertheless, it is possible that such multiple scattering processes do play an important role in the $s \gg |t|$ domain of Fig. 24 especially considering the fact that $\hat{t} \sim 1/9t$ is of order ~1 GeV^2. However, it is not clear why the ISR and FNAL data do not indicate a faster fall-off than t^{-8} considering the strong variation of $\alpha_s(t)$ and non-zero value of $\tilde{\xi}$ in Eq. (6.9).

7. The Phenomenology of pp Scattering at Large Angles

The overall features of $d\sigma/dt\,(pp \to pp)$ are sketched schematically in Fig. 25. The data at central angles $30^\circ \lesssim \theta_{cm} \lesssim 150^\circ$ fall rather uniformly as $s^{-10}f(\theta_{cm})$ and merges with a t^{-8} energy-independent "envelope" in the small θ_{cm} $s \gg |t|$ region. This s-independent small θ_{cm} envelope is consistent with multiple gluon exchange mechanisms which produce fixed $J = 1$ Regge behavior. The overall behavior of the <u>central</u> angle data is consistent with the quark

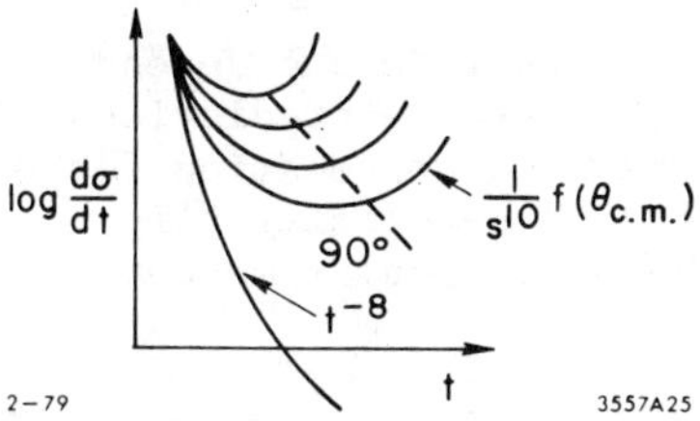

Fig. 25. Qualitative features of elastic pp scattering.

interchange[40] QCD diagrams; the observed shape of $f(\theta_{cm}) \sim (\sin\theta_{cm})^{-10 \text{ to } -14}$ is compatible with this ansatz. Furthermore, the roughly symmetrical shape of $d\sigma/dt$ ($np \to np$) around 90° and the large $pp/p\bar{p}$ ratio at 90° are all consistent with quark exchange or interchange hard scattering mechanisms.[40]

However, these mechanisms fail to account for either:

(a) the slow oscillation of the 90° cross section about the s^{-10} prediction (see Fig. 7), nor

(b) the striking, strongly varying spin correlation recently measured by A. Krisch and collaborators at Argonne.[14] (See Fig. 26.) At $s = 24$ GeV^2, $\theta_{cm} = 90^{\circ}$, it is ~4 times more likely for two protons to scatter with their spins aligned normal and parallel to the scattering plane than anti-parallel. In contrast, the quark interchange amplitude predicts this ratio should be close to 2 independent of angle.[44,45]

The possibility that non-perturbative (instanton) effects could be interfering with the quark interchange amplitude to produce large spin correlations has been investigated by Farrar, Gottlieb, Sivers, and Thomas.[45] More recently Brodsky, Carlson, and Lipkin[44] have considered another possibility: since the triple scattering pinch singularity contribution to pp scattering requires the $qq \to qq$ amplitude at $\hat{s} = 1/9s$ and the relatively low momentum transfer $\hat{t} \sim 1/9t \sim -1.1$ GeV^2, it is not unlikely that the qq scattering is dominated here by simple meson exchange or Reggeon diagrams. In fact a low energy pinch contribution computed with triple "σ" or "π" exchange, interfering with the quark interchange amplitude can

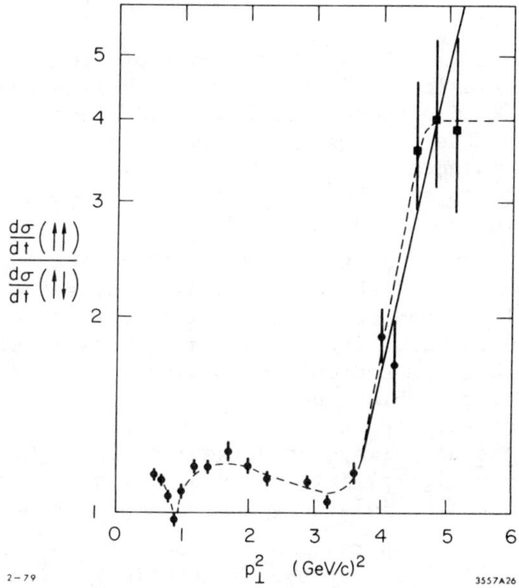

Fig. 26. Spin correlation for protons polarized normal to the scattering plane. (From Ref. 14.)

produce a spin-correlation with the observed magnitude and dependence on θ_{cm}. This model leads to a number of testable predictions including s-dependence, np/pp ratios, and the spin correlation for longitudinally polarized nucleons. A large double-helicity flip amplitude is predicted due to scalar and the pseudoscalar meson exchange.

8. Transverse Momentum Distributions and Inclusive Reactions

It is straightforward to extend the methods discussed here to calculate the $x \to 1$ and large $k_\perp$ behavior of each Fock component of hadronic wavefunctions. In general, one starts with the soft wavefunction $\Psi_\lambda(x,\vec{k}_\perp^2)$ and then uses the evolution equations to obtain the leading power-law and α_s dependence in the far-off-shell domain. In the case of the $q\bar{q}$ component of the (helicity zero) meson wavefunction, the asymptotic fall-off in $\vec{k}_\perp^2$ is given by

$$16\pi^2 \ \psi(\vec{k}_\perp,x) \ \equiv \ \frac{\partial}{\partial \vec{k}_\perp^2} \tilde{\phi}(x,\vec{k}_\perp^2) \tag{8.1}$$

$$\to \ \frac{3}{4} k_\pi \frac{(1-x)^2}{\vec{k}_\perp^2} \frac{C_F}{\beta} \alpha_s(\vec{k}_\perp^2) \cdot \left[\frac{\alpha_s(\vec{k}_\perp^2)}{\alpha_s(\lambda^2)}\right]^{-C_F/\beta}$$

where $k_\pi = f_\pi/\sqrt{n_c}$ in the case of pions. Thus the fall-off in the intrinsic wavefunction is slightly stronger than an inverse power of $\vec{k}_\perp^2$ due to the internal hard gluon interactions, and much slower than the exponential or Gaussian forms usually assumed. The dependence on λ^2 cancels in physical cross sections. This independence in the choice of λ^2 can be used to derive renormalization-group type equations.

The fact that the "tail" of the hadronic wavefunction can be computed at short distances removes the central uncertainty in the calculation of effects due to intrinsic transverse momentum fluctuations for reactions such as massive lepton pair production or high p_T hadronic processes - the hadronic wavefunction is no longer a "black box"! Since the wavefunction enters squared in the inclusive cross sections the spectator transverse momentum integrations are always convergent at large $\vec{k}_\perp^2$. This is in contrast to the $d\vec{k}_\perp^2/\vec{k}_\perp^2 \ \alpha_s(\vec{k}_\perp^2)$ integrations from gluon bremsstrahlung which lead to the scale-violations in the input structure functions.

The general analysis of large p_T inclusive reactions in QCD is complicated, but the asymptotic scaling behavior of the inclusive cross section to leading order in $\alpha_s(p_T^2)$ appears to be a tractable problem. Let us consider a given reaction $A+B \to C+X$ where C is detected at large CM angles with transverse momentum p_T and momentum fraction $x_R = |\vec{p}|/p_{max}$. Each hadron A, B, and C is represented by its Fock state components Ψ_λ.

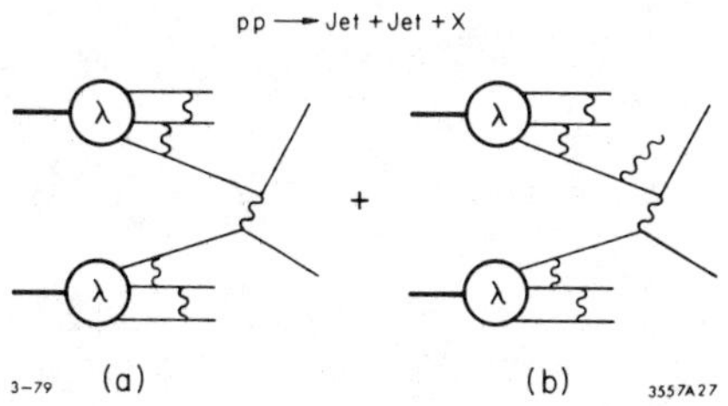

Fig. 27. Example of diagrams which contribute to high p_T pp + jet + X reactions.

We then identify[46] all hard scattering processes T_B (a+b → c+d) which can contribute to the final state high p_T trigger by the scattering of incident constituents (see Fig. 27a). The simplest subprocesses are qq → qq, qg → qg, gg → gg, etc., but for some kinematic regions we should also consider "high twist" subprocesses such as $q\bar{q} \to M\bar{M}$, Mq → Mq, Bq → Bq, etc. As in exclusive scattering, the nominal power-law fall off of T_B in p_T^2 reflects the number (n_{active}) of fundamental fields forced to change directions. By definition every intermediate state associated with T_B and its radiative corrections is off shell by at least λ^2; otherwise its contribution is already included in Ψ_λ.

Furthermore, as $x_R \to 1$, each constituent a,b,c (assuming $c \neq C$) requires the far-off-shell behavior of the Fock state wavefunctions for A,B,C. The leading power law behavior as $(1-x_R) \to 0$ is given by diagrams with the minimal number of gluons exchanged between the Fock state constituents which can transfer all of the hadronic momentum to the hard scattering subprocess. The evolution equations for the Fock state wavefunctions then gives anomalous logarithms $a_i[\alpha_s(k_{x_i}^2)]^{\gamma_i}$ where $k_{x_i}^2 = -(\vec{k}_{\perp i}^2 + \tilde{m}_i^2)/(1-x_i)$, as in the structure function calculations. Combining these factors we obtain at large p_T and $x_R \sim 1$

$$E \frac{d\sigma}{d^3p}(A+B \to C+X) \sim \left[\frac{\alpha_s(p_T^2)}{p_T^2}\right]^{n_{active}-2} \qquad (8.2)$$

$$\cdot\ (1-x_R)^{2n_s-1}\left[\alpha_s(k_{x_R}^2)\right]^{2n_s+\Gamma} f(\theta_{cm})$$

where n_s is the total number of spectators in A,B and C and Γ is the sum of the leading anomalous dimensions for the corresponding Fock states. (The spin complication noted in Section 5 will for simplicity be ignored here.)

Now let us consider the effects of gluon bremsstrahlung for the external lines a,b, and c of T_B, as in Fig. 27b. Taking into account the off-shell kinematics, the integration over gluon transverse momentum up to p_T^2 gives the additional factors

$$\prod_{i=a,b,c} (1-x_i)^{\xi_i(p_T^2,\, k_{x_i}^2)} \; P(\xi_i) \tag{8.3}$$

which can be incorporated into the structure functions $G_{a/A}(x_a\,,\,p_T^2)$, etc. Here

$$\xi_i = \frac{C_i}{\pi} \int_{k_{x_i}^2}^{p_T^2} \frac{d\ell^2}{\ell^2}\,\alpha_s(\ell^2) \tag{8.4}$$

where C_i is the color SU(3) Casimir operator for a,b or c:

$$C_i = \begin{cases} 0 & \text{color singlets} \\ 4/3 & 3 \text{ or } \bar{3} \\ 3 & \text{octet} \end{cases} \tag{8.5}$$

It is a striking fact that the scale-breaking corrections due to gluon bremsstrahlung vanish in the case of color singlets. Thus in the case of higher twist subprocesses such as $(q\bar{q}) + q \to M + q$, the dominant contribution to the $p_T^2 \to \infty$ asymptotic cross section at large x is obtained when the $(q\bar{q})$ system is in a color singlet.[1] In addition, if a,b,c or d is a color singlet composite system, then we also obtain anomalous logarithm factors from the evolution of the hadronic wavefunctions:

$$\prod_{i=a,b,c,d} \left[\sum_n a_n^{(i)} \left(\frac{\log\, p_T^2/\Lambda^2}{\log\, k_x^2/\Lambda^2} \right)^{-\gamma_n^{(i)}} \right] \tag{8.6}$$

In particular, the leading anomalous dimension γ_0 is zero for helicity zero mesons.

Notice that in the exclusive limit, fixed $\mathcal{M}^2 = (1-x_R)p_T^2$, $p_T^2 \to \infty$, we have $\log(k_x^2/\Lambda^2)/\log(p_T^2/\Lambda^2) \to 1$ and $\tilde{\xi} \to 0$; i.e.: because of the off-shell kinematics the anomalous logarithmic corrections to T_B do not appear in the exclusive limit. The asymptotic inclusive cross section in QCD thus takes the factorized form ($\hat{s} = x_a x_b s$, etc.)

$$E\frac{d\sigma}{d^3p}(AB \to CX) = \sum_{ab\to cd} \int_0^1 dx_a\, G_{a/A}(x_a,\xi_a) \int_0^1 dx_b\, G_{b/B}(x_b,\xi_b)$$

$$\int_0^1 \frac{dx_C}{x_C^2}\, G_{C/c}(x_C,\xi_C)\, \frac{d\hat{\sigma}}{d\hat{t}}(ab \to cd)\, \frac{\hat{s}}{\pi}\, \delta(\hat{s}+\hat{t}+\hat{u}) \tag{8.7}$$

where

$$G_{a/A}(x_a,\xi_a) \sim (1-x_a)^{2n_s^a - 1 + \xi_a(p_T^2,k_{x_a}^2)}\ P(\xi_a) \qquad (8.8)$$

is the structure function for finding constituent a in A with light-cone momentum fraction x_a. Equation (8.7) holds for subprocess $ab \to cd$ where a,b,c,d are each either quarks, gluons, or color singlets. The exclusive cross section $d\hat{\sigma}/d\hat{t}$ $(ab \to cd)$ includes the anomalous logarithms from Eq. (8.6) in the case of composite color singlets. [We have not considered the contribution of multiparticle scattering (pinch singularities) to the inclusive cross sections.] The fact that $\xi_a = 0$ for color singlets implies that structure functions obtained from higher Fock state components such as $G_{M/B}(x,\xi)$ are actually scale-invariant. It thus should be possible to obtain the normalization of these contributions from conventional Deck or Drell (meson-exchange) analyses of multiparticle exclusive processes.

Equation (8.7) is consistent with the usual factorization theorems for inclusive reactions, but it includes the effects of higher twist processes, off-shell effects, and predicts the $(1-x_R) \to 0$ behavior of the cross section. Aside from the computable logarithm corrections, the inclusive cross section is well characterized by the power-law formula (8.2). Since $\xi \to 0$ at $p_T^2 \to \infty$ and fixed $(1-x)p_T^2$, there is again a smooth connection with exclusive processes, and the spectator power law[3] $(1-x_R)^{2n_s-1}$ becomes precise in this limit.

Equation (8.7) is evaluated to leading order in $\alpha_s(p_T^2)$ and includes the effects of the k_T fluctuation due to both gluon radiation and the fall-off of the intrinsic wavefunction. Non-leading terms in m^2/p_T^2 and λ^2/p_T^2 are also obtained from the mass connection to T_B and the soft-wavefunction. Note that the Feynman amplitude $qq \to qqg$ contributes to both the $qq \to qq$ and $qg \to qg$ subprocesses in T_B, depending on whether the extra gluon's or quark's transverse momentum is integrated over. The region of phase space (as in Fig. 27b) where all three particles emerge at large p_T is of higher order in $\alpha_s(p_T^2)$. The use of the correct off-shell kinematics prevents anomalous or singular contributions from the integration over a gluon or quark pole.[46]

9. High p_T Phenomenology

In general, the inclusive cross section $Ed\sigma/d^3p$ $(A+B \to C+X)$ is given by a sum of contributions of the form (8.7). The cross section will be dominated at very high p_T by processes involving the minimum number (4) of active constituents, and it will be dominated at $x_R \to 1$, the edge of phase space, by processes involving the minimum number of Fock state spectators. In addition, for specific trigger particles or systems, specific channels or subprocesses may be anomalously suppressed. The most important effect, often referred to as "trigger bias"[47] greatly suppresses the contribution of processes requiring quark or gluon jet fragmentation.

As an example, consider the inclusive single particle trigger process $pp \to pX$. Hard scattering contributions such as $qq \to qq$ yield cross sections with the asymptotic large p_T, $x_R \to 1$ behavior

$$E \frac{d\sigma}{d^3p} (pp \to pX) \sim \frac{\alpha_s^2(p_T^2)}{p_T^4} (1-x_R)^{11+3\tilde{\xi}} \left[\alpha_s(k_{x_R}^2)\right]^{12+6\gamma_N} \quad (9.1)$$

Although this contribution should eventually dominate at very high p_T, it is suppressed in normalization by several orders of magnitude[47] because of the fact that only a finite fraction of the outgoing quark's momentum is transferred to the proton; the actual hard scattering subprocess thus occurs at a higher value of p_T where the cross section is much smaller. It is this effect that leads to the prediction of sizeable jet/single ratios in simple QCD subprocesses[48] measured at the same P_T.

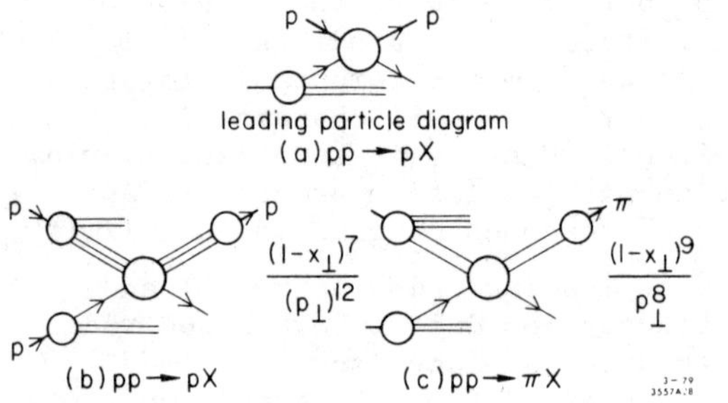

Fig. 28. High twist contributions to high p_T $pp \to pX$ and $pp \to \pi X$ inclusive reactions.

Alternatively we can consider processes where the trigger hadron emerges directly from the hard scattering reaction.[49,50] These are higher "twist" subprocesses in that more than the minimum number of active constituents are involved.

As we have discussed in Section 8, the leading contributions at large x occur when the composite systems are color singlets. The simplest such subprocess is the "leading particle" diagram of Fig. 28a where the hard scattering reaction is $pq \to pq$, i.e.: elastic proton quark scattering. The nominal order is

$$E \frac{d\sigma}{d^3p} (pp \to pX) \sim \frac{[\alpha_s(p_T^2)]^{6+2\gamma_N}}{p_T^{12}} (1-x)^{3+\tilde{\xi}} [\alpha_s(k_x^2)]^{4+2\gamma_N} \quad (9.2)$$

Such a contribution is expected to dominate near the exclusive boundary, but is strongly peaked toward forward angles because all of the beam energy is utilized. If we consider a high Fock-state component of the incident nucleon, e.g., $|qqqq\bar{q}\rangle$, then this problem is avoided and one obtains contributions with the form

$$E \frac{d\sigma}{d^3p}(pp \to pX) \sim \frac{[\alpha_s(p_T^2)]^{6+2\gamma_N}}{p_T^{12}} (1-x_R)^{7+\tilde{\xi}} [\alpha_s(k_x^2)]^{8+\Gamma} \quad (9.3)$$

from the $B+q \to p+q$ subprocess (see Fig. 28b). Similarly the $M+q \to \pi q$ subprocess (see Fig. 28c) leads to a contribution

$$E \frac{d\sigma}{d^3p}(pp \to \pi X) \sim \frac{\cdot\,\alpha_s^4(p_T^2)}{p_T^8} (1-x_R)^{9+\tilde{\xi}} [\alpha_s(k_x^2)]^{10+\Gamma} \quad (9.4)$$

Since these are the leading QCD contributions which produce hadrons at $\theta_{cm} = 90^o$ directly without the trigger bias suppression, these terms can dominate the inclusive cross section until very large p_T where the nominal p_T^{-4} terms take over.[51] Measurements at FNAL and ISR for $pp \to pX$ and $pp \to \pi X$ are in fact consistent with the predicted powers, Eqs. (9.3) and (9.4), respectively for $3 < p_T < 8$ GeV.[52] There are also indications from ISR measurements[53] of $pp \to \pi^o X$, that the fixed x_R power law fall-off changes from p_T^{-8} to $\sim p_T^{-5}$ for $8 \lesssim p_T \lesssim 12$ GeV.

Independent of these phenomenological questions, the important point is that higher twist subprocesses, including the scale-breaking corrections from wave-functions and bremsstrahlung processes (as in Fig. 29) can be systematically computed in QCD. It should also be possible to calculate the normalization of these processes directly from form factor normalizations, and relative normalization of different Fock state components. For example the important $(q\bar{q})+q \to \pi+q$ subprocesses for $pp \to \pi X$ requires the normalization of the $|qqqq\bar{q}\rangle$ Fock state component of the nucleon, which in turn can be computed from the intrinsic sea-quark distribution in the nucleon.

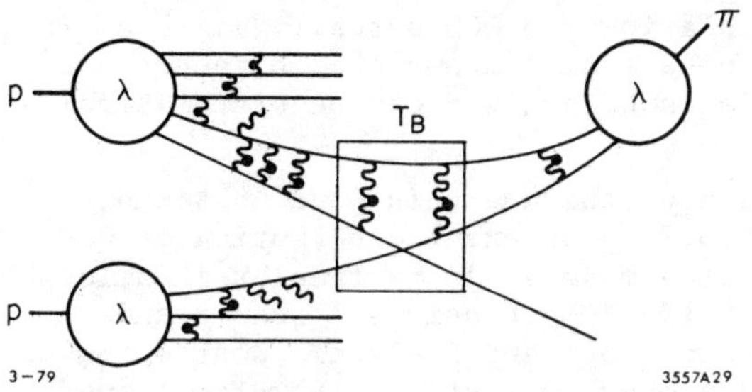

Fig. 29. Example of diagrams which contribute to scale-breaking and anomalous logarithm corrections to the $qM \to qM$ higher twist subprocess contribution to $pp \to \pi X$.

10. Summary

As we have discussed in this chapter, the testing ground of quantum chromodynamics can be extended to exclusive processes at large momentum transfer. The essential features which are required in the calculation of the cross sections are (a) the unambiguous separation of hard (far-off-shell) and soft regimes of each hadronic Fock component and (b) the derivation of evolution equations which determine the wavefunctions at short distances. The eigenvalues of the evolution equations yield the anomalous scale-breaking logarithms which are specific to each hadron and systematically correct the leading power behavior of large momentum transfer amplitudes. The dimensional counting rules, modulo calculable logarithmic corrections, thus emerge as predictions of perturbative QCD.

In particular we have obtained

(a) exact results for hadronic form factors at large Q^2 and hadronic structure functions for x near 1,
(b) the asymptotic behavior (including factorization theorems) for fixed angle amplitudes,
(c) the asymptotic suppression of Landshoff (pinch-singularity) contributions for exclusive processes,
(d) a smooth exclusive-inclusive connection in QCD.

The last result depends critically on recognizing the off-shell nature of subprocesses within hadrons, and the fact that the parameter ξ which controls scale-violations in QCD actually vanishes in the fixed $\mathcal{M}^2$, $Q^2 \to \infty$ limit. We have also shown that higher twist and constituent interchange model subprocesses can be systematically computed in QCD.

The Fock space decomposition of the hadronic wavefunctions, together with the axial gauge provides an exact description of QCD which is the analogue of the parton model. In particular it allows us to make a clean distinction between intrinsic sea-quarks and those evolved from the Q^2 dependence of deep inelastic scattering. In general, the lowest particle number Fock states $|B\rangle = |qqq\rangle$ and $M = |q\bar{q}\rangle$ dominate the power law behavior of large momentum transfer exclusive reactions and inclusive reactions at $x \to 1$. The existence of Fock states at $P \to \infty$ with a fixed number of constituents is a consequence of the fact that a hadron is a color singlet.

It is worth emphasizing that QCD predicts specific integral powers for the asymptotic form factors of hadrons:

$$F_\pi(t) \underset{t\to-\infty}{\Longrightarrow} \frac{\alpha_s(t)}{t^1} \quad , \quad F_N^{(1)}(t) \underset{t\to-\infty}{\Longrightarrow} \frac{[\alpha_s(t)]^{2+\gamma_N}}{t^2}$$

Similarly, the asymptotic form of the nucleon structure function is predicted

$$F_{2N}(x,Q^2) \Longrightarrow (1-x)^{3+\tilde{\xi}(Q^2,k_x^2)}$$

The integral powers in these expressions directly test both the scale-invariance of the internal interactions, and the SU(3) color prediction that the minimal Fock states which are color singlets are $|q\bar{q}\rangle$ and $|qqq\rangle$. In this sense, exclusive processes at large momentum transfer and the x near 1 dependence of structure functions provide the most direct and critical prediction of QCD dynamics and color SU(3) symmetry.

At this point, the study of exclusive reactions in QCD is just beginning, but there seems to be real hope that a microscopic description of a large range of hadronic physics will emerge. The anomalous logarithms which emerge from the evolution equation can play a dominant role in phenomenology and may be experimentally accessible from detailed comparisons of different reactions, e.g., $F_\pi(t)/F_K(t)$, meson photoproduction, Compton scattering, etc. The anomalous dimensions are fundamental parameters of each hadron and reflect the underlying symmetry properties of its wavefunction. In some cases the absolute normalization of amplitudes such as the meson form factors, $\gamma(q^2) + \gamma(k^2) \to \pi^0$ can be computed in the asymptotic limit. It also should be possible to compute the normalization of large angle scattering processes and higher twist (CIM) subprocesses in terms of the normalization of the meson and baryon form factors as well as their angular distributions. We are also analyzing the spin structure of baryon wavefunctions in the short distance limit. In principle, it should be possible to make a direct connection between the soft Fock space wavefunctions Ψ_λ and the wavefunctions used in hadronic spectroscopy, bag models, etc. The higher Fock state components also evidentally play an important role in the sea quark distribution, low momentum hadron exchange reactions, and inclusive reactions in the fast $(x_L \to 1)$ forward regime.[3]

Acknowledgements

We wish to thank our collaborators, Y. Frishman and C. Sachrajda for many discussions. We also thank R. Blankenbecler, G. Farrar, J. Gunion, R. Horgan, K. Wilson, and T. M. Yan for helpful conversations.

References

1. S. J. Brodsky and G. R. Farrar, Phys. Rev. Lett. 31, 1153 (1973); Phys. Rev. D11, 1309 (1975).

2. V. A. Matveev, R. M. Muradyan and A. V. Tavkheldize, Lett. Nuovo Cimento 7, 719 (1973).

3. The spectator rule holds for general structure functions $G_{a/A}(x)$ including the case where a is composite. See S. J. Brodsky and R. Blankenbecler, Phys. Rev. D10, 2973 (1974).

4. S. J. Brodsky, in Proceedings of the International Conference on Few Body Problems, Laval University, Quebec (1974), edited by R. Slobodrian et al.

5. S. J. Brodsky and B. Chertok, Phys. Rev. D14, 3003 (1976); Phys. Rev. Lett. 37, 269 (1976).

6. M. D. Mestayer, SLAC-Report No. 214 (1978).

7. R. Arnold et al., Phys. Rev. Lett. 35, 776 (1975).

8. R. Anderson et al., Phys. Rev. Lett. 30, 627 (1973).

9. M. A. Shupe et al., Tufts, Harvard, Cornell collaboration (to be published).

10. K. A. Jenkins et al., Phys. Rev. Lett. 40, 425 (1978).

11. P. V. Landshoff and J. C. Polkinghorne, Phys. Lett. 44B, 293 (1973).

12. J. L. Stone et al., Phys. Rev. Lett. 38, 1315 (1978); Nucl. Phys. B143, 1 (1978).

13. A. Hendry, Phys. Rev. D10, 2300 (1974).

14. D. G. Crabb et al., Phys. Rev. Lett. 41, 1257 (1978).

15. G. P. Lepage and S. J. Brodsky (in preparation). A preliminary report of this work is given in S. J. Brodsky, SLAC-PUB-2240 (December 1978), to be published in the Proceedings of the J. H. Weis Memorial Symposium on Strong Interactions (November 1978).

16. S. J. Brodsky, G. P. Lepage, Y. Frishman and C. Sachrajda (in preparation).

17. D. R. Jackson, Ph.D. Thesis, Cal Tech (1977). G. R. Farrar and D. R. Jackson (to be published).

18. A. V. Efremov and A. V. Radyushkin, paper submitted to the XIX International Conference on High Energy Physics (Tokyo, 1978) and to the International Seminar on High Energy Physics and Field Theory (Serpurkhov, July 1978). Note added: After our work was completed we received Dubna preprint E2-11983 in which Eq. (2.1) is also derived by these authors.

19. A. M. Polyakov, reporteur's talk, Lepton-Photon Symposium, SLAC (1975).

20. P. Menotti, Phys. Rev. D14, 3560 (1976); Phys. Rev. D11, 2828 (1975).

21. T. Appelquist and E. Poggio, Phys. Rev. D10, 3280 (1970).

22. M. L. Goldberger, A. H. Guth, D. E. Soper, Phys. Rev. D14, 1117 (1976).

23. M. Einhorn, Phys. Rev. D14, 3451 (1976); R. C. Brower, J. Ellis, M. G. Schmidt and J. H. Weis, Nucl. Phys. B128, 131 (1977); Phys. Lett. 65B, 249 (1976).

24. See S. Weinberg, Phys. Rev. 150, 1313 (1966); L. Susskind and G. Frye, Phys. Rev. 165, 1535 (1968); J. B. Kogut and D. E. Soper, Phys. Rev. D1, 2901 (1970); J. D. Bjorken, J. B. Kogut and D. E. Soper, Phys. Rev. D3, 1382 (1971). Renormalization theory is implemented in this framework by S. J. Brodsky, R. Roskies and R. Suaya, Phys. Rev. D8, 4575 (1978); and W. Caswell and G. P. Lepage, Phys. Rev. A18, 810 (1978) for the case of bound states.

25. We wish to thank R. P. Feynman for an illuminating discussion on this point.

26. W. Caswell and G. P. Lepage, Ref. 24.

27. S. J. Brodsky, R. Roskies and R. Suaya, Ref. 24.

28. See, e.g., G. Alterelli and G. Parisi, Nucl. Phys. B126, 298 (1977). For an excellent review see C. H. Llewellyn Smith, Acta. Physica Austriaca Suppl. XIX, 331-397 (1978).

29. Yu. L. Dokshitser, D. I. D'Yakanov and S. I. Troyan, SLAC-TRANS-183 (translation for Proceedings of the 13th Leningrad Winter School on Elementary Particle Physics (1978)).

30. See, e.g., D. J. Gross and S. B. Trieman, Phys. Rev. Lett. 32, 1145 (1974); and R. Coquereaux *et al.*, Ref. 43.

31. H. D. Politzer, Phys. Rev. Lett. 30, 1346 (1976); and Phys. Reports 14, 129 (1974).

32. D. J. Gross and F. Wilczek, Phys. Rev. Lett. 30, 1323 (1973); Phys. Rev. D8, 3633 (1973); and Phys. Rev. D9, 980 (1974).

33. S. J. Brodsky and J. F. Gunion, Phys. Rev. D19, 1005 (1979).

34. G. R. Farrar, Nucl. Phys. B77, 429 (1974); J. F. Gunion, Phys. Rev. D15, 3317 (1977), and D10, 242 (1974).

35. Z. F. Ezawa, Nuovo Cimento 23A, 271 (1974).

36. G. R. Farrar and D. R. Jackson, Phys. Rev. Lett. 35, 1416 (1975).

37. A. I. Vainshtain and V. I. Zakharov, Phys. Rev. Lett. 72B, 368 (1978).

38. E. L. Berger and S. J. Brodsky, SLAC-PUB-2247 (1979), to be published in Phys. Rev. Lett.

39. K. J. Anderson *et al*., Chicago-Princeton Report EFI-78-38.

40. R. Blankenbecler, S. J. Brodsky and J. F. Gunion, Phys. Rev. D8, 4117 (1973); Phys. Lett. 39B, 649 (1972).

41. P. V. Landshoff, Phys. Rev. D10, 1024 (1974). See also P. Cvitanovic, Phys. Rev. D10, 338 (1974); and S. J. Brodsky and G. Farrar, Ref. 1.

42. A. Donnachie and P. V. Landshoff, Preprint M/C 79/11 (January 1979).

43. This work on the asymptotic suppression of pinch singularities in QCD was done in collaboration with Y. Frishman.[16] The quark form factor in QCD is discussed by J. Cornwall and G. Tiktopolus, Phys. Rev. D13, 3370 (1976) and D15, 2937 (1977); R. Coquereaux and E. DeRafael, Phys. Lett. 74B, 135 (1978); and E. C. Poggio and H. R. Quinn, Phys. Rev. D12, 3279 (1975).

44. S. J. Brodsky, C. Carlson and H. Lipkin, SLAC-PUB (in preparation).

45. G. R. Farrar, S. Gottlieb, D. Sivers and G. H. Thomas, ANL preprint HEP-PR-78-43 (1978).

46. See S. J. Brodsky, R. R. Horgan and W. E. Caswell, Phys. Rev. D18, 2415 (1978).

47. S. D. Ellis, P. V. Landshoff and M. Jacob, Nucl. Phys. B108, 93 (1978); and P. V. Landshoff and M. Jacob, Nucl. Phys. B113, 395 (1976).

48. See, e.g., R. Field, R. P. Feynman and G. C. Fox, Phys. Rev. D18, 3320 (1978) and references therein.

49. R. Blankenbecler, S. J. Brodsky and J. F. Gunion, Phys. Rev. D18, 900 (1978) and references therein.

50. P. V. Landshoff and J. C. Polkinghorne, Phys. Rev. D10, 891 (1974) and references therein; M. K. Chase and W. J. Stirling, Nucl. Phys. B133, 157 (1978).

51. For a comprehensive discussion see D. Jones and J. F. Gunion, Phys. Rev. D19, 1032 (1979).

52. D. Antreasyan *et al.*, Phys. Rev. D19, 764 (1979) and references therein.

53. For a recent review see M. Jacob and P. V. Landshoff, Phys. Report 48, 285 (1978). For $7.5 < p_T < 14.0$ GeV/c, $\sqrt{s} = 53.1$ and 62.4 GeV, a good fit to the ISR-CCOR data is obtained with $Ed\sigma/d^3p\ (pp \to \pi^0 X) = p_T^{-5.1 \pm 0.4} f(x_T)$; CCOR collaboration, Phys. Lett. 79B, 505 (1978).

III. TWO-PHOTON COLLISIONS AND SHORT-DISTANCE TESTS OF QUANTUM CHROMODYNAMICS

In this chapter I will review the physics of two-photon collisions in $e^{\pm}$ storage rings with emphasis on the predictions of perturbative quantum chromodynamics for high transverse momentum reactions. Because of the remarkable scaling properties predicted by the theory, two-photon collisions may provide one of the cleanest tests of the QCD picture of short distance hadron dynamics. The contrasts between photon-induced and hadron-induced reactions at high transverse momentum are remarkable and illuminating. Most of the work reported here was done in collaboration with T. DeGrand, J. Gunion, and J. H. Weis.[1] After this short survey of two photon collisions we will go on to the more complicated physics of hadron-hadron collisions.

The photon plays a unique role in strong interaction dynamics because of its elementarity and its direct interactions with the hadronic constituents. Although it is well-known that highly virtual photons have asymptotically scale-free interactions with the quark current in QCD, it is perhaps not sufficiently emphasized that the interactions of real on-shell photons also become dominantly point-like in large momentum transfer (short-distance) processes. The predictions by Bjorken and Paschos[2] for deep inelastic Compton scattering, and dimensional counting predictions[3] for exclusive and inclusive processes involving real photons are all based on the existence of direct $\gamma q\bar{q}$ perturbative couplings, and imply the breakdown of the vector meson dominance description[4] of the photon's hadronic interactions at short-distances and large momentum transfer. As a general rule, VMD can only be valid in QCD for low momentum transfer, nearly on-mass-shell processes where perturbation theory in α_s is invalid. Whenever a photon couples to far-off shell quarks (as in $\gamma q \to \gamma q$) the net real and virtual gluon radiative corrections are of order $\alpha_s(p_T^2) \sim O(\log^{-1}(p_T^2/\Lambda^2))$, and the pointlike Born amplitude are expected to dominate in the asymptotic limit.

The production of hadrons in the collisions of two photons should provide an ideal laboratory for testing many features of the photon's hadronic interactions, including its short distance aspects. It is well known that photon-photon inelastic collisions in e^+e^- storage rings become an increasingly important source of hadrons as the center-of-mass energy $\sqrt{s} = 2E_e$ is raised.[5] The dominant part of the cross section for $e^+e^- \to e^+e^-$ + hadrons arises from the annihilation of two nearly on-shell photons emitted at small angles to the beam (see Fig. 1). The resulting cross section increases logarithmically with energy ($m_e^2/s \to 0$,

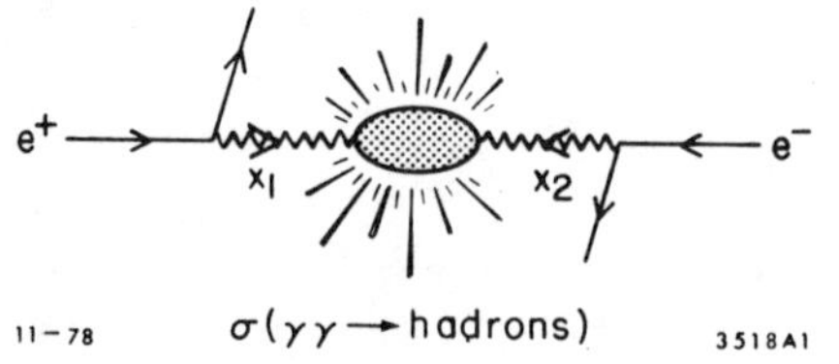

Fig. 1. Two-photon annihilation into hadrons in e^+e^- collisions.

$s \gg m_H^2$):

$$\frac{d\sigma_{e^+e^-\to e^+e^-X}(s)}{dm_H^2} \cong \frac{\alpha^2}{\pi^2} \log^2\left(\frac{s}{m_e^2}\right) \frac{\sigma_{\gamma\gamma}(m_H^2)}{m_H^2} \log\left(\frac{s}{m_H^2}\right) \quad (1)$$

where m_H is the invariant mass of the produced hadronic system. In contrast, the e^+e^- annihilation cross section decreases quadratically with energy. For example, at the beam energy of $E_e = 15$ GeV, the standard vector dominance estimate for $\sigma_{\gamma\gamma}(m_H^2)$ gives $\sigma(e^+e^- \to e^+e^-$ hadrons) $\cong 15$ nb for $m_H \geq 1$ GeV, compared to the annihilation cross section $\sigma_{e^+e^-\to\gamma\to \text{hadrons}} \equiv R\sigma_{e^+e^-\to\gamma\to\mu^+\mu^-} \cong (0.1\ \text{nb})R$.

The event rate can be large because of (1) the relatively large efficiency for an electron to emit a photon: ($x \equiv (k_0 + k_3)/(p_0 + p_3) \cong \omega/E$)

$$xG_{\gamma/e}(x) = x\frac{dN_{\gamma/e}}{dx} \cong \left(\frac{\alpha}{2\pi}\log\frac{s}{m_e^2}\right)\left(1 + (1-x)^2\right)$$

$$\cong .051 \quad (\sqrt{s} = 30\ \text{GeV},\ x \to 0) \quad (2)$$

(2) the factor of $\log s/m_H^2$ from the integration over the nearly flat rapidity distribution of the produced hadronic system, and (3) the fact that the cross section is dominated by low-mass hadronic states. For untagged leptons, the cross section for $ee \to eeX$ in the equivalent photon spectrum takes the general form[6]

$$d\sigma_{e^+e^-\to e^+e^-X}(s,t,u) = \int_0^1 dx_1 \int_0^1 dx_2\, G_{\gamma/e}(x_1)\, G_{\gamma/e}(x_2)$$

$$\times\, d\sigma_{\gamma\gamma\to X}(\hat{s} = x_1x_2s,\ \hat{t} = x_1t,\ \hat{u} = x_2u) \quad (3)$$

where $G_{\gamma/e}(x)$ is the equivalent photon energy spectrum and $d\sigma_{\gamma\gamma\to X}$ is the differential cross section for the scattering of two oppositely directed unpolarized photons (of energy $x_1\sqrt{s}/2$, $x_2\sqrt{s}/2$ in the e^+e^- c.m. system) into a final state X. If the scattered lepton kinematics are measured, then the photon momenta are determined and the full range of hadronic $\gamma\gamma$ physics analogous to pp colliding ring physics becomes accessible.

A large-scale experimental investigation of two-photon physics is now planned at PEP and PETRA. Among the areas of interest are[7]

(a) the production of heavy leptons[8] ($\gamma\gamma \to \tau^+\tau^-$, etc.).
(b) The production of even charge conjugation states and hadronic resonances ($\gamma\gamma \to \eta_c$, etc.).
(c) The measurement of the total $\sigma_{\gamma\gamma}(s)$ cross section, including heavy quark thresholds.
(d) Measurements of the $\pi-\pi$ and K^+-K^- phase shifts via $\gamma\gamma \to M\bar{M}$ and unitarity, as well as checks of dimensional-coupling scaling laws for the crossed Compton amplitude at large s and t.
(e) Deep inelastic scattering on a photon target,[9] via electrons or positrons tagged at large momentum transfer $e\gamma \to e'X$.

In each case the spacelike mass of each photon can be individually tuned by tagging the scattered $e^{\pm}$. The photon linear polarization is determined by the lepton scattering plane. We also note that the $e^{\pm}$ circular polarization of an incident lepton is transferred to the emitted photon with 100% efficiency as $x_\gamma \to 1$.

A. Large p_T Two Photon Reactions

Perhaps the most interesting application of two photon physics is the production of hadrons and hadronic jets at large p_T. The elementary reaction $\gamma\gamma \to q\bar{q} \to$ hadrons yields an asymptotically scale-invariant two-jet cross section at large p_T proportional to the fourth power of the quark charge. The $\gamma\gamma \to q\bar{q}$ subprocess[10] implies the production of two non-colinear, roughly coplanar high p_T (SPEAR-like) jets, with a cross section nearly flat in rapidity. Such "short jets" will be readily distinguishable from $e^+e^- \to q\bar{q}$ events due to missing visible energy, even without tagging the forward leptons. It is most useful to determine the ratio,

$$R_{\gamma\gamma} \equiv \frac{d\sigma(e^+e^- \to e^+e^-q\bar{q} \to e^+e^- + \text{jets})}{d\sigma(e^+e^- \to e^+e^-\mu^+\mu^-)} \tag{4}$$

since experimental uncertainties due to tagging efficiency and the equivalent photon approximation tend to cancel. In QCD, with 3-colors, one predicts[1]

$$R_{\gamma\gamma} = 3 \sum_{q=u,d,s,c,\ldots} e_q^4 \left(1 + O\left[\frac{\alpha_s(4p_T^2)}{\pi}\right]\right) \tag{5}$$

where p_T is the total transverse momentum of the jet (or muon) and $\alpha_s(Q^2) \to 4\pi/(\beta \log Q^2/\Lambda^2)$, $\beta = 11-2/3\, n_f$ for n_f flavors. Measurements of the two-jet cross section and $R_{\gamma\gamma}$ will directly test the scaling of the quark propagator $\not{p}^{-1}$ at large momentum transfer, check the color factor[11] and the quark fractional charge. The QCD radiative corrections are expected to depend on the jet production angle and acceptance. Such corrections are of order $\alpha_s(p_T^2)$ since there are neither infrared singularities in the inclusive cross section, nor quark mass singularities at large p_T to give compensating logarithmic factors. The onset of charm and other quark thresholds can be studied once again from the perspective of $\gamma\gamma$-induced processes. The cross section for the production of jets with total hadronic transverse momentum ($p_T > p_{Tmin}$) from the $\gamma\gamma \to q\bar{q}$ subprocess alone can be estimated from the convenient formula,[1,12]

$$\sigma_{e^+e^- \to e^+e^- \mathrm{Jet} + X}(s, p_T^{jet} > p_T^{min}) \equiv R_{\gamma\gamma}\,\sigma_{e^+e^- \to e^+e^-\mu^\pm\mu^\mp}(s, p_T^{\mu\pm} > p_T^{min})$$

$$\cong R_{\gamma\gamma}\,\frac{32\pi\alpha^2}{3}\left(\frac{\alpha}{2\pi}\log\frac{s}{m_e^2}\right)^2 \frac{\left(\log\frac{s}{p_{Tmin}^2} - \frac{19}{6}\right)}{p_{Tmin}^2}$$

$$\cong \frac{0.5\ \mathrm{nb\ GeV}^2}{p_{Tmin}^2} \quad \text{at} \quad \sqrt{s} = 30\ \mathrm{GeV} \quad . \tag{6}$$

where we have taken $R_{\gamma\gamma} = 3\sum_q e_q^4 = 34/27$ above the charm threshold. For $p_{Tmin} = 4$ GeV, $\sqrt{s} = 30$ GeV, this is equivalent to 0.3 of unit of R; i.e., 0.3 times the $e^+e^- \to \mu^+\mu^-$ rate. We note that at $\sqrt{s} = 200$ GeV, the cross section from the $e^+e^- \to e^+e^-q\bar{q}$ subprocess with $p_{Tmin} = 10$ GeV is 0.02 nb, i.e., about 9 units of R! At such energies e^+e^- colliding beam machines are more nearly laboratories for $\gamma\gamma$ scattering then they are for e^+e^- annihilation! A useful graph[12] of the increase in R from the $\gamma\gamma \to q\bar{q}$ process for various $x_{Tmin} = 2p_{Tmin}/\sqrt{s}$ is shown in Fig. 2. The $\log s/p_{Tmin}^2 - 19/6$ in Eq. (6) arises from integration over the nearly flat rapidity distribution of the $\gamma\gamma$ system. The final state in high p_T $\gamma\gamma \to q\bar{q}$ events in the $\gamma\gamma$ center of mass should be very similar in

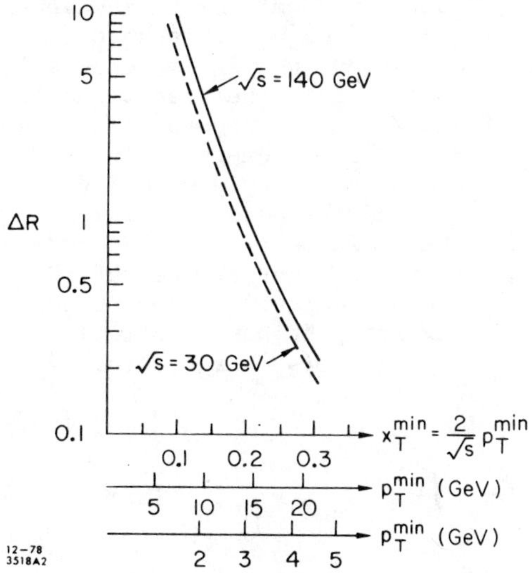

Fig. 2. The contribution to R from $\gamma\gamma \to q\bar{q}$ two jet processes at $\sqrt{s} = 30$ and 140 GeV (from Ref. 12).

multiplicity and other hadronic properties as $e^+e^- \to \gamma^* \to q\bar{q}$, although $u\bar{u}$ and $c\bar{c}$ events should be enhanced relative to $d\bar{d}$ and $s\bar{s}$ due to the e_q^4 dependence. Monte Carlo studies of SPEAR events at $s = 4p_T^2$ distributed uniformly in rapidity would be useful in order to learn how to identify and trigger $\gamma\gamma \to q\bar{q}$ events.

Although the above prediction for $R_{\gamma\gamma}$ is one of the most straightforward consequences of perturbative QCD, it should be noted that from the perspective of photon physics of 10 years ago, the occurence of events with the structure $\gamma\gamma \to \text{jet} + \text{jet}$ at high p_T could only be regarded as revolutionary. From the VMD standpoint, a real photon acts essentially as a sum of vector mesons; however, it is difficult to imagine an inelastic collision of two hadrons producing two large p_T jets without energy remaining in the beam direction!

On the other hand, if the $\gamma\gamma \to$ two jet events are not seen at close to the predicted magnitude with an approximately scale invariant cross section, then it would be hard to understand how the perturbative structure of QCD could be applicable to hadronic physics. In particular, unless the pointlike couplings of real photons to quarks are confirmed, then the analogous predictions for perturbative high p_T processes, involving gluons such as $gg \to q\bar{q}$ are probably meaningless.

B. Multi-Jet Processes and the Photon Structure Function

In addition to the two-jet processes, QCD also predicts 3- and 4-jet events from subprocesses such as $\gamma q \to gq$ (3-jet production where one photon interacts with the quark constituent of the other photon) as well as the conventional high p_T QCD subprocesses $qq \to qq$ and $q\bar{q} \to gg$ (which lead to jets down the beam direction plus jets at large p_T) (see Fig. 3). The structure of these events are very similar to that for hadron-hadron collisions. The cross section for $E d\sigma/d^3p_J$ ($\gamma\gamma \to \text{jet} + X$ or $ee \to ee\ \text{jet} + X$) can be computed in the standard way from the hard scattering expansion ($\hat{s} = x_a x_b s$, etc.)[13]

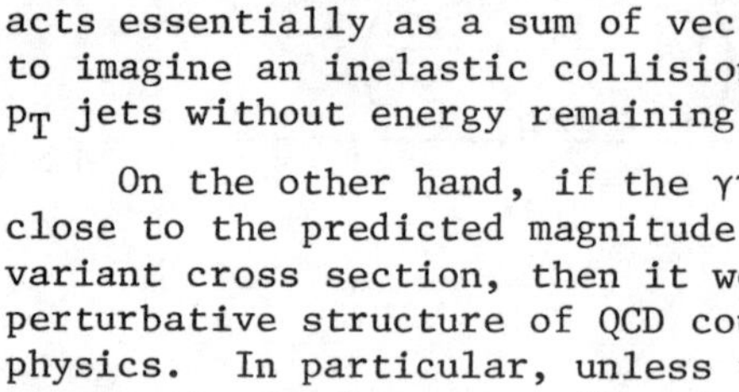

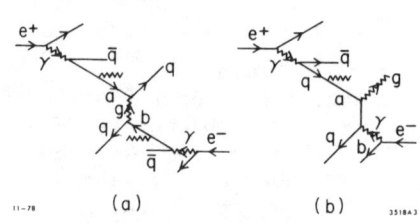

Fig. 3. Contributions from QCD subprocesses to (a) 4-jet and (b) 3-jet final states.

$$E \frac{d\sigma}{d^3p}(AB \to CX) = \sum_{abd} \int_0^1 dx_a \int_0^1 dx_b\, G_{a/A}(x_a)\, G_{b/B}(x_b)$$

$$\frac{d\sigma}{d\hat{t}}(ab \to cd)\bigg|_{s,t,u} \frac{\hat{s}}{\pi}\, \delta(\hat{s} + \hat{t} + \hat{u}) \qquad (7)$$

where the hard scattering occurs in $ab \to cd$ and the fragmentation function $G_{a/A}(x_a)$ gives the probability of finding constituent a with light-cone fraction $x_a = (p_a^0 + p_a^3)/(p_A^0 + p_A^3)$. In general, $G_{a/A}$ has a scale-breaking dependence on $\log p_T^2$ which arises from the constituent transverse momentum integration when gluon bremsstrahlung or pair production is involved.[14]

However, there is an extraordinary difference between photon and hadron induced processes. In the case of proton-induced reactions, $G_{q/p}(x,Q^2)$ is determined from experiment, especially deep inelastic lepton scattering. In the case of the photon, the $G_{q/\gamma}$ structure function required in Eq. (7) has a perturbative component which can be predicted from first principles in QCD. This component, as first computed by Witten,[15] has the asymptotic form at large probe momentum Q^2

$$G_{q/\gamma}(x,Q^2) \Longrightarrow \frac{\alpha}{\alpha_s(Q^2)} f(x) + O(\alpha^2) \tag{8}$$

i.e.: aside from an overall logarithmic factor, the $\gamma \to q$ distribution Bjorken scales; $f(x)$ is a known, calculable function. Unlike the proton structure function which contracts to $x = 0$ at infinite probe momentum $Q^2 \to \infty$, this component of the photon structure function <u>increases</u> as $\log Q^2$ independent of x. This striking fact is of course due to the direct $\gamma \to q\bar{q}$ perturbative component in the photon wavefunction. (The apparent violation of probability conservation when $\alpha_s(Q^2) < \alpha$ should be cured when higher order terms in α are taken into account.) In addition to the perturbative component, one also expects a nominal hadronic component due to intermediate vector meson states.

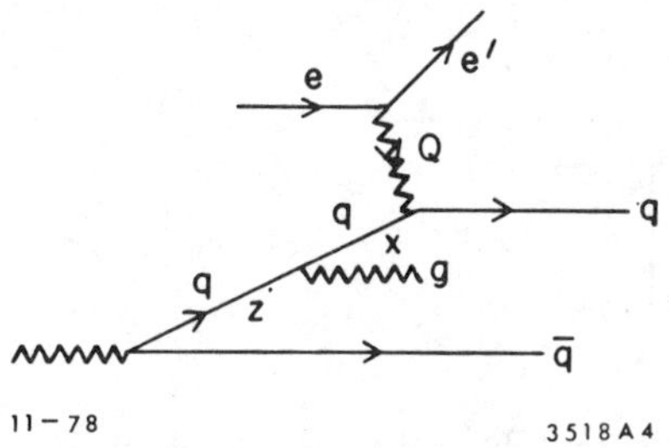

Fig. 4. Representation of the QCD photon structure function in deep inelastic scattering on a photon target. Real and virtual gluon corrections to all orders are included in the analytic results.

The calculation of the photon structure function is straightforward if we keep only leading logarithms in each order of perturbation theory. The leading contribution can be written as a simple convolution: (see Fig. 4)[14]

$$G_{q/\gamma}(x,Q^2) = \frac{3\alpha}{2\pi} e_q^2 \int_{\mu^2}^{Q^2} \frac{dk^2}{k^2} \int_x^1 \frac{dz}{z} \left[z^2 + (1-z)^2\right] G_{q/q}\left(\frac{x}{z}, Q^2, k^2\right) \tag{9}$$

where $G_{q/q}(x/z,Q^2,k^2)$ is the standard non-singlet distribution due to gluon bremsstrahlung for quarks in a target quark of mass k^2 being probed at four-momentum squared Q^2. The factor of three includes the sum over quark colors. In addition one can include smaller sea quark contributions form $g \to q\bar{q}$ processes. The region $k^2 < \mu^2$ can be identified with the VDM contribution to $G_{q/\gamma}$.

Taking moments, we have[1]

$$G_{q/\gamma}(j,Q^2) = \frac{3\alpha}{2\pi} e_q^2 \int_{\mu^2}^{Q^2} \frac{dk^2}{k^2} f(j)\, G_{q/q}(j,Q^2,k^2) \qquad (10)$$

where

$$G(j) \equiv \int_0^1 dx\, x^{j-1}\, G(x) \qquad (11)$$

$$f(j) = \int_0^1 dz\, z^{j-1} \left(z^2 + (1-z)^2 \right)$$

$$= \frac{1}{j} - \frac{2}{j+1} + \frac{2}{j+2} \qquad (12)$$

and

$$G_{q/q}(j,Q^2,k^2) = \left[\frac{\alpha_s(k^2)}{\alpha_s(Q^2)} \right]^{\gamma_{j-1}} \qquad (13)$$

The γ_j are the standard valence anomalous dimensions, as defined in (II.2.18). Performing the k^2 integral in (10) yields

$$G_{q/\gamma}(j,Q^2) = \frac{3}{2\pi} e_q^2 \frac{\alpha}{\alpha_s(Q^2)} \left[\frac{4\pi f(j)}{\beta(1-\gamma_{j-1})} \right] \qquad (14)$$

This exhibits the remarkable scaling features of the photon structure function discussed above.

It is easy to invert the moment equation via the method of Yndurian.[16] A graph of $xG_{q/\gamma}(x)$ calculated in valence approximation in QCD and in the parton model is given in Fig. 5. Good agreement is obtained with the (valence plus singlet) results of Llewellyn Smith[5] over nearly the entire range of x.

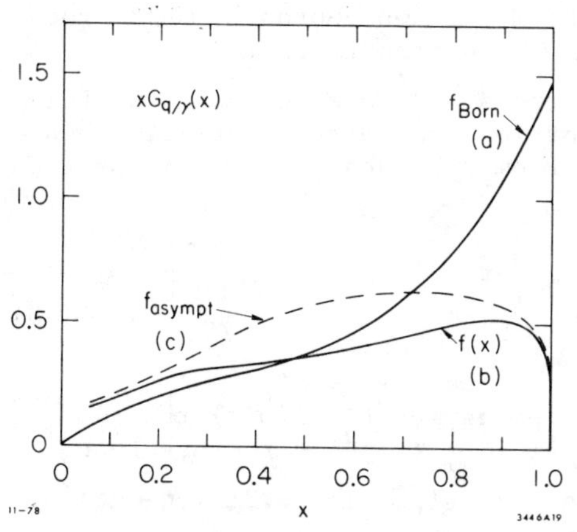

Fig. 5. The valence photon structure function $G_{q/\gamma}(x)$ as calculated in (a) Born approximation, (b) to all orders in QCD, and (c) the $x \to 1$ limit (Eq. (17)). An overall factor proportional to $\log Q^2/\Lambda^2$ is factored out (from Ref. 1).

The x near 1 behavior of $G_{q/\gamma}(x)$ can be obtained more directly from a direct integration of (10), using the $x \to 1$ form for the quark structure function[14]

$$G_{q/q}(x,Q^2,k^2) = \frac{\exp\left[(3-4\gamma_E)\xi C_2\right](1-x)^{4C_2\xi - 1}}{\Gamma(4C_2\xi)} \tag{15}$$

where $\gamma_E = 0.577\ldots$ is Euler's constant, $C_2 = (N^2-1)/2N = 4/3$, and

$$\xi = \frac{1}{\beta} \ln \frac{\alpha_s(k^2)}{\alpha_s(Q^2)} \tag{16}$$

One then obtains[14]

$$G_{q/\gamma}(x,Q^2) \underset{x\to 1}{=} \frac{3}{2\pi} e_Q^2 \frac{\alpha}{\alpha_s(Q^2)} \frac{4}{\beta - (3-4\gamma_E)C_2 + 4C_2 \ln \frac{1}{1-x}} \tag{17}$$

This result is numerically accurate only for $x \gtrsim 0.97$ but is off by no more than a factor of 2 for $x > 0.1$ (see Fig. 5).

It is interesting to note that for fixed $\mathcal{M}^2$, $Q^2 \to \infty$, this expression for $G_{q/\gamma}(x,Q^2)$ approaches a constant. This implies, via the Drell-Yan relation, perfect power-law scaling for the $\gamma \to \pi^0$ transition form factor. [See Eq. (2.26), Chapter II.]

Compared to meson distributions which fall as a power at $x \to 1$, the photon structure function is nearly flat in x, again due to the underlying $\gamma q\bar{q}$ pointlike vertex. In principle the photon structure can be determined experimentally from the two photon $e\gamma \to e'X$ process, i.e.; deep inelastic scattering from a photon target.[9]

Returning to the high p_T jet cross sections, we note the following striking fact: in each contribution to the four-jet cross section the two factors of $\alpha_s(p_T^2)$ from the subprocess cross section, e.g.,

$$\frac{d\sigma}{dt}(qq \to qq) \sim \frac{2}{9}\frac{4\pi\alpha_s^2(t)}{t^2} \tag{18}$$

(see Fig. 3a) actually cancel (in the asymptotic limit) the two inverse powers of $\alpha_s(p_T^2)$ from the two $G_{q/\gamma}(x,p_T^2)$ structure functions.[1] Similarly the single power of $\alpha_s(p_T^2)$ in $d\sigma/dt\ (\gamma q \to gq)$ cancels the single inverse power of $\alpha_s(p_T^2)$ structure function in the 3-jet cross section. (See Fig. 3b.) Thus miraculously all of these jet trigger cross sections obey exact Bjorken scaling

$$E\frac{d\sigma}{d^3p}(\gamma\gamma \to \text{Jet} + X) \underset{p_T^2 \to \infty}{\Longrightarrow} \frac{1}{p_T^4} f(x_T, \theta_{cm}) \tag{19}$$

when the leading QCD perturbative corrections to all orders are taken into account.[17] Furthermore, the asymptotic cross sections are even independent of $\alpha_s(p_T^2)$! The asymptotic prediction thus has essentially zero parameters.

Quite detailed numerical predictions can be made for the $ee \to ee$ Jet + X cross sections by computing $G_{q/e}$ (from the convolution of the equivalent photon approximation $G_{\gamma/e}$ and the photon structure function $G_{q/\gamma}$), and then summing in Eq. (7) over all 2-2 hard scattering QCD processes, including all quark colors and flavors. In our calculations[1] we have found it useful to display approximate analytic forms which have the correct power-law dependence at large p_T and at the edge of phase space ($x_R = 2p_J/\sqrt{s} \to 1$). The analytic forms usually agree with the numerically integrated results to within 20%. For the analytic calculations, we have used the simplified form

$$xG_{q/e}(x) = e_q^2\left(\frac{\alpha}{2\pi}\log\eta\right)\frac{\alpha}{2\pi}F_Q(1-x) \tag{20}$$

for each quark flavor and color, where $\log\eta = \log s/4m_e^2$ if the scattered electron is not tagged. The factor F_Q which is $\sim \log s/4m_q^2$ if we use the Born approximation for $\gamma \to q\bar{q}$, becomes of order $1/\alpha_s(Q^2)$ when the QCD radiative corrections are taken into account. We have found empirically that the value $\alpha_s(Q^2)F_Q \cong 0.8$ gives a good characterization of the QCD normalization. [For $x \to 1$ $G_{q/e}$ actually falls as $(1-x)\log 1/(1-x)$.] Note also that for $x \to 1$, the quark and electron tend to have the same helicity.

For the 4-jet cross section, the sum over all types of jet triggers near 90° gives

$$E \frac{d\sigma}{d^3p_J}(e^+e^- \to e^+e^- \text{ Jet} + X) \cong \left(\frac{\alpha}{2\pi} \log \eta\right)^2 \left[\frac{\alpha}{2\pi} F_Q \alpha_s(p_T^2)\right]^2$$

$$\cdot \left[80\left(\sum_f e_f^2\right)^2 + \frac{52}{9}\left(\sum_f e_f^4\right)\right] \frac{(1-x_R)^3}{p_T^4}$$

$$= \quad 0.8 \times 10^{-2} \text{ nb GeV}^2 \frac{(1-x_R)^3}{p_T^4} \quad [\sqrt{s} = 30 \text{ GeV}]. \qquad (21)$$

The sum f is over contributing quark flavors. The subprocesses include $qq \to qq$, $q\bar{q} \to q\bar{q}$, and $q\bar{q} \to gg$.

For the 3-jet events, the subprocesses $\gamma q \to gq$ and $\gamma\bar{q} \to g\bar{q}$ yield the cross section

$$E \frac{d\sigma}{d^3p_J}(e^+e^- \to e^+e^- \text{ Jet} + X) \cong \alpha\left(\frac{\alpha}{2\pi} \log \eta\right)^2 \left[\frac{\alpha}{2\pi} F_Q \alpha_s(p_T^2)\right]$$

$$\cdot \left[40 \sum_f e_q^4\right] \frac{(1-x_R)^2}{p_T^4}$$

$$\cong \quad 2.5 \times 10^{-2} \text{ nb GeV}^2 \frac{(1-x_R)^2}{p_T^4} \quad [\sqrt{s} = 30 \text{ GeV}]. \qquad (22)$$

The corresponding result for the two jet cross section from $\gamma\gamma \to q\bar{q}$ is

$$E \frac{d\sigma}{d^3p_J}(e^+e^- \to e^+e^- \text{ Jet} + X) \cong 3 \times 10^{-2} \text{ nb GeV}^2 \frac{(1-x_R)}{p_T^4}, \qquad (23)$$

i.e.: in general, $\sigma(2 \text{ jet}) > \sigma(3 \text{ jet}) > \sigma(4 \text{ jet})$. It is clear that there is no double counting of cross sections here since each type of jet cross section has a distinctive topological structure and different pattern of q, $\bar{q}$ and g jets. A graph of these cross sections is shown in Fig. 6.

To remind ourselves how critical the pointlike photon couplings are to these results, let us estimate the contribution to high p_T jet production when both photons are meson dominated. We have

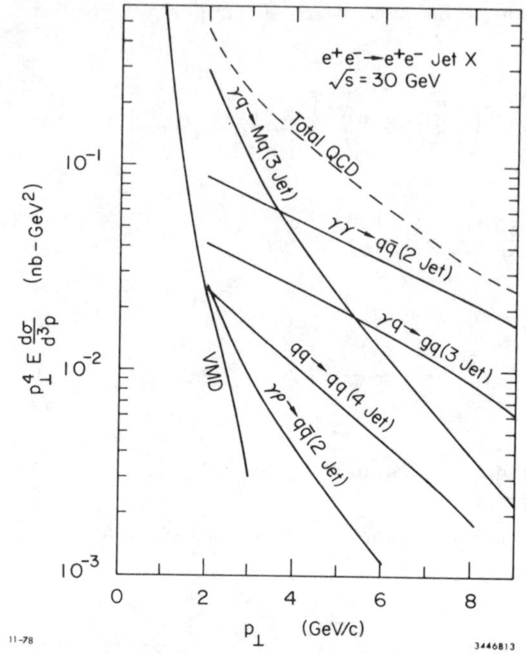

Fig. 6. QCD (and VMD) contributions to the $e^+e^- \to e^+e^-$ Jet + X. The 4-jet cross section includes the contributions from $qq \to qq$, $q\bar{q} \to q\bar{q}$, and $q\bar{q} \to gg$ (from Ref. 1).

$(f_\rho^2/4\pi \cong 2)$

$$d\sigma^{VDM}(\gamma\gamma \to \text{Jet} + X) = \left(\frac{e}{f_\rho}\right)^4 d\sigma\,(\rho\rho \to \text{Jet} + X) \cong \left(\frac{4\pi\alpha}{f_\rho^2}\right)^2 \left(\frac{2}{3}\right)^2 \times \frac{d\sigma(pp \to \text{Jet} + X)}{(1-x_R)^4} \qquad (24)$$

since we expect $G_{q/p} \sim (1-x)^2\, G_{q/\rho}$. If we take $E d\sigma\,(pp \to \text{Jet} + X)/d^3p \sim 300 \times E d\sigma/d^3p\,(pp \to \pi X) \sim 1.1\ \text{nb GeV}^6\,(1-x_R)^9\, p_T^{-8}$, then the convolution over photon momentum distributions yields the rough estimate ($\theta_{cm} \cong 90^\circ$, $\sqrt{s} = 30$ GeV):

$$E\,\frac{d\sigma^{VDM}}{d^3p_J}\,(e^+e^- \to e^+e^-\ \text{Jet} + X) \cong 1.4\ \text{nb GeV}^6\ \frac{(1-x_R)^7}{p_T^4}\ , \qquad (25)$$

which is negligible compared to the pointlike contributions for $p_T > 2$ GeV (see Fig. 6). We have also checked explicitly that the QCD ($q\bar{q} \to q\bar{q}$ hard scattering) contributions from processes such as $\gamma\rho \to q\bar{q}q\bar{q}$ or $\rho\rho \to q\bar{q}q\bar{q}$, where one or both photons are meson dominated, are also small.

The overall scaling properties of QCD cross sections due to specific subprocesses can be easily determined from counting rules:[18]

$$E\,\frac{d\sigma}{d^3p}\,(A+B \to C+X) \cong \frac{f(\theta_{cm})}{(p_T^2)^{n_{active}-2}}\,(1-x_R)^{2n_{spect}^{bnd} + n_{fm} - 1} \qquad (26)$$

where n_{active} is the number of elementary fields (q,e,γ,g, etc.) participating in the hard scattering subprocess,[3] n^{bnd}_{spect} is the number of bound spectators, i.e.: the number of constituent fields which do not interact (and thus "waste" the incident energy),[19] and n_{fm} are the number of unbound spectator fermions (q,e) from pair production or bremsstrahlung scattering processes, as in the equivalent photon approximation. In Eqs. (21)-(23) the number of active fields in each case is 4; $n^{bnd}_{spect} = 0$, and $n_{fm} = 4,3$, and 2 respectively. The counting rules have small corrections due to logarithmic scale-breaking effects and the log (1/1-x) behavior of $G_{q/\gamma}$.

C. High p_T Meson Production in γγ and pp Collisions

We have also considered in some detail background contributions to the $\gamma q \to \text{Jet} + X$ cross section from (higher "twist") subprocesses that involve more than the minimum number of active fields in the hard scattering subprocess.[20] The most significant background comes from subprocesses of the form (see Fig. 7)

$$\gamma q \to Mq$$

where a photon from one beam photoproduces a meson at large p_T on a quark constituent of the other beam. The meson trigger, the recoil quark jet, and the spectator $\bar{q}$ jet together provides a background to the $\gamma\gamma \to$ 3-jet events.

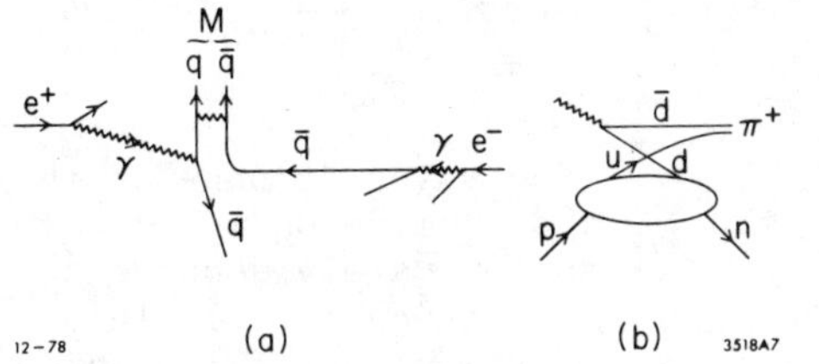

Fig. 7. Contribution of the $\gamma q \to Mq$ subprocesses to (a) $e^+e^- \to e^+e^-\pi^+X$ and (b) $\gamma p \to \pi^+ n$.

The normalization of the $\gamma q \to Mq$ amplitude can be inferred in a straightforward way from $\gamma p \to \pi^+ n$ photoproduction at large momentum transfer: (see Fig. 7(b))

$$\frac{d\sigma}{dt}(\gamma p \to \pi^+ n) \propto F_p^2(t)\,\frac{d\hat{\sigma}}{dt}(\gamma q \to \pi q) \qquad (27)$$

The 90° exclusive cross section[21] falls as $s^{-7.3 \pm 0.4}$ in agreement with the s^{-7} behavior predicted by Eq. (27), and dimensional counting.[3] [See Chapter II.] The net result is

$(n_{active} = 5)\ E\,\frac{d\sigma}{d^3p}\,(e^+e^- \to e^+e^-\ \text{Jet} + X) = 1.1\ \text{nb GeV}^4\ (1-x_R)^2/p_T^6$ where the sum over all pseudo-scalar and vector meson $q\bar{q}$ bound states in the 35 + 1 representation of SU(3) constitutes the "jet" trigger. As shown in Fig. 6, this contribution falls faster in p_T but at $\sqrt{s} = 30$ GeV dominates the $\gamma q \to gq$ 3-jet cross section until $p_T^{jet} \sim 6$ GeV, and (though distinguishable by topology) it even dominantes the $\gamma\gamma \to q\bar{q}$ contribution until $p_T^{jet} \gtrsim 4$ GeV.

It is possible that the normalization of the $\gamma q \to Mq$ subprocess has been overestimated; nevertheless this amplitude must occur at some level, producing a characteristic p_T^{-6} $f(x_R, \theta_{cm})$ cross section. The most important check of its contribution will come from single particle production at large p_T, such as $e^+e^- \to e^+e^-\pi^+X$. In the case of hard scattering processes such as $\gamma\gamma \to q\bar{q}$, $\gamma q \to gq$, and $qq \to qq$, the final state fragmentation $G_{\pi/q} \sim (1-x)$ leads to a strong suppression ($\sim 10^{-2}$) in the π^+/Jet ratio, since the quark jet must be produced at higher momentum than the trigger particle (the "trigger bias" effect).[22] For example, the leading $\gamma\gamma \to q\bar{q}$ subprocess gives

$$E \frac{d\sigma}{d^3p} (e^+e^- \to e^+e^-\pi^+X) = 6 \times 10^{-4} \text{ nb GeV}^2 \frac{(1-x_R)^3}{p_T^4} . \qquad (28)$$

On the other hand, the $\gamma q \to \pi^+ q$ subprocess produces a pion at high p_T, without suppression from fragmentation:

$$E \frac{d\sigma}{d^3p} (e^+e^- \to e^+e^-\pi^+_{\text{prompt}}X) = 3 \times 10^{-2} \text{ nb GeV}^2 \frac{(1-x_R)^2}{p_T^6} . \qquad (29)$$

(Inclusion of non-"prompt" π's from resonance decay approximately doubles the production rate.) This contribution is thus predicted to dominate single pion production in the $\gamma\gamma$ process until very high p_T. With the above normalization, and in the absence of electron or positron tagging, the two-photon reaction provides a significant background to the 90° inclusive π^+ spectrum from $e^+e^- \to \gamma^* \to \pi^+ + X$ for $x_T \lesssim 0.15$ at $\sqrt{s}$ = 30 GeV. (See Fig. 8.)

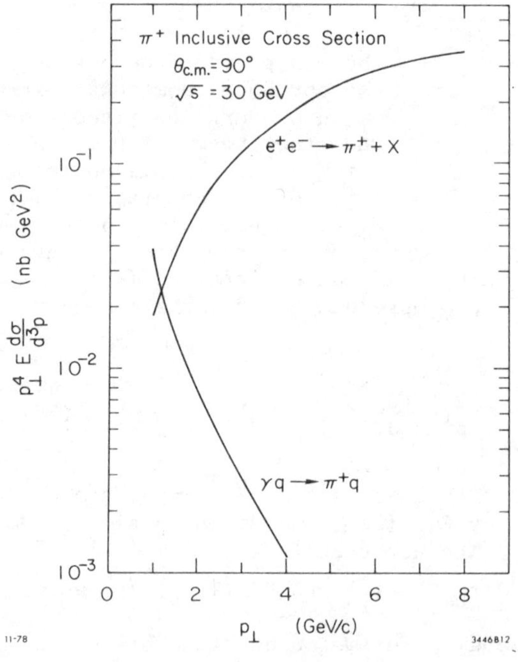

Fig. 8. Leading contributions to inclusive pion contributions from e^+e^- annihilation and $e^+e^- \to e^+e^-\pi^+X$ (from Ref. 1).

It will be extremely interesting to verify the normalization and especially the power law of the $\gamma\gamma \to \pi^+ + X$ cross section. The p_T^{-6} power is derived directly from the lowest order diagram for $\gamma q \to (q\bar{q})q$ where the $q\bar{q}$ system is at fixed mass; higher order QCD corrections can only modify the result by an overall logarithmic factor. The fact that the single hadron trigger is produced directly in the hard scattering subprocess rather than by quark or gluon fragmentation also is an important feature in hadron-hadron collision. In this case, as described in the constituent interchange model (CIM),[20] dominant subprocess contributing to the $pp \to \pi^+ X$ and $pp \to pX$ cross sections for $p_T < 8$ GeV are expected to be the prompt hard-scattering reactions such as $qM \to qM$ and $qB \to qB$, respectively. These subprocesses immediately explain why the observed power law for $E d\sigma/d^3p$ at fixed x_T and θ_{cm} are close to p_T^{-8} (meson production) and p_T^{-12} (proton production) for data below $p_T = 8$ GeV. The CIM approach also can account for the observed angular distributions, same side momentum correlations, and charge correlations (flavor transfer) between opposite sides.[23] We will discuss the central issues for hadron collisions in the next chapter.

In summary, it becomes evident that two photon collisions can provide a clean and elegant testing ground for perturbative quantum chromodynamics. The occurrence of $\gamma\gamma$ reactions at an experimentally observable level implies that the entire range of hadronic physics which can be studied, for example, at the CERN-ISR can also be studied in parallel in $e^{\pm}e^{-}$ machines. Although low p_T $\gamma\gamma$ reactions should strongly resemble meson-meson collisions, the elementary field nature of the photon implies dramatic differences at large p_T. We have especially noted the sharp contrasts between hadron-and photon-induced reactions due to the photon's pointlike coupling to the quark current and the ability of a photon to give nearly all of its momentum to a quark. The large momentum transfer region can be a crucial testing ground for QCD since not only are a number of new subprocesses accessible ($\gamma\gamma \to q\bar{q}$, $\gamma q \to gq$, $\gamma q \to Mq$, deep inelastic scattering on a photon target) with essentially with no free parameters, but most important, one can make predictions for a major component of the photon structure function directly from QCD. We also note that there are open questions in hadron-hadron collisions, e.g., whether non-perturbative effects (instantons, wee parton interactions) are important for large p_T reactions.[24] Such effects are presumably absent for the perturbative, pointlike interactions of the photon. We also note that the interplay between vector-meson-dominance and pointlike contributions to the hadronic interactions of photon is not completely understood in QCD, and $\gamma\gamma$ processes may illuminate these questions.

Footnotes and References

1. S. J. Brodsky, T. A. DeGrand, J. F. Gunion and J. H. Weis, Phys. Rev. Lett. 41, 672 (1978); and SLAC-PUB-2199 (submitted to Phys. Rev.).

2. J. D. Bjorken and E. A. Paschos, Phys. Rev. 185, 1975 (1969).

3. S. J. Brodsky and G. Farrar, Phys. Rev. Lett. 31, 1153 (1973); Phys. Rev. D11, 1309 (1975); V. A. Matveev, R. M. Muradyan and A. N. Tavkheldize, Lett. Nuovo Cimento 7, 719 (1973).

4. For a comprehensive review of the vector meson dominance model, see T. H. Bauer, R. D. Spital, F. M. Pipkin and D. R. Yennie, Rev. Mod. Phys. 50, 261 (1978).

5. S. J. Brodsky, T. Kinoshita and H. Terazawa, Phys. Rev. D4, 1532 (1971). For reviews see V. M. Budnev *et al.*, Phys. Reports 15C (1975); H. Terazawa, Rev. Mod. Phys. 45, 615 (1973); and the reports of S. J. Brodsky, H. Terazawa and T. Walsh in the Proceedings of the International Colloquium on Photon-Photon Collisions, published in Supplement au Journal de Physique, Vol. 35 (1974). See also, G. Grammer and T. Kinoshita, Nucl. Phys. B80, 461 (1974); R. Bhattacharya, J. Smith and G. Grammer, Phys. Rev. D15, 3267 (1977); J. Smith, J. Vermaseren and G. Grammer, Phys. Rev. D15, 3280 (1977).

6. See S. J. Brodsky *et al.*, Ref. 5, and F. Low, Phys. Rev. 120, 582 (1960). In the case of tagged leptons $\log s/m_e^2 \to \sim\log \theta^2_{max}/\theta^2_{min}$. Derivations and more precise formula are given in Ref. 5. An excellent discussion of the experimental considerations is given by J. Field, LEP Summer Study/1-13, October 1978.

7. For discussion and references, see S. J. Brodsky, Ref. 5.

8. For detailed calculations see J. Smith *et al.*, Ref. 5, and J. Field, Ref. 6. The $e^+e^- \to e^+e^-\tau^+\tau^-$ cross section is comparable to the $e^+e^- \to \tau^+\tau^-$ cross section at $\sqrt{s} = 30$ GeV.

9. S. J. Brodsky, T. Kinoshita and H. Terazawa, Phys. Rev. Lett. 27, 280 (1971); T. F. Walsh, Phys. Lett. 36B, 121 (1971).

10. The $\gamma\gamma \to q\bar{q}$ process for single hadron production at large p_T in e^+e^- collisions was first considered in the pioneering paper by J. D. Bjorken, S. Berman and J. Kogut, Phys. Rev. D4, 3388 (1971). A discussion of this process for virtual γ reactions has been given by T. F. Walsh and P. Zerwas, Phys. Lett. 44B, 195 (1973).

11. In the case of integrally-charged Han-Nambu quarks, locality of the $\gamma\gamma \to q\bar{q}$ matrix element at high p_T implies for the u,d first generation quarks

$$\mathcal{M}_{\gamma\gamma} \propto \langle 0|j^2_{em}(0)|X\rangle = \begin{cases} \sum\limits_{c=R,Y,B} u_c^\dagger u_c\left(\frac{2}{3}\right) + d_c^\dagger d_c\left(\frac{1}{3}\right) & \left(\begin{array}{c}\text{below}\\ \text{color}\\ \text{threshold}\end{array}\right) \\ u_R^\dagger u_R + u_B^\dagger u_B + d_Y^\dagger d_Y & \left(\begin{array}{c}\text{above}\\ \text{color}\\ \text{threshold}\end{array}\right) \end{cases}$$

which aside from a sign change for the down quark is identical to the $\langle 0|j_{em}(0)|X\rangle$ matrix element. Thus, we have the identity

$$R^{HN}_{\gamma\gamma} = R$$

both below and above the color threshold. In particular, $R^{HN}_{\gamma\gamma}$ = 5/3 × (number of flavor generations) below color threshold, and $R^{HN}_{\gamma\gamma}$ = 3 × (number of flavor generations) above color threshold, compared to 17/27 × (number of flavor generations) for the standard QCD model. See also M. Chanowitz in "Color Symmetry and Quark Confinement," Proceedings of the 12th Rencontre de Moriond, 1977, edited by Tran Thanh Van, and P. V. Landshoff, LEP Summer Study/1-13, October 1978.

12. Equation (6) has also been derived in an equivalent form by K. Kajantie, University of Helsinki reprint, September 1978.

13. For a review see D. Sivers, R. Blankenbecler and S. J. Brodsky, Phys. Reports 23C:1 (1976). For a discussion of the validity of the hard scattering expansion in field theory, see W. E. Caswell, R. R. Horgan and S. J. Brodsky, Phys. Rev. D18, 2415 (1978). Note that processes where one electron balances the transverse moment of the high p_T jet trigger also occur in (3).

14. Yu. L. Dokshitser, D. I. D'Yakanov and S. I. Troyan, Stanford Linear Accelerator Center translation SLAC-TRANS-183, translated for Proceedings of the 13th Leningrad Winter School on Elementary Particle Physics, 1978.

15. E. Witten, Nucl. Phys. B120, 189 (1977). The results of Witten have been rederived by summing ladder graphs by W. Frazer and J. Gunion, University of California at Davis preprint 10P10-194 (1978) and by C. H. Llewellyn Smith, Oxford preprint 67/78. See also Ref. 14.

16. F. J. Yndurian, Phys. Lett. 74B, 68 (1978).

17. This remarkable scaling property was first pointed out by C. H. Llewellyn Smith, Oxford preprint 56/78 (1978).

18. See Refs. 13, R. Blankenbecler, S. J. Brodsky and J. F. Gunion, Phys. Rev. D18, 900 (1978), and Ref. 1. If the spin of a constituent a does not match that of the projectile A, then one expects a suppression factor $(1-x)^{2|S_z^a - S_z^A|}$ in the leading scaling term. S. J. Brodsky, J. F. Gunion and M. Scadron (to be published). Logarithmic QCD corrections to the power-law behavior are discussed in Chapter II.

19. S. J. Brodsky and R. Blankenbecler, Phys. Rev. D10, 2973 (1974).

20. R. Blankenbecler, S. J. Brodsky and J. F. Gunion, Phys. Rev. D18, 900 (1978); P. V. Landshoff and J. C. Polkinghorne, Phys. Rev. D8, 927 (1973), Phys. Rev. D10, 891 (1974).

21. R. Anderson *et al*., Phys. Rev. Lett. 30, 627 (1973). For a comparison of Compton scattering and photoproduction, see M. A. Shupe *et al*., Cornell preprints (1978). [See Chapter II.]

22. S. D. Ellis, M. Jacob and P. V. Landshoff, Nucl. Phys. B108, 93 (1976). See also J. D. Bjorken and G. R. Farrar, Phys. Rev. D9, 1449 (1974).

23. For recent discussions see S. J. Brodsky, SLAC-PUB-2217, October 1978, and D. Jones and J. F. Gunion, Phys. Rev. D9, 1032 (1979). Also see Chapters II and IV.

24. I wish to thank J. Ellis for conversations on this point.

IV. HADRON AND PHOTON PRODUCTION AT LARGE TRANSVERSE MOMENTUM AND THE DYNAMICS OF QCD JETS

1. Introduction

The most direct tests of the interactions of quarks and gluons at short distances involve the production of single hadrons, hadronic jets, and photons at large transverse momentum. In this chapter we will review several areas of hadronic phenomenology which test predictions of quantum chromodynamics calculated from perturbation theory, including:

(a) The production of direct photons at large transverse momentum in hadron-hadron collisions.[1,2] In perturbative QCD, the ratio of gluon jet and direct photon cross sections is directly calculable, and leads to important phenomenological constraints.
(b) The multiplicity and distribution of hadrons in inclusive reactions may be related to color separation of the initiating subprocesses.[3,4] The consequences of this ansatz for gluon and quark jets are discussed. We also review other possible discriminants of jet parentage.
(c) The hadronic decay of the upsilon via three gluon jet[5,6] or a photon plus two gluon jets[6] could provide some of the most definite tests of QCD.
(d) Gluon jets may be "oblate" with principal axes correlated with the gluon polarization.[7]
(e) The gluon distribution of a hadron is connected with the size of the source due to coherent effects and is not determined solely by the quark distribution.[8]

At present, the most controversial area of QCD phenomenology concerns the production of single hadrons at large transverse momentum in proton-proton collisions.[9,10] We shall begin our discussion with a short review of the current issues.

2. Production of Large Transverse Momentum Particles in Hadron-Hadron Collisions

There are currently two main approaches to large p_T phenomena -- <u>both</u> based on perturbative QCD and a "hard scattering expansion."

(A) Quark, gluon scattering models. The basic collision subprocesses responsible for the large momentum transfer are assumed to be $qq \to qq$, $qg \to qg$, and $gg \to gg$, as calculated in Born approximation QCD.[11] Violation of scale-invariance occurs through the running coupling constant $\alpha_s(Q^2)$, the quark and gluon structure functions, and the transverse momentum (k_T) distributions of the constituents in the hadronic wavefunctions. The calculations automatically include those parts of higher particle number subprocesses such as $qq \to qqg$ which contribute to logarithmic scaling violations in the structure functions.

(B) The Constituent Interchange Model.[12] In addition to all of the contributions listed in (A), QCD also predicts "higher twist" subprocesses[13] where more than the minimal number of quark and gluon fields participate in the hard scattering reaction, such as $qM \to qM'$, $qB \to qB'$, $gq \to Mq$, $q\bar{q} \to M\bar{M}$, etc. Here "M" and "B" indicate $q\bar{q}$ and qqq clusters of fixed mass relative to p_T. The cross sections for these subprocesses are readily computed from minimal QCD diagrams.[13,14] As in (A) logarithmic scaling violations occur.[15] By definition, higher twist subprocesses are responsible for all large p_T exclusive reactions involving hadrons.

The basic distinction between these two approaches for an inclusive reaction such as $pp \to \pi X$ is simply whether (a) the high p_T trigger meson is formed after the hard scattering (e.g., $q_1 q_2 \to q_1 q_2$ with $q_1 \to q_1 + \pi$) or (b) formed before the collision and then scattered (e.g., $\pi q \to \pi q$). Obviously both types of subprocesses contribute to the cross section at some level -- it is a question of kinematics where each dominates: for fixed $x_T = 2p_T/s$ and θ_{cm}, the Born contributions clearly dominate at $p_T \to \infty$ since

$$\frac{\frac{d\sigma}{dt}(\pi q \to \pi q)}{\frac{d\sigma}{dt}(qq \to qq)} \sim F_\pi^2(p_T^2) \sim O\left[\frac{1\ \text{GeV}^4}{p_T^4}\right] . \tag{2.1}$$

On the other hand, the necessity for final state fragmentation in any quark or gluon scattering reaction implies a numerical suppression of the cross section by 2 to 3 orders of magnitude! This crucial factor (called "trigger bias" by Ellis, Jacob, and Landshoff[16]) results because a quark typically gives 75% of its momentum to the trigger particle due to its rapidly falling fragmentation function $G_{\pi/q}(z)$ at $z \sim 1$. The $qq \to qq$ subprocess then occurs at an effectively higher p_T where the cross section is orders of magnitude smaller. (It is this effect that yields large jet/single ratios, since the (quark or gluon) jet trigger is not suppressed by this effect.) On the other hand, if the pion trigger emerges directly from the subprocess (as in the CIM $Mq \to \pi q$ subprocesses) then there is no trigger bias suppression. Thus for some range of p_T, the "higher twist" QCD-CIM subprocesses will be numerically important. Ignoring (logarithmic) scale-violating effects (see Chapter II) the cross sections have the representative forms (see Fig. 1)

$$\text{QCD-Born:} \quad \frac{d\sigma}{d^3p/E}(pp \to \pi X) \sim \frac{\alpha_s^2}{(100)p_T^4}(1 - x_T)^9 \tag{2.2}$$

versus

$$\text{QCD-CIM:} \quad \frac{d\sigma}{d^3p/E}(pp \to \pi X) \sim \frac{\alpha_M^2}{p_T^8}(1 - x_T)^9 \tag{2.3}$$

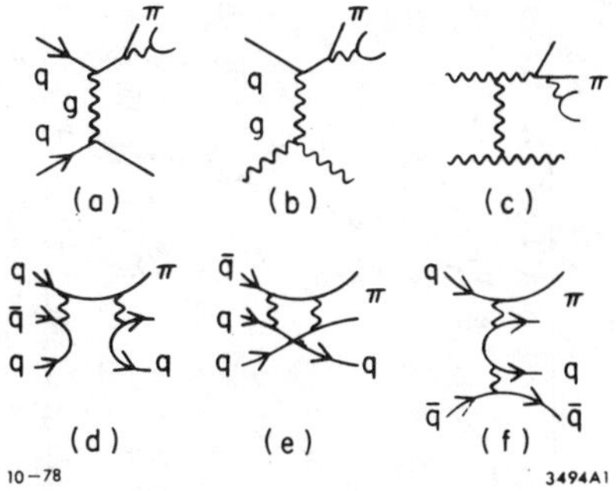

Fig. 1. QCD hard scattering subprocesses for $pp \to \pi X$. In (a), (b) and (c) the π is formed after the hard scattering via quark or gluon fragmentation. In (d) and (e), the "higher twist" CIM contributions, the meson is formed before the hard scattering. Diagram (f) represents the "fusion" CIM $q\bar{q} \to \pi + M$ contribution.

The critical question is determining the magnitude of each contribution.

In principle, it is straightforward to determine the normalization of the $2 \to 2$ QCD subprocess contributing to the inclusive cross section, since α_s and the structure functions are to a large extent determined (although there is some uncertainty in determining the gluon distributions in hadrons). The effect of the k_T distributions of the hadronic constituents is controversial. An essential point ignored in many model calculation is that the interacting constituents are <u>always</u> off the mass shell and spacelike:[17]

$$k^2 = -\frac{\vec{k}_T^2 + \tilde{m}^2}{1-x} \qquad (2.4)$$

where $\tilde{m}^2$ is a linear combination of squares of spectator and incident hadron masses (see Chapter II). The off-shell kinematics ensure that the gluon pole in the $qq \to qq$ amplitude never occurs in the physical region, and serve to damp out the effects of large k_T. In practice, one finds that k_T fluctuations <u>do not</u> increase the inclusive cross section by more than a factor of 2 for $p_T \geq 2$ GeV, even if we assume very large mean $k_T \sim$ 850 MeV Gaussian smearing.[18] A representative calculation is shown in Fig. 2. (If one uses on-shell kinematics, the cross section can be increased by an arbitrary amount depending on a cut-off.) Off-shell kinematics are of course required whether one uses covariant Feynman amplitudes or time-ordered perutrbation theory.

The cross section for subprocesses such as $qM \to qM$ has the form[19]

$$\frac{d\sigma}{dt}(qM \to qM) = \frac{\pi\alpha_M^2}{su^3} \qquad (2.5)$$

corresponding to the QCD amplitude shown in Fig. 1(d). The $qM \to qM$ amplitude falls as s^{-1} at fixed u because of the exchanged fermion in the u-channel. The power fall-off at fixed center-of-mass angle agrees with the dimensional counting rules $d\sigma/dt \propto s^{-(n-2)}$ where n (= 6 here) is the number of active fields in the initial and final state.[20] The constant α_M is proportional to $\alpha_s(p_T^2)$ times the meson wavefunction at the origin. It can be fixed phenomenologically (to

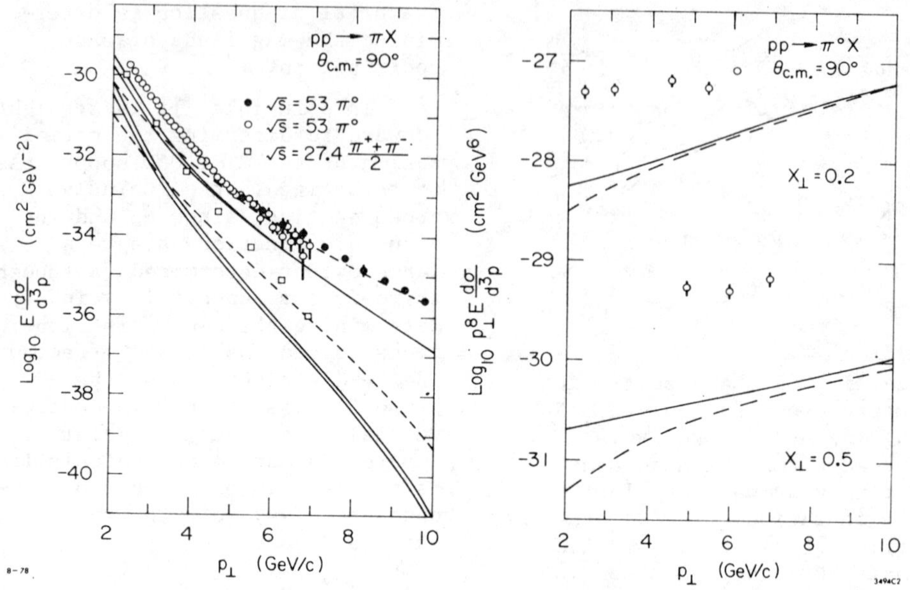

Fig. 2. (a) Data and QCD contributions for $Ed\sigma/d^3p$ ($pp \to \pi X$) at $\theta_{cm} = 90^o$. The dotted line has no scale violations or k_T fluctuations. The lower solid curve indicates scale violations in the structure functions and α_s. The upper solid curve indicates scale violations plus k_T fluctuations calculated with off-shell kinematics. (b) QCD results for $p_T^8\ Ed\sigma/d^3p$ ($pp \to \pi^o X$). The dashed curves indicate scale violations. The solid curves indicate scale violations plus off-shell k_T fluctuations. (From Horgan and Scharbach, Ref. 18.) The sum of QCD plus CIM diagrams give a good fit to the data. See Figs. 5 and 8.

within a factor of ~2), since the $qM \to qM$ amplitude enters directly in the meson elastic form factor and meson-proton elastic scattering at large momentum transfer (see Fig. 3). In a recent paper, Blankenbecler, Gunion and I have found that within errors of order ±50%, $\alpha_M \cong 2\ GeV^2$.[19]

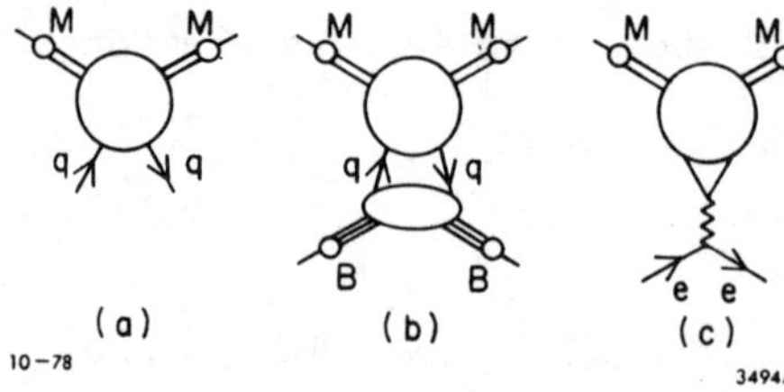

Fig. 3. Contribution of $qM \to qM$ amplitude (a) to meson-baryon scattering (b) and the meson form factor (c).

In order to determine the size of the contribution of the $Mq \to Mq$ subprocess to the $pp \to \pi X$ (see Fig. 4) we also need the normalization of $G_{M/q}(x)$, the distribution of virtual $q\bar{q}$ states in the proton. (The same normalization enters virtual meson-

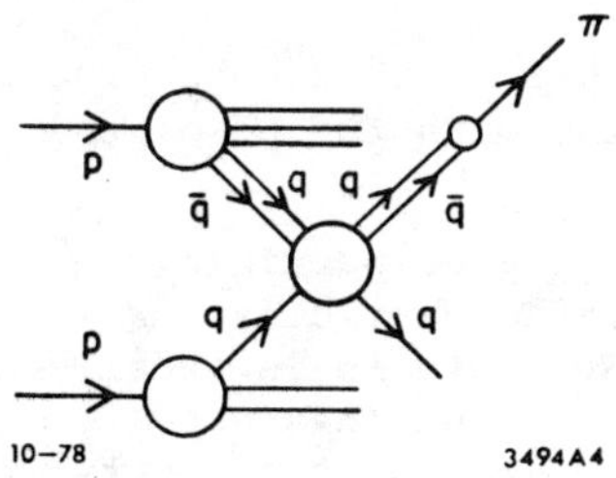

Fig. 4. The CIM $Mq \to \pi q$ contribution to $pp \to \pi X$ at large p_T. The virtual meson M is a $q\bar{q}$ component of the nucleon.

induced reactions, such as Deck or Drell diagrams in low t hadronic physics and the height of the meson plateau in forward reactions.) We have assumed a normalization such that ~1/2 of the $\bar{q}$ sea can be identified as constituents of the virtual $q\bar{q}$ states.

With these normalizations, we find that contributions (B) are in fact consistent with the normalization of FNAL[21] and ISR data[22] for $pp \to \pi X$ up to $p_T \sim 8$ to 10 GeV. At that point we predict the $2 \to 2$ QCD -- Born subprocesses contributions (A) will cross over and dominate the inclusive cross section.[19,22] (See Fig. 5.) Moreover, we note the following:

(1) The best power-law fit to the Chicago-Princeton[21] FNAL data is

$$E\frac{d\sigma}{d^3p}(pp \to \pi^+X) = \frac{1}{p_T^{8.2 \pm .5}}(1 - x_T)^{9.0 \pm 0.5} \qquad (2.6)$$

is agreement with the predicted CIM powers.

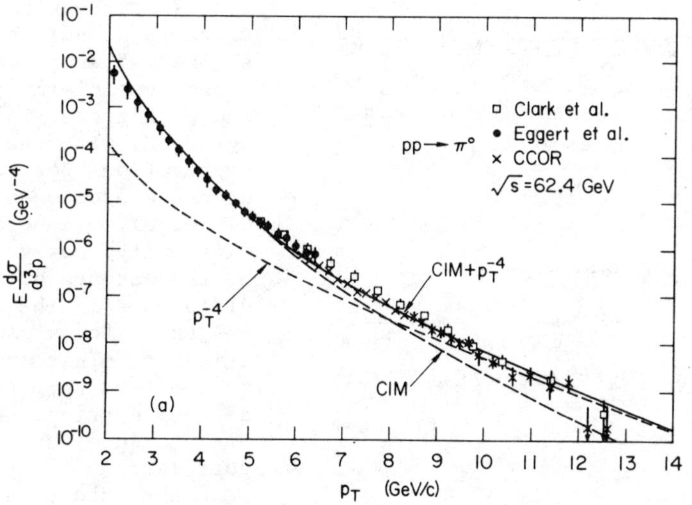

Fig. 5. Comparison with data of CIM plus QCD (p_T^{-4}) contributions to the $pp \to \pi^0X$ cross section. Scale-breaking is neglected and $\alpha_s = .15$ in the QCD term. (From Jones and Gunion, Ref. 23.)

(2) The best fit[24] to the angular distribution of the subprocess in $pp \to \pi X$ is $d\sigma/dt \propto 1/su^3$ or $1/st^3$ in agreement with the predicted CIM form.

(3) The CIM mechansim predicts that the trigger particle usually emerges alone without same-side correlated particles, or from the decay of resonances, especially the ρ. This is in excellent agreement with the results of the British-French-Scandinavian[25] group's experiment at the ISR, who find that in ~85% of the events with a 4 GeV trigger, the trigger particle is unaccompanied by same-side charged particles (aside from the usual low momentum background). The small growth of the same-side momentum with the trigger p_T observed in the experiment indicates that on average more than 90% of the trigger momentum is carried by the trigger pion -- much larger than the ~75% expected from q or g jet fragmentation.[26] The BSF data clearly does not support the hypothesis that the same-side jet is a quark or gluon jet.

(4) The $qM \to qM$ subprocesses implies that flavor is generally exchanged in the hard scattering reaction.[27] For example, consider the quark interchange and $q\bar{q} \to M\bar{M}$ fusion contributions to $pp \to K^{\pm}X$ shown in Fig. 6. The average charge of the recoil quark is slightly positive for the K^+ trigger and $> +1/3$ for the case of the K^- trigger. Thus the charge and flavor of the away-side jet in the CIM can be correlated with the flavor quantum numbers of the trigger. In contrast, gluon exchange diagrams predict very small[26] flavor correlations between the away-side and same-side systems. The data from the BSF-ISR group (see Fig. 7) for various charge triggers at 90° show striking flavor correlations, especially for K^- and $\bar{p}$ triggers, in general, agreement with the above expectations for the quark exchange processes of the CIM model. (A possible difficulty, however, may be the absence of a strong difference in the away-side +/- ratio for π^+ and π^- triggers. This may be due to the fact that resonance decays, particularily $\rho^0 \to \pi^+\pi^-$, dilute the charge correlations.) It should be emphasized that the CIM terms are not maximal for back to back configurations because of the difference in q and M distributions. [This could explain why charge correlations are

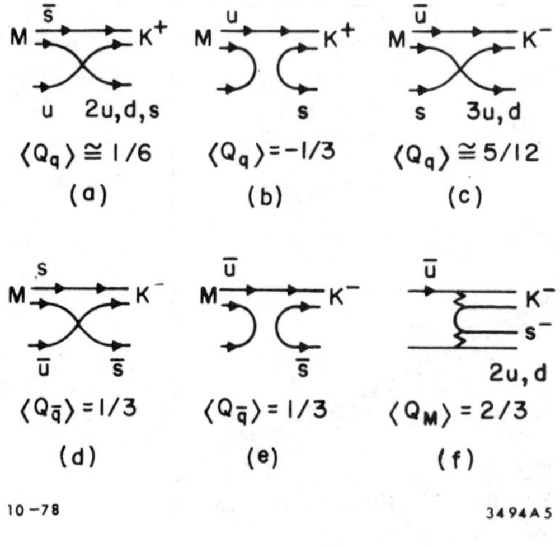

Fig. 6. Analyses of charge flow in CIM diagrams for $pp \to K^{\pm}X$. Quark exchange in the subprocess implies charge correlations between the trigger and away-side jet.

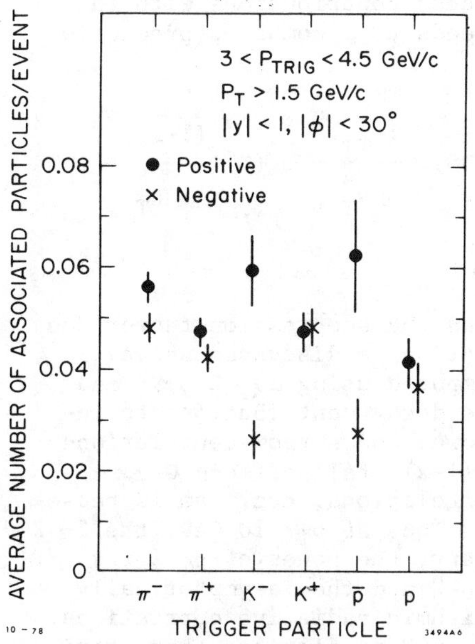

Fig. 7. Number of fast positive and negative particles on the side-away from a 90° trigger for various trigger type. (From Ref. 25.) The gluon exchange QCD diagrams give an away-side jet nearly independent of the trigger type. See R. Field, Ref. 10.

strongest away from zero rapidity on the away-side in the BSF-ISR[25] experiment and why only minimal flavor correlations are observed in the FNAL experiment of R. J. Fisk et al.,[28] who only look at particles directly opposite a 90° trigger. The correlations will also be reduced because of the nuclear target.]

In each case we would expect that these charge correlations will disappear at very high p_T when the $2\to 2$ QCD -- Born subprocesses become dominant. It is interesting to note that for K^- and $\bar{p}$ triggers, the cross-over point is predicted by Jones and Gunion[23] to occur (for pp collisions) at a relatively small p_T (~ 4 to 5 GeV at ISR energies) due to the rapid fall-off of the CIM terms as $x_T \to 1$ for these triggers. Thus there is a rich, dynamical structure controlled by the p_T and x_T kinematics which can be unraveled by quantum number correlations.

(5) In the case of $pp\to pX$, the dominant CIM subprocess is the $qB\to qp$ subprocess. The theoretical prediction is $Ed\sigma/d^3p\ (pp\to pX) \propto p_T^{-12}(1-x_T)^7$. The Chicago-Princeton[21] fit at 90° in fact gives $p_T^{-11.7}(1-x_T)^{6.8}$ at FNAL energies, $p_T < 7$ GeV with uncertainties in the exponent of order ±0.5. We emphasize that a successful model for single particle production must account for both high p_T meson and baryon data. There does not seem any way to account for the $pp\to pX$ scaling behavior in terms of $2\to 2$ QCD subprocesses without enormous scale-breaking in the $q\to p$ distribution function; we note that data from DESY for $e^+e^-\to \bar{p}X$ appears to be reasonably consistent with scale-invariance. On the other hand, we find that the normalization of the $Bq\to Bq$ subprocess required here is consistent with elastic $pp\to pp$ scattering and the proton form factor.[19] In addition, at $\theta_{cm} = 90^o$, $x_T > 0.6$, we predict that the direct scattering process $pq\to pq$ (where the incident proton itself scatters in the subprocess) should become dominant, leading to $p_T^{-12}(1-x_T)^3$ behavior. The direct scattering contribution to inclusive $pp\to pX$ connects smoothly to elastic scattering $pp\to pp$, in agreement with the Bjorken-Kogut "correspondence principle" arguments.[29]

Combining the QCD 2-2 Born subprocess contributions with the CIM (higher twist QCD) contributions leads to a combined prediction for $pp \to \pi^+ X^-$ of the form ($\theta_{cm} = 90^o$)[19]

$$E \frac{d\sigma}{d^3p} (pp \to \pi^+ X^-) = \alpha_s(p_T^2)(0.035) \frac{(1-x_T)^9}{p_T^4} + (9) \frac{(1-x_T)^9}{p_T^8} , \tag{2.7}$$

in GeV units. The 0.035 factor includes the suppression factor due to trigger bias[16] from $q \to M+q$ fragmentation as discussed above. (The factor of 9 in the CIM term is computed using $\alpha_M = 2$ GeV^2 and an estimated factor of 2 from resonance decay contributions to inclusive π^+ production.) The $(1-x_T)^9$ power comes from convolutions of valence distributions $G_{q/p}(x)$ with $(1-x)^3$ fall-off and $G_{\pi/q} \sim (1-x)^5$. Asymptotic freedom,[5] spin correlations, etc. can increase the effective power to $(1-x_T)^{10}$ or ^{11}. Thus at $p_T \sim 10$ GeV, the $2 \to 2$ subprocesses are predicted to be dominant, the power of p_T for $\pi^{\pm,0}$, K^+, and production should decrease to p_T^{-6} and then asymptotically approach p_T^{-4} scaling, modulo QCD logarithmic radiative corrections. At these values of p_T all the canonical QCD predictions characteristic of the Born diagrams should hold; in particular the same-side system will cease to be dominated by single particles, and flavor correlations between the trigger and away-side system will tend to zero. An important prediction of QCD is the eventual dominance of gluon jet recoil.[11,26]

We note that recent ISR data[22] for the $pp \to \pi^0 X$ cross section for $6 < p_T < 12$ GeV are indeed consistent with a sum of terms of the form of Eq. (2.7) (see Fig. 8). For $p_T < 8$ GeV, the experimental data are consistent with dominance of the CIM terms. We emphasize that the predicted QCD $2 \to 2$ Born contributions alone are at least a factor of 5 below the data for $p_T \sim 4$ GeV, even allowing for a factor of 2 from k_T smearing corrections and uncertainties in the effective value of α_s; in any event these contributions are inconsistent with all of the features of the data, (1) through (5) discussed above.

An important theoretical question is how to systematically include the effects of higher particle number hard-scattering subprocesses $2 \to n$ and even $m \to n$. In a recent paper by Casewell, Horgan, and myself[18] we showed that for ϕ^3 field theory, the inclusive cross section for $A+B \to C+X$ can be computed systematically in terms of a sum of incoherent hard scattering contributions, as expected by parton-model considerations. In the ϕ^3 model all effects associated with large k_T in the incident wavefunction are automatically included when the higher order subprocesses are taken into account. Subprocesses with higher number of active fields suffering the large momentum transfer give higher powers of p_T fall-off.

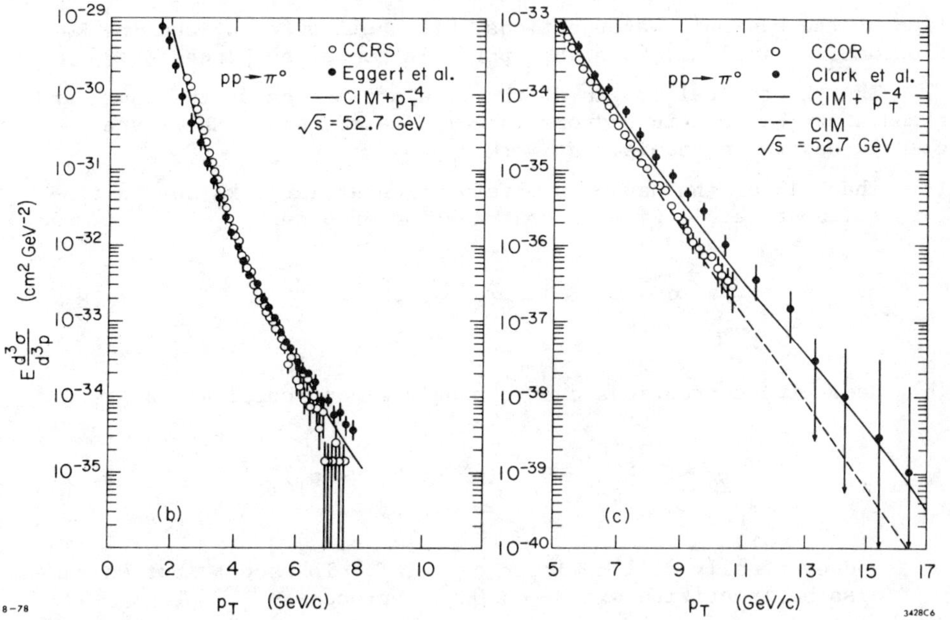

Fig. 8. Comparison with data of CIM plus QCD (p_T^{-4}) contributions to the $pp \to \pi^0X$ cross section. Scale-breaking is neglected and α_s = .15 in the QCD term. (From Jones and Gunion, Ref. 23.) The data may include contributions from direct photons, $pp \to \gamma X$.

The situation in QCD is best illustrated by an example (see Fig. 9). A Feynman diagram which corresponds to $qq \to qq$ scattering with gluon bremsstrahlung yields contributions to both the $qq \to qq$ hard scattering subprocess (when the emitted gluon g_1 is parallel to q_1) and to the $qg \to qg$ subprocess (when the exchange gluon g_2 is at low k_T relative to q_2). The contribution where the q_3, q_4, and g all emerge at different θ_{cm} is suppressed by a power of $\log p_T^2$. Note that (1) off-shell kinematics are required in order to obtain the correct contribution to the $gq \to gq$ subprocess; (2) it would be double-counting to include both k_T fluctuations to $qq \to qq$ scattering plus the $gq \to gq$ subprocess; and (3) the leading logarithmic corrections to the $qq \to qq$ scattering are already included when the measured $G_{q/p}$ structure function is used. A consistent treatment in QCD requires simultaneous considera-

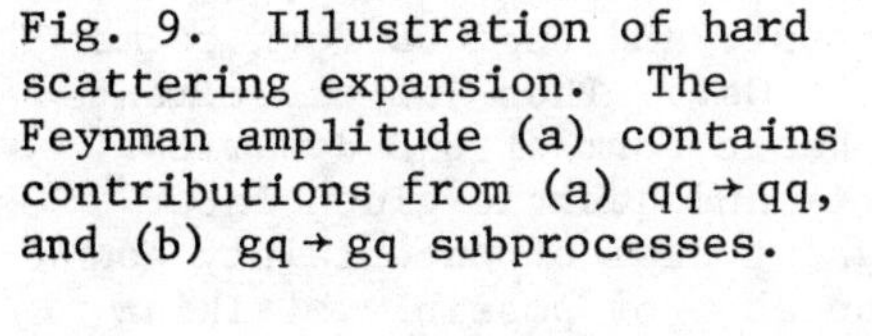

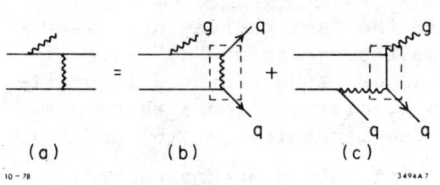

Fig. 9. Illustration of hard scattering expansion. The Feynman amplitude (a) contains contributions from (a) $qq \to qq$, and (b) $gq \to gq$ subprocesses.

tion of the hadronic wavefunctions, off-shell effects, and the k_T fluctuations implicit in higher particle number subprocesses.

The theoretical origin of the k_T distribution of the quark and gluon distributions in hadrons is complicated, since there are clearly several mechanisms at work:

(a) The tail of the hadronic wavefunction at large k_T due to constituent recoil gives a contribution of order

$$\frac{dN}{dk_T^2} \sim \frac{\alpha_M}{k_T^4} \qquad (m^2 << k_T^2 << p_T^2) \tag{2.8}$$

(b) Radiative corrections due to single gluon recoil gives

$$\frac{dN}{dk_T^2} \sim \frac{\alpha_s(k_T^2)}{k_T^2} \qquad (m^2 << k_T << p_T^2) \tag{2.9}$$

and eventually will dominant over (a). This contribution can also be identified with $2 \rightarrow 3$ QCD subprocesses.

(c) In any inclusive process in which color is virtually separated the radiated soft gluons taken together give an effective k_T distribution. According to the analysis of Dokshitser, D'yakanov, and Troyan[30] for the Drell-Yan process, the effective distribution has a computable Gaussian-like shape.

(d) The intrinsic k_T distribution of the hadronic wavefunction due to binding and other non-perturbative effects. The recent bubble chamber measurements of the final state hadron distribution in deep inelastic neutrino-proton scattering reported recently by Vander Velde[31] shows that the intrinsic k_T of the constituents are in fact small; the fast hadrons near $x_F \cong -1$ in the $W^{\pm}$ - proton cm frame (from the spectator "qq" jet) have $\langle k_T^2 \rangle \cong 0.1$ GeV^2. The large values of k_T observed in Drell-Yan and large p_T reactions (from p_{out} distributions) thus must be attributed to a combination of the mechanics (a),(b) and (c).

As we discussed, the CIM (higher twist QCD) diagrams can temporarily dominate the $2 \rightarrow 2$ Born subprocess contributions because of the trigger bias in single particle high p_T reactions. In the case of jet triggers, the trigger bias is absent, and the QCD Born terms are expected to be dominant even at $p_T \sim 4$ GeV. Thus jet experiments can provide a direct tool to check the basic form of QCD dynamics, verify the form and magnitude of the tri- and quartic-gluon interactions, etc. At present, there is a great deal of uncertainty how to define a jet trigger, particularly because of possibly striking differences in the structure of gluon and quark jets. The study of jet production in two photon physics and the recoil system in deep inelastic scattering should be helpful for establishing workable definitions for jet triggers.

The CIM-QCD approach to large p_T dynamics, combined with dimensional counting rules for determining the leading power behavior, makes a large number of phenomenological predictions (see Refs. 19, 23). Thus far, I am not aware of any serious conflicts with data. In particular, the observed particle ratios such as $pp \to K^- X / pp \to K^+ X$ and beam ratios $\pi p \to \pi X / pp \to \pi X$ are not inconsistent with the CIM (although in the latter case, the situation is complicated by the presence of several competing subprocesses). It is very interesting that corrections to scaling can now be systematically evaluated in perturbative QCD for the higher twist/CIM subprocesses (see Chapter II).

3. Photon Production at Large P_t

In addition to $\gamma\gamma$ collisions (see Chapter III), other photon-induced reactions such as $\gamma p \to \pi X$, $\gamma p \to \gamma X$, and $\gamma p \to$ jet X are sensitive to "direct" QCD reactions such as $\gamma q \to Mq$, and $\gamma q \to \gamma q$, where the incident photon participates in the hard scattering subprocess (and no forward hadrons are produced)[32] as well as standard QCD or CIM subprocesses such as $qq \to qq$, $qM \to qM$, and $qM \to \gamma M$, where the perturbation QCD "anti-scaling" structure function of the incident photon is important.

Photon production at large p_T can also be used as an important probe of the underlying hard scattering subprocesses.[1,2] Discarding photons which are produced from hadron decay ($\pi^0 \to \gamma\gamma$, $\eta^0 \to \gamma\gamma$, etc.), we can distinguish several mechanisms in QCD:

(a) QCD Born contributions with quark fragmentation, e.g.: $e^+e^- \to q\bar{q}$, $qq \to qq$, $gq \to gq$ with $q \to \gamma q$. The $G_{\gamma/q}(x,Q^2)$ fragmentation distribution has the Witten[33] anti-scaling form and is nearly flat in x until x very close to 1. If a QCD 2-2 subprocess dominates both π and γ production, then

$$\frac{\dfrac{d\sigma}{d^3p/E}(pp \to \gamma X)}{\dfrac{d\sigma}{d^3p/E}(pp \to \pi X)} \sim \frac{G_{\gamma/q}(x_T)}{G_{\pi/q}(x_T)} \sim \frac{\alpha}{1 - x_T} \qquad (3.1)$$

at $\theta_{cm} = 90^o$, independent of p_T.

(b) <u>Direct</u> QCD Born contributions, from subprocesses such as:

$$gq \to \gamma q \qquad \text{and} \qquad q\bar{q} \to \gamma g$$

In these processes the photon is produced in the subprocess itself. Since there are no accompanying trigger jet hadrons one can easily distinguish such reactions from fragmentation processes.[2] These reactions can also be important for producing massive lepton pairs at large transverse momentum.

(c) CIM-type subprocesses, such as $Mq \to \gamma q$, where the incident meson M is a correlated $q\bar{q}$ pair in the incident hadron wavefunction.

Both processes (b) and (c) can dominate over (a) for moderate p_T because of the absence of trigger bias suppression. The nominal scaling laws are[2]

$$E\frac{d\sigma}{d^3p}(pp\to\gamma X) \sim \begin{cases} \alpha_s^2\,\alpha_\gamma\,\dfrac{\epsilon^8}{p_T^4} & \text{(a) } qq\to qq, \text{ etc.} \\ \alpha_s\,\alpha\,\dfrac{\epsilon^8}{p_T^4} & \text{(b) } gq\to\gamma q \\ \alpha_M\,\alpha\,\dfrac{\epsilon^9}{p_T^6} & \text{(c) } Mq\to\gamma q \end{cases} \qquad (3.2)$$

where $\alpha_\gamma \sim \alpha/\alpha_s(p_T^2) \sim \alpha \log p_T^2/\Lambda^2$ and $\epsilon = 1-x_T$. A systematic discussion of these contributions and their relative magnitude is discussed in Ref. 2. We found that with conventional parameterizations, the CIM contributions (c) exceed the QCD (a)+(b) terms until $p_T^\gamma \sim 8$ GeV at $\sqrt{s} = 33$ GeV, and until $p_T^\gamma \sim 9$ GeV at $\sqrt{s} = 61$ GeV. The ratio of γ to pion production (parameterized as ϵ^9/p_T^8) is shown in Fig. 10. For $p_T \lesssim 8$ GeV where the CIM subprocesses dominate both pion and photon production, we predict at 90° the cross section ratio:[19,2]

$$\frac{\gamma}{\pi} \cong 0.007\ p_T^2/\text{GeV}^2$$

roughly independent of s. This dependence of p_T and s should be readily distinguishable from the $\gamma/\pi \sim \alpha_\gamma/(1-x_T)$ dependence characteristic of conventional QCD calculations. We also note that the predictions of Fontannaz[34] and Blankenbecler et al.[35] which are based on the $Mq\to\gamma^* q$ subprocess appear to account for a large share of the p_T distribution of messive lepton pairs.[36]

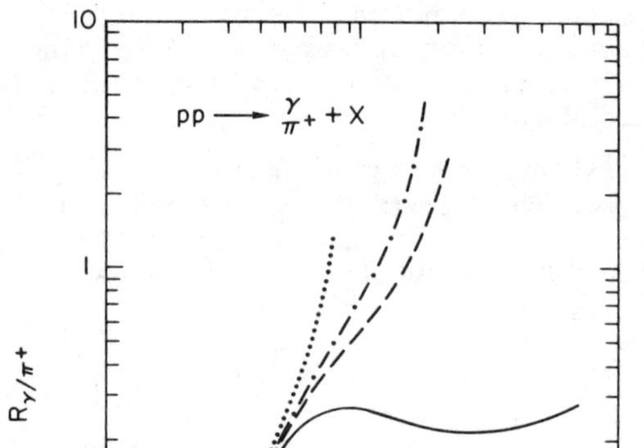

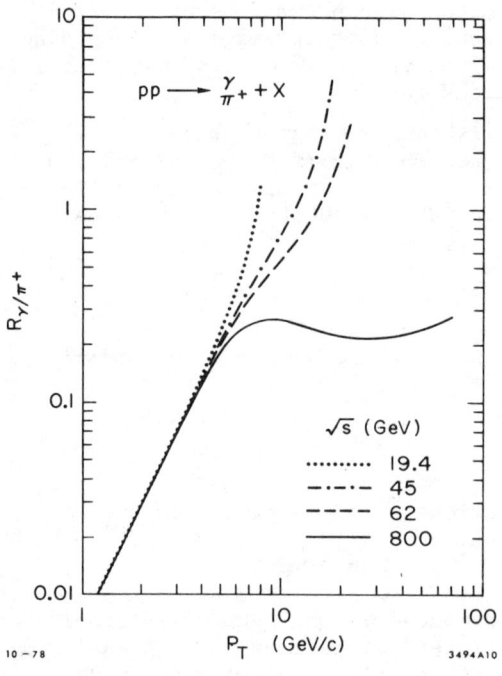

Fig. 10. Predicted ratio from QCD plus CIM contributions for γ/π in pp collisions. (From R. Rückl et al., Ref. 2.)

4. Photon and Gluon Jet Production

It should be emphasized that direct large p_T photon production at the magnitude discussed here is an essential prediction of the hard-scattering approach to hadron dynamics. In particular, since photons and gluons enter subprocesses in a similar manner, there is a close relationship between gluon jet and direct photon production. For example, consider the subprocesses[11]

$$\frac{d\sigma}{dt}(gq \to gq) = \frac{\pi\alpha_s^2}{s^2}\left[-\frac{4}{9}\left(\frac{s}{u}+\frac{u}{s}\right)+\frac{s^2+u^2}{t^2}\right] \tag{4.1}$$

and

$$\frac{d\sigma}{dt}(gq \to \gamma q) = \frac{\pi\alpha\alpha_s e_q^2}{3s^2}\left[-\left(\frac{s}{u}+\frac{u}{s}\right)\right] \tag{4.2}$$

At 90°, this implies

$$\frac{\frac{d\sigma}{d^3p/E}(pp \to \gamma X)}{\frac{d\sigma}{d^3p/E}(pp \to gX)} = \left(\frac{1}{22}\right)\frac{\alpha}{\alpha_s(p_T^2)} \tag{4.3}$$

from these subprocesses alone. Direct photons and gluon jets from these contributions have the same scaling laws, independent of structure functions, k_T smearing, etc. We note that the $gq \to gq$ subprocess gives ~1/4 of the total jet production cross section from all QCD $2 \to 2$ subprocesses.[26] Therefore we have a lower bound

$$\frac{\frac{d\sigma}{d^3p/E}(pp \to \gamma X)}{\frac{d\sigma}{d^3p/E}(pp \to \pi X)} \geq \left(\frac{\mathrm{Jet}}{\pi}\right)_{\mathrm{expt}} \cdot \frac{\alpha}{88\alpha_s(p_T^2)} \tag{4.4}$$

For example, if the jet/π ratio is of order 300 (to 600) as in the FNAL E260 experiment[37] ($p_T \sim 4.5$ GeV, $E_{Lab} = 200$ GeV), then the γ/π lower bound is 12.5% (to 25%). Conversely, the experimental upper bound for γ/π of (.55 ± .92)% as reported by J. H. Cobb *et al.*[38] at $2 < p_T < 3$ GeV, $\sqrt{s} = 55$ GeV implies an upper bound for jet/π production of order 30, which is in severe disagreement with QCD expectations and the trend of experimental results. Thus the production of direct photons may provide one of the most important constraints on QCD subprocesses.

5. Color and Hadron Multiplicity

One of the most intriguing problems in QCD is how to unravel the mechanisms which control the development of hadron multiplicity in large momentum transfer reactions. The "inside-outside" space-time development of hadron production as discussed by Casher, Kogut, and Susskind[39] and Bjorken[40] for $e^+e^- \to q\bar{q} \to$ hadron is consistent with causality and confinement. This picture implies that the fastest hadrons (which contain the valence quarks) are formed last, and the slow polarization cloud first. Weiss and I,[41] building on earlier work,[42] have shown that in such a picture, the charge of a quark jet (on the average) is equal to the charge of parent quark plus the average charge of anti-quarks in the sea:

$$Q_{jet} = Q_q + \langle Q_{\bar{q}} \rangle_{sea} \tag{5.1}$$

Here Q_{jet} is obtained by integrating the charge density in the jet starting from y_0 (anywhere in the central region) to Y_{max}. Gluon jets have $Q_{jet} = 0$. These results hold for all conserved quantum numbers Q.

The inside-outside description of jet dynamics leads to the following ansatz for QCD:[3] Soft hadron production in a hard scattering reaction depends only on the effective color separation. Accordingly, two reactions which initially separate any two 3 and $\bar{3}$ systems (q, $\bar{q}$, $\overline{qq}$, qq, etc.) will have the same distribution of hadrons in the central region. (Only the fragmentation region discriminates the flavor and composition of the jet.) Thus we expect the same multiplicity distributions (e.g., plateau height) in the central region for the hadron system X in $e^+e^- \to X$, $\gamma^* p \to X$, and $pp \to \mu^+\mu^- + X$ (Drell-Yan mechanism), given the same rapidity separation of the 3 and $\bar{3}$ systems. For large p_T reactions, the subprocess $qq \to qq$ leads to four 3 or $\bar{3}$ jets. The multiplicity and associated coherence effects associated with these jets can be computed in analogy with the soft-photon production formulae of QED for the corresponding charge separation reaction, positronium + positronium $\to e^+ + X$ [$e^+e^\pm \to e^+e^\pm$ subprocesses]. The net multiplicity corresponds to 4 quark jets, with coherent enhancement in the interference zone.

An important consequence of the color separation ansatz is that gluon (color 8) jets must have a different soft hadron spectrum than quark jets. In fact, for $N_c \to \infty$, the color separation for a gluon jet is the same as two incoherent quark jets. More generally, the number of soft gluons bremsstrahled from a gluon source compared to a quark source is given by the ratio of Casimir operators for the adjoint and fundamental representation:[3,4]

$$\frac{\langle n^{soft}\rangle_g}{\langle n^{soft}\rangle_q} = \frac{2}{1 - N_c^{-2}} = \frac{9}{4} \quad \text{for color SU(3)} \tag{5.2}$$

Thus we expect that the plateau height for soft gluons (or sea quarks) in the gluon jet is 9/4 that of quark jets (i.e., a color octet has 3/2 the "color charge" of the color triplet). If we assume the density of produced hadrons is linearly related to the sea-quark density, then gluon jets will have more than twice as many soft hadrons in the central region compared to quark jets. Further,[43,44] the energy of a gluon jet will be contained in a larger solid angle due to its increased "straggling" -- again due to the 9/4 color factor. The leading particle distribution in a gluon jet will also be depleted more strongly by soft gluon radiation.

On the other hand the dependence of hadron multiplicity on soft gluon or quark production may not be as strong as linear. For example, the lower density of $g \to q\bar{q}$ pairs in a color triplet jet implies that the average cluster (singlet $q\bar{q}$) mass will be of higher mass than clusters due to the more copious bremsstrahlung from the color octet jet. Since the heavier clusters decay with a higher multiplicity, the net difference between quark and gluon multiplicities may not be as severe as indicated by QCD perturbation theory.[45] Nevertheless, taking into account their different structure at the short distance level, it would be very surprising if the hadron distribution from quark and gluon jets turned out to be identical.

QCD and "Hole" Partons

Several years ago Bjorken[46] postulated the concept of a "hole" parton to describe the development of the final state multiparticle distribution after a deep inelastic lepton reaction. It is an interesting question whether this parton model ansatz has an analogue in QCD.

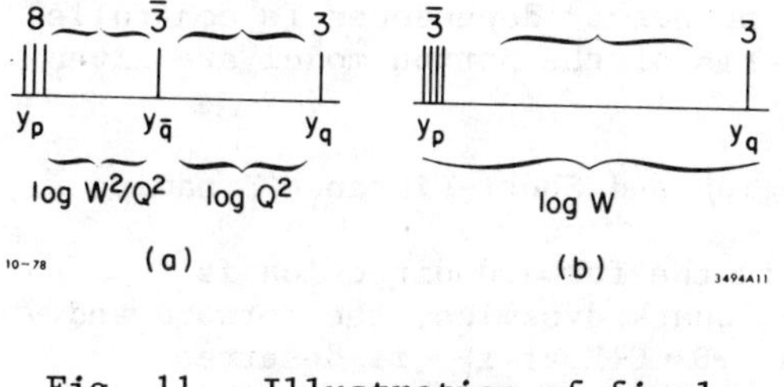

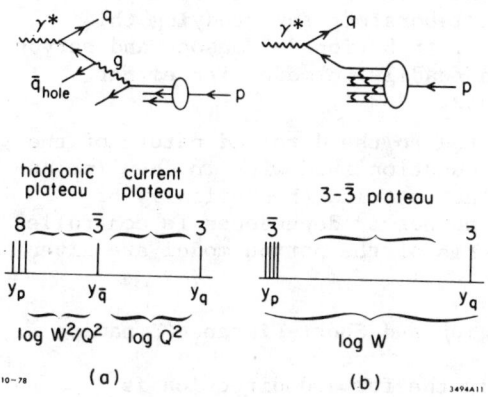

Fig. 11. Illustration of final state hadron distribution deep inelastic lepton scattering on a sea quark arising from (a) gluon bremsstrahlung or (b) a $(q\bar{q}\ qqq)$ Fock state.

A common phenomenological assumption is that sea quarks in a hadron arise as low-mass pair states created from gluon bremsstrahlung. If this perturbative picture is correct, then after a sea quark with rapidity y_0 is struck by a deep inelastic γ or W, the spectator system consists of (1) an antiquark (hole parton) at $y \sim y_0$ with quantum numbers opposite to those of the struck quark, and (2) a leading particle system with the rapidity of the target hadron, but with color 8 (see Fig. 11(a)). There are thus two rapidity regions created from the color neutralization:

(a) a "current" plateau region of length log Q^2 between the 3 and $\bar{3}$, and (b) an "hadronic" plateau of length log (W^2/Q^2) between the 8 and hole parton $\bar{3}$. The density of soft gluons created in the neutralization of the 8-8 system will be 9/4 that of the 3-$\bar{3}$ separated system; thus we expect the height of the hadronic plateau to be higher than that of the current plateau; i.e., the hadronic multiplicity will be a function of both W^2 and Q^2. Despite these expectations, data from deep inelastic electron and neutrino reactions indicate that the current and hadron multiplicity plateaus have equal heights. We note that dual string picture also predicts that the "hadronic" plateau should be twice as high as the "current" plateau.

There is however an alternative description of the proton gluon and quark distribution, which requires giving up a simple perturbative picture of the $q\bar{q}$ sea.[47] The hadronic state is evidentally a complicated coherent color state: all constituents tend to have the same rapidity in order that the system remains a coherent singlet over the semi-infinite time before collision. The virtual gluon, quark, and antiquark states are thus continually exchanging momentum. When a virtual sea quark is struck at y_0, the remaining state is that of a coherent $\bar{3}$ at the original rapidity Y of the target. Because of the exchange of momentum in the initial state, there is no special reason for a $\bar{q}$ with opposite quantum numbers to be at the struck quark rapidity, and there is no "hole" parton (see Fig. 11(b)). Furthermore, there is no separate current or hadronic plateaus; the multiplicity should only depend on log W^2, in agreement with data.

The question of the color and quantum number content of the hadronic state before and after a deep inelastic reaction is a fascinating subject, which deserves much more theoretical and experimental attention. The associated multiplicity in massive lepton pair production events could be an ideal laboratory for studying this problem since both valence and sea distributions of mesons and baryons can be probed, and a comparison can readily be made with either normal events or low-mass pair events.

Another important problem related to the detailed nature of the hadronic wavefunction concerns the question shadowing in deep inelastic events on nuclei. It is still not settled theoretically or experimentally whether the nucleon number A^{α} dependence is controlled by Bjorken x or q^2. Analyses in terms of the parton model are given in Refs. 40 and 48.

6. The Forward Fragmentation Region and Short-Distance Dynamics

Although hadronic scattering in the forward direction is normally not regarded as a probe of quark dynamics, the forward and backward fragementation regions in $A+B \to C+X$ at $x_L^C \sim \pm 1$ deserves special attention. In order for C to have nearly all the momentum of A or B, there must be the exchange of large momentum transfer between constituents which are far off-shell. The forward systems produced in low p_T reactions can be regarded, in a general sense, as hadronic jets and many of their properties (multiplicity, k_T

distributions, quantum number correlations) are not dissimilar from jets in e^+e^- annihilation or large p_T reactions.[3] Blankenbecler and I[49] have emphasized the unity and continuity of physics throughout the Peyrou plot; in particular, the dynamics at the quark and gluon level for large p_T reactions at $x_R = p_C/p_C^{max} \sim 1$ at fixed θ_{cm} must connect smoothly with forward reactions at $x_L \sim 1$ as $\theta_{cm} \to 0$ or π. In particular, Ochs[50] has noted the phenomenological similarity between particle ratios at $\theta_{cm} = 90^o$ and 0° in pp collisions.

The first suggestion that the behavior of the forward fragmentation region in inclusive reactions can be related to the quark distributions in hadrons is due to H. Goldberg.[51] However, the simplest implementation of this idea fails: For example, for the reaction $pp \to \pi^+X$, one can imagine that either before or after an initial soft scattering, a u-quark in the proton, with the distribution $G_{q/p}(x)$ (obtained from deep inelastic lepton scattering) fragments to the fast pion with the distribution $G_{\pi^+/u}$ (obtained from e^+e^- annihilation) (see Fig. 12(a)). Although this ansatz can account for the observed particle ratios in the forward direction, it predicts a too-small and too-steeply falling distribution,

$$\frac{1}{\sigma}\frac{d\sigma}{dx_L}(pp \to \pi^+X) \propto (1-x_L)^5 \qquad \text{(prediction)}^{49} \qquad (6.1)$$

vs.

$$\frac{1}{\sigma}\frac{d\sigma}{dx_L}(pp \to \pi^+X) \propto (1-x_L)^{3.1 \pm 0.5} \qquad \text{(experiment)}^{52} \qquad (6.2)$$

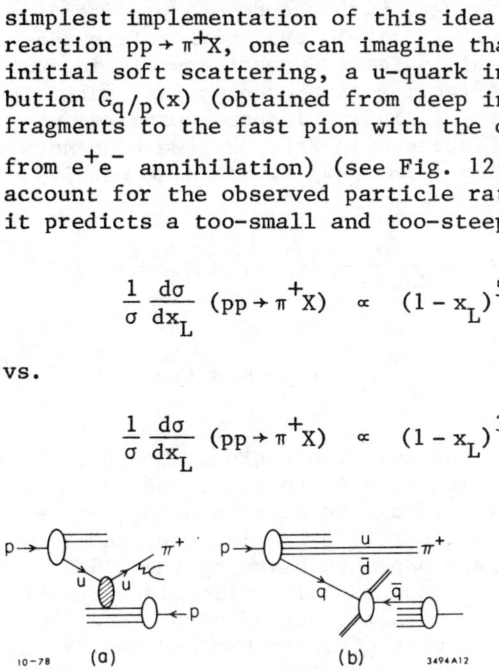

Fig. 12. Production of high energy, low p_T pions in $pp \to \pi^+X$ arising from (a) diffractive or gluon exchange processes or (b) $q\bar{q}$ annihilation of sea quarks.

The prediction (6.1) can also be derived in QCD if one assumes that soft hadronic interactions are represented by gluon exchange.

There is however another possibility:[53] consider the five quark $|uud\bar{q}_{sea}q_{sea}\rangle$ component of the proton wavefunction. The sea quark has a flat distribution in rapidity and can be exchanged or annihilated in the target, giving a constant total cross section [see Fig. 12(b)]. (This is the QCD analogue of Feynman's "wee parton" mechanism for high energy interactions.) The distribution of meson systems $u\bar{q}_{sea}$ in the remaining 4-quark state is

$$\frac{1}{\sigma}\frac{d\sigma}{dx_L}(pp \to \pi^+X) \propto G_{\pi^+/uud\bar{d}}(x) \sim (1-x_L)^3 \qquad (6.3)$$

where we have used the QCD-based spectator counting rule,[49,54]

$$G_{a/A}(x) \propto (1-x_L)^{2n_s-1} \qquad (x \to 1) \qquad (6.4)$$

where n_s is the number of bound spectators which are required to stop as $x = (k_0^a + k_3^a)/(p_0^A + p_3^A) \to 1$. Notice that for the gluon exchange mechanism, there are 3 spectators for $pp \to \pi^+ X$, versus 2 spectators for the wee-quark exchange case.

A large number of forward reactions have been measured; the results are generally in good agreement with the powers predicted by the q-exchange mechanisms.[53] It is likely that both soft q and g exchange mechanisms are important in forward reactions; it is just that sea quark exchange is more effective in producing fast particles. Other consequences of this picture, including induced correlations between particles at $x_L = \pm 1$ are discussed in Ref. 53. We also note that two particle correlations at $x_L(1) + x_L(2) \to 1$ are also readily predicted:

$$\frac{dN}{dx_1\,dx_2}(pp \to \pi^+\pi^+X) \sim \frac{dN}{dx_1\,dx_2}(pp \to \pi^+\pi^-X)$$

$$\gg \frac{dN}{dx_1\,dx_2}(pp \to \pi^-\pi^-X) \qquad (6.5)$$

where we utilize the two valence quarks in the proton. Tests of these ideas can illuminate the multi-quark correlations in hadronic wavefunctions. A recent test of the quark spectator rule for the distribution of fast forward particles in large p_T reactions in correlation with various triggers has been given by the CCHK group.[55] There are also a number of successful applications of this rule to nuclear-induced reactions. An alternative parton model for forward-fragmentation processes has been given by Das and Hwa.[56] A comparison between these approaches and applications to Drell-Yan processes is given in Ref. 57.

7. Gluon Jets

The essential property of QCD which distinguishes it from a generalized quark-parton model, is the prediction of jets derived from the initial creation of a gluon quantum. Gluon jets are predicted in e^+e^- annihilation (3-jet decay from $e^+e^- \to q\bar{q}g$) and in deep inelastic scattering ($eq \to eqg$). The identification of multi-jet events corresponding to such subprocesses is not completely straightforward because of severe backgrounds from subprocesses such as $e^+e^- \to \pi q\bar{q}$; the relatively large $q \to Mq$ coupling dominates the $q \to gq$ process until quite large p_T. Selection of events with high multiplicity could be used to favor gluon jet production.

By comparing the processes $q\bar{q} \to \gamma + g$ and $q\bar{q} \to \mu^+\mu^-$ for high p_T γ or μ^+ production, we can obtain a prediction for gluon jet production which is independent of the initial state:[58]

$$\frac{\dfrac{d\sigma}{d^3p/E}(AB \to \gamma X)}{\dfrac{d\sigma}{d^3p/E}(AB \to \mu^+ X)} = \frac{\frac{4}{3}\alpha_s(p_T^2)}{\alpha}\,\frac{4}{\langle \sin^2\hat{\theta}\rangle} \tag{7.1}$$

Here $\sin^2\hat{\theta}$ is the subprocess center-of-mass production angle. Low mass $\gamma^* \to \ell\ell^-$ pairs can be used here to avoid backgrounds.

The decay of heavy quark systems $Q\bar{Q}$ such as the Υ into 3 gluons[5] or γ+2 gluons[6] could provide the cleanest test of QCD gluon jet phenomenology. The standard perturbation formulae for positronium decay, updated for color factors gives the branching ratio[59,60]

$$\frac{\Gamma(\Upsilon \to \gamma gg)}{\Gamma(\Upsilon \to ggg)} \cong \frac{36}{5}\left(\frac{e_Q}{e}\right)^2 \frac{\alpha}{\alpha_s(M_\Upsilon^2)} \tag{7.2}$$

where $Q^2 \cong M_\Upsilon^2$ is the effective off-shell value to be used in the running coupling constant. If $e_Q = 1/3$, the branching ratio is ~3%. Predictions for the angular distributions of the Υ decay plane relative to the beam axis and decay distributions are given in Ref. 6. The $\Upsilon \to \gamma gg$ two-jet channel is particularly interesting since the $gg \to X$ mass can be varied and a direct comparison with SPEAR $q\bar{q}$ jets at the same energy can be made. The predicted spectrum based on perturbation theory as a function of $M_{gg}^2 = M_\Upsilon^2(1-x_\gamma)$, $x_\gamma = p_\gamma/p_{max}$, is shown in Fig. 13. Resonances with the gg quantum numbers (η, η', η_c, glueballs) can be expected to modulate the perturbative prediction over a local region if we assume local duality.

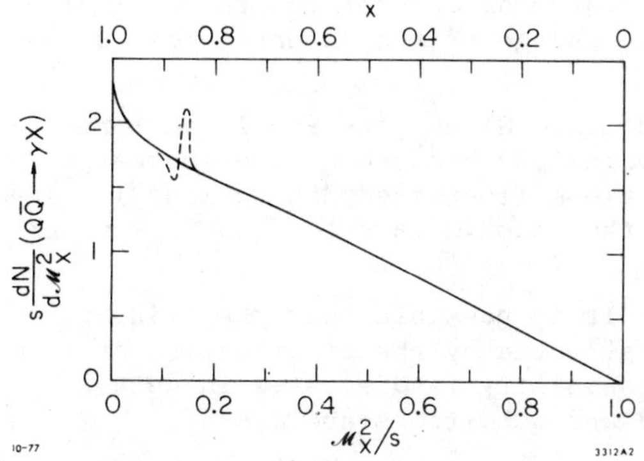

Fig. 13. Decay distributions for $\Upsilon \to \gamma + X$ in $x = 2\omega/M_\Upsilon$ and $\mathcal{M}_X^2/s$ from the simpliest QCD diagrams $\Upsilon \to g+g+\gamma$, and massless gluons. The modulation of a singlet resonance at fixed $\mathcal{M}_X^2$ is shown schematically.

It should be noted that all of these predictions for gluon jet production treat the gluon as strictly massless. Although this is evidently correct for QCD matrix elements, the fact that the gluon "decays" to a massive jet may indicate that we should include mass spectrum effects and thresholds in the phase space calculations. Such effects could distort simple QCD predictions; e.g., the $\Upsilon \to \gamma gg/ggg$ ratio will be enhanced. We also note that higher order (in α_s) channels $\Upsilon \to gq\bar{q}$ and $q\bar{q}$ could be relatively more important than indicated by perturbation theory if the gluon jet has an effectively heavier mass spectrum than the quark jet.[61]

To summarize, let us list the discriminants which could distinguish quark and gluon jets:

(a) Multiplicity. As discussed in Section 6, color octet separation leads to multiplicity of soft gluons and sea quarks 9/4 as large as color triplet separation.[3,4] If this translates into higher hadron multiplicity, then $\Upsilon \to 3g$ decay events with low sphericity will have a higher rapidity plateau in the central region with respect to the $g + (gg)$ jet axis.

(b) Leading particles. If we trust lowest order QCD perturbation theory, then the distribution of charged particles as $x \to 1$ falls off faster in gluon jets compared to hadron jets. A simple form which has the predicted $x \to 0$ and $x \to 1$ limiting behavior is[3,62]

$$D_{H^\pm/g}(x) = \frac{9}{8}\left[D_{H/q}(x) + D_{H/\bar{q}}(x)\right](1-x) \qquad (7.3)$$

Gluon jets, however, may have enhanced number of $I = 0$ states at $x \to 1$ which have a strong gluon component, e.g., $g \to \eta, \omega, \psi$, etc.[63]

(c) Quantum numbers. The total charge of the jet in its fragmentation region is related to the charge of the parent as discussed in Section 6.

(d) Transverse momentum distribution. Gluon jets should be more diffuse (large $\langle k_T^2 \rangle$) than quark jets because of the increased number of soft gluon interactions (increased "straggling").[43,44] This effect also results if the gluon decays to $q\bar{q}$ before color neutralization occurs.

(e) Gluon jets may be "oblate." It is possible that the (linear) polarization of a gluon is reflected by the distribution of hadrons in the jet. This possibility is discussed in detail in a recent paper by DeGrand and Schwitters and myself.[7]

For example, suppose that hadrons are produced from gluon jets after the decay $g \to q\bar{q}$. Then by convolution,

$$G_{H/g}(z,\phi) \cong \int_z^1 \frac{dx}{x}\, G_{H/q}\left(\frac{z}{x}\right) G_{q/g}(x,\phi) + (q \to \bar{q}) \ . \qquad (7.4)$$

In lowest order perturbation theory spin 1/2 quarks from $g \to q\bar{q}$ are aligned with respect to the gluon linear polarization:

$$G_{q/g}(x,\phi) \propto [1 - 4 \cos^2\phi \; x(1-x)]$$

$$\cos \phi = \hat{q} \cdot \hat{\varepsilon} \qquad (7.5)$$

This then implies a sum rule for the momentum weighted distribution of hadrons:

$$\frac{d\varepsilon}{d\phi} \equiv \sum_H \int_0^1 dz \; z \; D_{H/g}(z,\phi)$$

$$= \frac{1}{4\pi} (1 + 2 \sin^2\phi) \quad . \qquad (7.6)$$

In this model, hadrons are 3 times more likely to be produced orthogonal rather than parallel to $\hat{\varepsilon}$, thus producing a non-cyclindrical "oblate" jet. Oblateness can be determined experimentally by finding the principal axes of $\sum_H p_i^H p_j^H$ as in sphericity analyses.

Equation (7.6) should be regarded as an upper limit to the oblateness effect in QCD, since (1) not all hadrons arise from the q and $\bar{q}$ decay products, and (2) the "straggling" from $g \to g_{soft} + g$ due to soft gluon emission depolarizes the gluon. The latter effect is of order $\alpha_s(s)$ and can probably be diminished by selecting events with fast hadrons. The main problem is that gluons are not produced 100% linearly polarized in a given direction.

For example, in $\eta(Q\bar{Q}) \to g+g \to q\bar{q} + q\bar{q}$ (pseudoscalar decay analogue of π^0 double Dalitz decay), the correlation between gluon polarizations is

$$\frac{dN}{d\psi} = \frac{1}{9\pi} (4 + \sin^2\psi) \qquad (7.7)$$

and

$$\frac{d\varepsilon}{d\psi} \equiv \sum_{H_a, H_b} \int_0^1 dz_a \int_0^1 dz_b \; \frac{dN}{dz_a \, dz_b \, d\psi} \, z_a \, z_b$$

$$= \frac{15 + 2 \sin^2\psi}{32\pi} \qquad (7.8)$$

gives the summed correlation between hadrons of the two jets. The maximal effect is only 13%. Similarly in $\Upsilon \to 3q$, the polarization of each gluon is correlated with the normal to the decay plane. Summing over hadrons (from $g \to q\bar{q} \to \text{hadrons} + q + \bar{q}$) gives

$$\frac{d\varepsilon}{d\chi} = (X_1^2 + X_2^2 + X_3^2) + \frac{1}{4}(1 - 2\cos^2\chi)\, X_1 X_2 X_3 \qquad (7.9)$$

where $\cos\theta_{23} = 1 - X_1$ is the cosine of the angle between the gluon jets, and $\cos\chi = p_{\hat{H}} \cdot \hat{n}$ is the projection of the hadron direction with the decay plane normal. The maximal effect occurs for $\theta_{ij} = 120^\circ$ ("tripod" configuration), where we predict that a hadron is 9/7 more likely to be aligned in the plane than normal to the plane.

Finally, for $e^+e^- \to q\bar{q}g$ or $eq \to eq'g$ events, the distribution of gluon polarization is given by

$$z \frac{dN}{dz\, d\phi\, d^2k_T} = zG_{g/q}(z,\phi,\vec{k}_T^2)$$

$$\sim \frac{4}{3}\frac{\alpha_s}{\pi^2}[z^2 + 4(1-z)\cos^2\phi]\frac{\vec{k}_T^2}{(\vec{k}_T^2 + z^2m_q^2)^2} \qquad (7.10)$$

where $\cos\phi = \hat{\varepsilon}\cdot\hat{n}$ and n is in the plane of $q\bar{q}$ or qq'. The average over ϕ gives the standard $1 + (1-z)^2$ distribution.

Although these model calculations give only a first estimate, it seems likely that the spin-1 nature of the gluon in QCD will be reflected in the oblateness of the distribution of its decay products.

8. The Gluon Distribution in Hadrons[8]

Another important question concerning the hadron wavefunction is the nature of its gluon distribution. In QCD, there are three essentially different sources of gluons within a meson or baryon, as discussed in Chapter II.

Often only gluons from quark bremsstrahlung (type (a)) have been taken into account in the standard QCD phenomenological analyses: one assumes that at some Q_0^2 the proton only consists of valence quarks, and that the gluons and sea quarks can be generated by QCD evolution equations for $Q^2 \neq Q_0^2$. In such an approach the probability distribution for quarks is sufficient to determine the probability distribution for gluons. Only "diagonal" terms are computed; off-diagonal diagrams involving two quarks are not considered. This analysis reproduces the q^2-dependent QCD moments for structure functions.[64]

However, for x and $\vec{k}_T \to 0$ (the long wavelength limit) the gluon only "sees" a color singlet source; thus there must be coherent cancellations between the different quark currents. The diagonal approximation can only be accurate for large transverse momentum gluons (see Fig. 14).

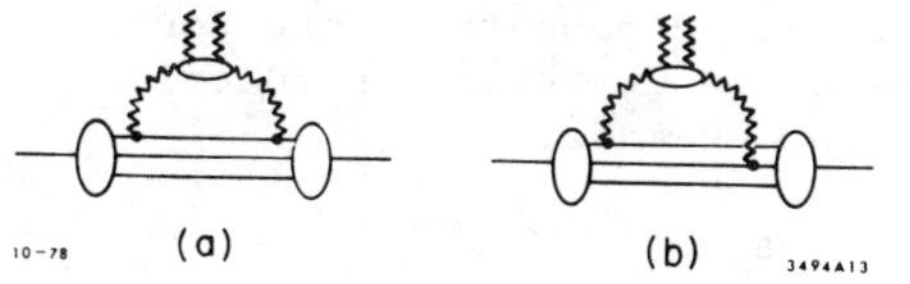

Fig. 14. Contributions to quark sea from (a) diagonal, and (b) off-diagonal gluon contributions. Only (a) is considered in usual analyses.

Gunion and I[8] have recently considered a simple gauge theory model of the meson which preserves gauge invariance and allows a detailed study of the color coherence effects. (The same physics also occurs in QED when one determines the photon distribution in a neutral atom, such as positronium). In the model, the gluon distribution can be computed in lowest order analytically for all x and k_T. For small x, we find

$$G_{g/M}(x,\vec{k}_T^2) = \frac{8}{3}\frac{\alpha_s}{\pi^2}\frac{1}{x}\frac{1}{\vec{k}_T^2}\left[1 - F_M\left(\frac{\vec{k}_T^2}{(1-\bar{x}_q)^2}\right)\right] \tag{8.1}$$

where $F_M(Q^2) \cong 1/(1+Q^2/M_V^2)$ is the meson electromagnetic form factor. The first term in the bracket is the usual (diagonal) contribution obtained from the convolution of $G_{q/M}$ or $G_{\bar{q}/M}$ with $G_{g/q}$. The F_M term from the (off-diagonal) coherence of the q and $\bar{q}$ distributions is only unimportant at large $k_T^2 >> (1-\bar{x}_q)^2 M_V^2$, where $\bar{x}_q \cong 1/2$ is the average momentum fraction of the quark in the meson, and M_V sets the scale of the electromagnetic form factor and hadron size. The coherence of the color singlet bound state eliminates the usual infrared divergences at $\vec{k}_T^2 \to 0$. In this simple model, the standard denominator for a quark target $(\vec{k}_T^2 + x^2 m_q^2)^{-1}$ is replaced by $(\vec{k}^2 + M_V^2)^{-1}$; i.e., there is no quark mass singularity for $m_q \to 0$.

The most important consequence for phenomenology is the fact that the gluon distribution in a hadron reflects its size and constituency. The gluon momentum and sea quark fractions will be bigger the larger the size (λ_H) of the hadron $xG_{g/H}(x) \sim \log\left[(1 + \lambda_H^2\, k_{T\max}^2)\right]$. In addition, the gluon distribution in a hadron clearly tends to increase with the number of quark constituents. Eventually, at large enough log q^2 the QCD radiative corrections will cause the structure functions to contract to the $x \sim 0$ region, and the gluon and quark momentum fractions will reach an asymptotic equilibrium independent of the nature of the target. However, in the preasymptotic domain, target effects are important for determining the gluon and sea-quark distributions. A number of applications are discussed in Ref. 7 including the prediction that the gluon momentum fraction in mesons at present q^2 is appreciably smaller than in nucleons. This prediction can be tested in reactions such as ψ production or gluon jet production in hadronic collisions, where a gluon-induced subprocess is expected to play a major role.[65]

The models used thus far for the gluon distribution in hadrons are primitive and can only take into account perturbative effects. Non-perturbative claculations which can account for final state interaction effects, and higher Fock state components in the bound state wavefunction will be required before a definitive prediction of the gluon distribution in a hadron can be made.

9. Conclusions

Although there are tantalizing hints of success, there is as yet no convincing quantitative evidence that inclusive hadronic reactions are described by perturbative quantum chromodynamics. A great deal of experimental and theoretical work will be required to provide bona fide tests of the theory at even the 10-20 percent level. Among the outstanding problems:

(1) The production cross section for jets in hadron-hadron collisions is not known to within a factor of 2 or 3, let alone its scaling properties at fixed x_T and θ_{cm}. If the combined QCD plus CIM description given here is correct, the jet/π cross section should increase as $\sim p_T^2$ at fixed x_T and θ_{cm} for $p_T \gtrsim 4$ GeV/c. Scale breaking due to QCD radiative corrections are discussed in Chapter II. The k_T smearing effect for $p_T > 4$ GeV/c changes the predictions by less than a factor of 2 if off-shell kinematics are used.[18]

(2) The existence of charge correlations between the trigger and away side hadrons, as observed by the BFS collaboration[25] evidentally eliminate 2 to 2 QCD Born subprocesses as the dominant hard scattering mechanism for single hadron production in the region up to $p_T > 4$ GeV. The extension of these measurements to higher p_T and x_T is critical. Nuclear targets tend to obscure flavor correlations because of charge averaging and final state interactions.

(3) Cross sections for hadron pairs at large $\mathcal{M}^2$ tend to be insensitive to the controversial k_T smearing effect. It is particularly interesting to compare hadron pairs and muon pairs at the same kinematics. One predicts[66]

$$\frac{\dfrac{d\sigma}{d\mathcal{M}^2 dy}(pp \to H^+H^-X)}{\dfrac{d\sigma}{d\mathcal{M}^2}(pp \to \mu^+\mu^-X)} = \left(\frac{1}{\mathcal{M}^2}\right)^k f\left(y, \frac{\mathcal{M}^2}{s}\right) \qquad (9.1)$$

where $k = 2$ for meson pairs and $k = 4$ for baryon pairs. If 2 to 2 QCD diagrams are dominant, then $k = 0$, and there are only minor scale violations from the relevant structure functions and an overall factor of $[\alpha_s(\mathcal{M}^2)/\alpha]^2$.

(4) It is very important that QCD predictions for direct high p_T photon reactions be tested, starting with the original Bjorken-Paschos[67] inelastic Compton reaction $\gamma p \to \gamma X$ and inclusive photoproduction $\gamma p \to \pi X$ (reactions without forward hadrons) to direct photon production $pp \to \gamma X$, two photon processes $\gamma\gamma \to X$, $e^+e^- \to \gamma + \pi^\pm + X$ (charge asymmetry), and $e^\pm p \to e^\pm \gamma X$ ($e^\pm$ asymmetry).[1,2] The photon is the only non-colored elementary field that directly participates in QCD dynamics at short distances; unless its pointlike couplings to quarks are confirmed, predictions for perturbative processes involving gluons are probably meaningless. The close relationship between photon production to gluon and quark jet production is discussed in Section 5. We also note the remarkable fact that the asymptotic photon structure function is scale-invariant up to an overall factor of $\alpha_s^{-1}(p_T^2)$, and photon-induced cross sections such as $ed\sigma/d^3p$ ($\gamma\gamma \to$ Jet + X) are asymptotically scale free and independent of $\alpha_s(p_T^2)$ when perturbative contributions to all orders are included (see Chapter III).

(5) The complete picture of quark and gluon distributions in hadrons will require attention to coherent effects and multiparticle correlations, as discussed in Section 8. Measurements of the final states in deep inelastic processes and massive lepton pair production processes, together with comparisons with low q^2 and low $\mathcal{M}^2$ events, can give detailed information on the evolution of multiquark and gluon jets, including the effect of color separation, "hole" parton production, and the influence of nuclear targets.

(6) Perhaps the most convincing evidence for underlying scale-invariant quark interactions comes from large momentum transfer exclusive measurements such as the form factors at large t and hadron scattering and photoproduction at large t and u.

As we have discussed in Chapter II, this is potentially the most important testing ground of the dynamics and symmetry properties of QCD.

References

1. C. O. Escobar, DAMTP 78/9 (1978); G. R. Farrar and S. C. Frautschi, Phys. Rev. Lett. 36, 1017 (1976); and J. D. Bjorken et al., Phys. Rev. D4, 3388 (1971). An indication that the interactions of large transverse moment real photon have point-like interactions is given by the measurements of the $e^+ - e^-$ asymmetry is deep inelastic bremsstrahlung by D. L. Fancher et al., Phys. Rev. Lett. 38, 800 (1977). See also S. J. Brodsky, J. F. Gunion and R. L. Jaffe, Phys. Rev. D6, 2487 (1972).

2. R. Rückl, S. J. Brodsky and J. Gunion, SLAC-PUB-2115 (to be published in Phys. Rev.); F. Halzen and D. Scott, Phys. Rev. Lett. 40, 1117 (1978); and H. Fritzsch and P. Minkowski, Phys. Lett. 63B, 99 (1976).

3. S. J. Brodsky and J. F. Gunion, Proceedings of the 7th International Collisions on Multiparticle Practions, Munich (1976); and Phys. Rev. Lett. 37, 402 (1976).

4. K. Konishi, A. Ukawa and G. Veneziano, CERN-TH-2509 (1978).

5. T. DeGrand, Y. J. Ng and S. H. H. Tye, Phys. Rev. D16, 3251 (1977); K. Koller, H. Krasemann and T. F. Walsh, DESY 78137 (1978); K. Koller and T. Walsh, 781/6 (1978); and A. DeRujula, J. Ellis, E. G. Floratas and M. K. Gaillard, CERN preprint TH-2455 (1978).

6. S. J. Brodsky, D. G. Coyne, T. A. DeGrand and R. R. Horgan, Phys. Lett. 73B, 203 (1978).

7. S. J. Brodsky, T. A. DeGrand and R. F. Schwitters, SLAC-PUB-2160 (1978) (to be published in Phys. Lett.).

8. S. J. Brodsky and J. F. Gunion, SLAC-PUB-2163 (1978).

9. D. Sivers, R. Blankenbecler and S. J. Brodsky, Phys. Reports 23C:1 (1976); and M. Jacob and P. V. Landshoff (to be published in Phys. Reports).

10. R. D. Field, CALT-68-683 (presented at the XIX International Conference on High Energy Physics, Tokyo, 1978, and references therein).

11. R. Cutler and D. Sivers, Phys. Rev. D16, 679 (1977) and Phys. Rev. D17, 196 (1978); and B. L. Combridge, J. Kripfganz and J. Ranft, Phys. Lett. 234 (1977).

12. R. Blankenbecler, S. J. Brodsky and J. F. Gunion, Phys. Rev. D18, 900 (1978); and P. V. Landshoff and J. C. Polkinghorne, Phys. Rev. D8, 927 (1973), and Phys. Rev. D10, 891 (1974).

13. We cavalierly refer to subprocesses which involve more than the minimum number of external fields as "high twist".

14. S. J. Brodsky and G. Farrar, Phys. Rev. D11, 1309 (1975).

15. Note, however that the standard QCD corrections to the structure functions $G_{a/A}(x,Q^2)$ depend on the color Casmir operators of a and thus vanish if the constituent is a color singlet (see Chapter II).

16. S. D. Ellis, M. Jacob and P. V. Landshoff, Nucl. Phys. B108, 93 (1976). See also J. D. Bjorken and G. R. Farrar, Phys. Rev. D9, 1449 (1974).

17. P. V. Landshoff, J. C. Polkinghorne and R. D. Short, Nucl. Phys. B28, 222 (1971); and S. J. Brodsky, F. E. Close and J. F. Gunion, Phys. Rev. D8, 3678 (1973).

18. For an extensive discussion see W. E. Caswell, R. R. Horgan and S. J. Brodsky, SLAC-PUB-2106 (to be published in Phys. Rev.). Detailed calculations for QCD subprocesses are given by R. R. Horgan and P. Scharbach, SLAC-PUB-2188 (1978). The importance of off-shell kinematics in calculating k_T fluctuations has also been discussed by K. Kinoshita and Y. Kinoshita, KYUSHU-78-HE-6 (1978); M. Chase, DAMTP 77/29 (1977); and R. Raitio and R. Sosnowski, HU-TFT-77-22 (1977).

19. R. Blankenbecler, S. J. Brodsky and J. F. Gunion, Ref. 12, and Phys. Rev. D6, 2652 (1972).

20. S. J. Brodsky and G. Farrar, Phys. Rev. Lett. 31, 1/53 (1973); and V. A. Matveev, R. M. Muradyan and A. N. Tavkheldize, Lett. Nuovo Cimento 7, 719 (1973).

21. D. Antreasyan et al., preprint EFI 78-29 (1979), and Phys. Lett. 38, 112 (1977); and J. W. Cronin et al., Phys. Rev. D11, 3105 (1975). See also M. Shochet, Proceedings of the XVIII International Conference on High Energy Physics, Tblisi (1976).

22. A. G. Clark et al., Phys. Lett. 74B, 267 (1978). CERN-Columbus-Oxford-Rockefeller Expt., reported by L. Di Lella in Workshop on Future ISR Physics, September (1977), edited by M. Jacob.

23. D. Jones and J. F. Gunion, SLAC-PUB-2157 (1978).

24. R. P. Feynman, R. D. Field and G. C. Fox, Nucl. Phys. B128, 1 (1977); R. D. Field and R. P. Feynman, Phys. Rev. D15, 2590 (1977); R. Baier, J. Cleymans, K. Kinoshita and B. Peterson, Nucl. Phys. B118, 139 (1977); and J. Kripfganz and J. Ranft, Nucl. Phys. B124, 353 (1977).

25. M. G. Albrow et al., NBI preprints (1978); R. Moller, Symposium on Jets in High Energy Collisions (1978); and K. H. Hansen, presented to the XIX International Conference on High Energy Physics, Tokyo (1978), and Nucl. Phys. B135, 461 (1978).

26. R. P. Feynman, R. D. Field and G. C. Fox, CALT-68-651 (1978). See also A. P. Contogouris, R. Gaskell and S. Papadopoulos, Phys. Rev. D17, 2314 (1978); A. P. Contogouris, McGill preprint (1978); J. F. Owens and J. D. Kimel, FSU-HEP-780330 (1978); J. F. Owens, FSU-HEP-780609 (1978), and Phys. Lett. 76B, 85 (1978); J. Ranft and G. Ranft, preprints KMU-HEP-7806 and 7805 (1978), and Lett. Nuovo Cimento 20, 669 (1979); and M. Fontannaz, Nucl. Phys. B132, 452 (1978).

27. A simplified analysis of charge correlations is given in S. J. Brodsky, Proceedings of the VII International Symposium on Multiparticle Dynamics, Kaysersberg, France, June (1977). An analysis of the QCD Born subprocesses is given by R. P. Feynman et al., Ref. 26.

28. R. J. Fisk et al., Phys. Rev. Lett. 40, 984 (1978).

29. J. D. Bjorken and J. B. Kogut, Phys. Rev. D8, 1341 (1973).

30. Y. L. Dokshitser, D. I. Dyakanov and S. I. Troyan, SLAC-TRANS-0183, Trans. from Proceedings of the 13th Leningrad Winter School on Elementary Particle Physics (1978); Leningrad preprint (1978).

31. J. C. Vander Velde, presented at the Symposium on High Energy Collisions, Niels Bohr Institute and Nordita, July (1978) (to be published in Physica Scripta); and J. S. Bell et al., UMBC-9 (1978).

32. This point has been emphasized by W. Ochs and L. Stodolsky (unpublished).

33. F. Witten, Nucl. Phys. B120, 189 (1977); and W. Frazer and J. F. Gunion, UCSD preprint 10P10-194 (1978).

34. M. Fontannaz, Phys. Rev. D14, 3127 (1976).

35. M. Duong-van, K. V. Vasavada and R. Blankenbecler, Phys. Rev. D16, 1389 (1977).

36. C. M. Debeau and D. Silverman (to be published) have given a combined QCD + CIM calculation of the massive lepton pair transverse momentum distribution.

37. C. Bromberg et al, Phys. Rev. Lett. 38, 1447 (1977), and Nucl. Phys. B134, 189 (1978); and M. D. Corcoran et al., presented at the XIX International Conference on High Energy Physics, Tokyo (1978).

38. J. K. Cobb et al., CERN preprint 78-0925 (1978).

39. A. Casher, J. Kogut and L. Susskind, Phys. Rev. Lett. 31, 792 (1973).

40. J. Bjorken, SLAC-PUB-1756, Lectures given at the International Summer School in Theoretical Physics, DESY (1975).

41. S. J. Brodsky and N. Weiss, Phys. Rev. D16, 2325 (1977).

42. G. R. Farrar and J. Rosner, Phys. Rev. D7, 2747 (1973); J. L. Newmeyer and D. Sivers, Phys. Rev. D9, 2592 (1974); and R. Cahn and E. Colglazier, Phys. Rev. D9, 2592 (1974).

43. K. Shizuya and S. H. H. Tye, Fermilab-PUB-78/54 (1978).

44. M. Einhorn and B. Weeks, SLAC-PUB-2164 (1978).

45. A. Casher, H. Neuberger and S. Nussinov, Tel Aviv preprint TAUP-694-78 (1978).

46. J. Bjorken, Phys. Rev. D7, 282 (1973).

47. See Refs. 3, 19, and 41.

48. S. J. Brodsky, J. F. Gunion and J. Kühn, Phys. Rev. Lett. 39, 1120 (1977).

49. S. J. Brodsky and R. Blankenbecler, Phys. Rev. D10, 2973 (1974).

50. W. Ochs, MPI-PAE/PTH 33/77 (1977), Proceedings of the 12th Rencontre de Moriond, Flaine (1977), and Nucl. Phys. B118, 397 (1977).

51. H. Goldberg, Nucl. Phys. B44, 149 (1972).

52. J. C. Sens et al., CHLM Colloboration (unpublished).

53. S. J. Brodsky and J. F. Gunion, Phys. Rev. D17, 848 (1978).

54. If the spin of the leading particle does not match that of the projectile, we expect an additional suppression factor $(1-x)^{2|S_z^a - S_z^A|}$. S. J. Brodsky, J. Gunion, N. Fuchs and M. Scadron (to be published).

55. E. E. Kluge et al. (CCHK Collaboration), presented at the Symposium on Jets in High Energy Collisions, Niels Bohr Institute and Nordita, July (1978) (to be published in Physica Scripta); and D. Drijard et al., CERN/EP/Phys. 78-14 (1978).

56. K. P. Das and R. C. Hwa, Phys. Lett. 68B, 459 (1977), and 73B, 504 (1978).

57. T. A. DeGrand and H. I. Miettinen, Phys. Rev. Lett. 40, 612 (1978); and T. A. DeGrand, SLAC-PUB-2182 (1978).

58. S. J. Brodsky, W. Caswell and R. Horgan (unpublished). See S. J. Brodsky, SLAC-PUB-1937, Proceeding of the 12th Recontre de Moriond, Flaine (1977).

59. M. Chanowitz, Phys. Rev. D12, 918 (1975); and L. Okun and M. Voloshin, ITEP-95 (1976).

60. V. A. Novikov *et al.*, Phys. Report 41C, 1 (1978).

61. S. J. Brodsky, M. Dine and E. Fahri (in progress).

62. S. J. Brodsky, Ref. 58.

63. H. Fritzsch and K. H. Streng, CERN-TH-2520 (1978).

64. See, e.g., A. Buras and K. Gaemer, Nucl. Phys. B132, 249 (1978). For a detached discussion of the diagonal approach see F. Martin, SLAC-PUB-2192 (1978).

65. S. Ellis and M. Einhorn, Phys. Rev. D12, 2007 (1975), and Phys. Rev. Lett. 36, 1263 (1976); and C. Carlson and R. Suaya, Phys. Rev. D14, 3115 (1976), and Phys. Rev. D18, 760 (1978).

66. R. Blankenbecler and S. J. Brodsky (unpublished).

67. J. D. Bjorken and E. Paschos, Phys. Rev. 185, 1975 (1969).

Acknowledgements

Parts of these lectures overlap with material presented at the Symposium on Jets in High Energy Collisions (Copenhagen, July 1978; SLAC-PUB-2217), the Joseph M. Weis Memorial Symposium on Strong Interactions, University of Washington, November 1978; SLAC-PUB-2240), and the Workshops on High Energy Physics (Cal Tech, February 1979; SLAC-PUB-2294).

Chapter II was written in collaboration with G. Peter Lepage. The new applications of QCD to exclusive processes are based on work done with Lepage and also with Y. Frishman and C. Sachrajda, and I am grateful to them for many helpful conversations.

A large part of these lectures are based on collaborations with many other colleagues, including E. Berger, R. Blankenbecler, C. Carlson, W. Caswell, G. Farrar, T. DeGrand, J. Gunion, R. Horgan, H. Lipkin, R. Rückl, R. Schwitters, and N. Weiss. I wish to thank them and the participants at the La Jolla Workshops for many helpful discussions.

I am also grateful to William Frazer, the organizer of these workshops, for his assistance and hospitality.

APPLICATIONS OF QUANTUM CHROMODYNAMICS

R.D. Field
California Institute of Technology

I.	INTRODUCTION	98
II.	THE EFFECTIVE COUPLING $\alpha_s(Q^2)$	100
III.	CORRECTIONS TO R IN e^+e^- HADRONS	108
IV.	THE QUARK AND GLUON DISTRIBUTIONS	112
V.	ANALYSIS OF CHARGED CURRENT NEUTRINO AND ANTINEUTRINO PROCESSES	135
VI.	QCD PERTURBATION THEORY	149
VII.	LARGE-MASS MUON-PAIR PRODUCTION	183
VIII.	QUARK AND GLUON "JETS": FRAGMENTATION FUNCTIONS	199
IX.	LARGE $p_\perp$ MESON AND "JET" PRODUCTION IN HADRON-HADRON COLLISIONS	202
X.	THE SEARCH FOR THREE-JET EVENTS	227
XI.	SUMMARY AND CONCLUSIONS	233

ISSN: 0094-243X/79/50097-146$1.50

APPLICATIONS OF QUANTUM CHROMODYNAMICS†*

R. D. Field
California Institute of Technology, Pasadena, California 91125

ABSTRACT

Perturbative applications of the theory of Quantum Chromodynamics (QCD) are examined and compared with experimental data. Particular emphasis is placed on understanding the similarities and differences between the QCD results and the expectations of the naive parton model.

I. INTRODUCTION

During the last several years, a new framework to describe strong interaction physics has emerged: quantum chromodynamics (QCD). It is the simplest field theory which incorporates a color-dependent force among quarks. The forces among the colored quarks are generated by the exchange of colored vector gluons which are coupled to the quarks in a gauge-invariant manner. The theory is closely related to the most successful quantum field theory: QED. The only (but very important) difference between QED and QCD is the gauge group involved. QED is an Abelian gauge theory (the photons do not couple to each other); QCD is a non-Abelian gauge theory [gauge group SU_3(color)]. (The gluons carry color and thus are coupled to each other.) The formal Lagrangian for QCD is given by[1,2]

$$\mathcal{L}_{QCD}(x) = -\frac{1}{4} F^a_{\mu\nu} F^{a\mu\nu} + i\bar{\psi}_{q_i}(x)\gamma_\mu D^\mu_{ij}\psi_{q_j}(x), \tag{1.1a}$$

where $F^a_{\mu\nu}$ is related to the vector potential A^a_μ according to

$$F^a_{\mu\nu} = \partial_\mu A^a_\nu - \partial_\nu A^a_\mu + g f_{abc} A^b_\mu A^c_\nu \tag{1.1b}$$

and

---

†Lectures given at La Jolla Institute Summer Workshop, July 31 - August 18, 1978.

* Work supported in part by the U.S. Department of Energy under Contract No. EY76-C-03-0068.

$$D^{\mu}_{ij} = \partial^{\mu}\delta_{ij} - igA^{\mu}_{a}T^{a}_{ij}. \tag{1.1c}$$

The structure constants (for the group SU(3) of color) are given by

$$[\lambda^a, \lambda^b] = 2if_{abc}\lambda^c, \tag{1.2a}$$

where λ^a_{ij} are the eight SU(3) matrices introduced by Gell-Mann[3] and

$$T^a = \frac{1}{2}\lambda^a. \tag{1.2b}$$

Quantizing in a covariant gauge leads to the Feynman rules summarized in Fig. 1.1[1]. Notice that all flavors of quarks as well as gluons couple with a universal strength g.

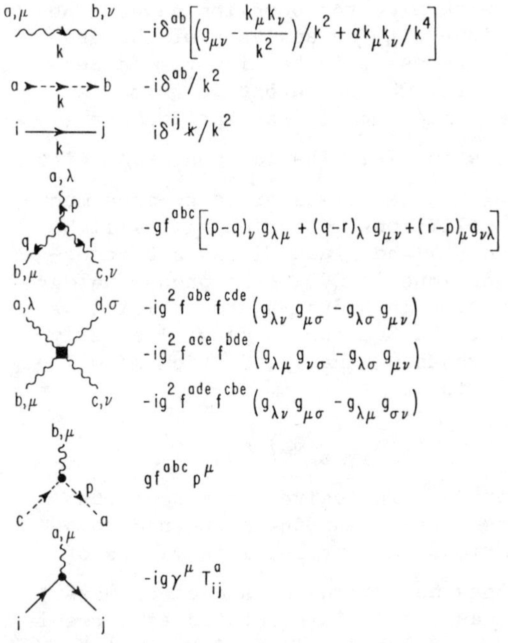

Fig. 1.1. Feynman rules for a gauge theory with fermions (from Ref. 1). The solid and wavy lines are fermions and vector gluons, respectively. The dotted lines are (non-existent) "ghost" particles which are introduced in the theory to aid in doing calculations that involve gluons.

Although the theory is well defined, precisely what it predicts is not yet clearly known. For example, it is not known if the theory actually confines quarks and gluons within hadrons nor has the spectrum of hadronic states been calculated. At present, the mathematical complexities are still too great. However, at very high energy or momentum transfer, Q, the theory is asymptotically free; the effective coupling between quarks and gluons decreases with increasing Q^2. As emphasized by Politzer, this permits calculation of those parts of a process involving high Q^2 by the use of perturbation theory and the Feynman rules in Fig. 1.1. Yet most real processes involve both high <u>and</u> low Q^2 together and precisely how to separate these parts is just becoming clear.

In these lectures, we will examine many of the present day applications of QCD to processes involving large momentum transfers. Some of these applications are rather crude and involve ideas that are rather phenomenological in nature. Nevertheless, comparisons with data are quite encouraging, although many of the most dramatic (and definitive) tests are yet to come. I will not have time to explain or derive everything from the beginning; however, I will attempt to explain the "physics" behind the formulas. In many ways, QCD provides a justification of the parton picture, but the predictions differ from the naive parton model due, for example, to the Bremsstrahlung radiation of gluons. I will try and emphasize the similarity and differences between the QCD results and the naive parton model expectations.

I will begin in Section II with a discussion of the QCD "effective" coupling $\alpha_s(Q^2)$ and a comparison with the electrodynamic results. In Section III, the gluon corrections to $R = \sigma(e^+e^- \to \text{hadrons})/\sigma(e^+e^- \to \mu^+\mu^-)$ will be examined. We will then proceed to investigate how QCD affects the quark and gluon distributions within hadrons (Section IV) and compare with electroproduction data. The confrontation with data will continue with an analysis of charged current neutrino and antineutrino processes in Section V. In Section VI, I will write down the general QCD perturbative formalism and we will examine how the naive parton model expectations are altered by order $\alpha_s(Q^2)$ corrections in QCD. The confrontation with data will resume in Section VII with an analysis of large-mass muon-pair production in hadron-hadron collisions. Section VIII will be devoted to a short discussion of quark and gluon "jets" and an examination of the fragmentation functions in QCD. The phenomenology of large $p_\perp$ meson and "jet" production in hadron-hadron collisions will be investigated in Section IX. In Section X, we will briefly examine the search for three-jet events. Finally, Section XI will be reserved for summary and conclusions.

II. THE EFFECTIVE COUPLING $\alpha_s(Q^2)$

The theory of QCD does not produce inclusive cross sections that "scale." One cannot use dimensional counting arguments to determine the behavior of cross sections (at intermediate values of Q^2). This is because the theory has an intrinsic "scale" or mass parameter μ (the renormalization mass) that is generated as a result of the interaction between quarks and gluons. These interactions result in an effective strong interaction coupling $\alpha_s(Q^2)$ that decreases logarithmically with increasing Q^2, where Q is some characteristic momentum in a collision. I would like to discuss for a moment the origin of the Q^2 dependence of α_s. The following discussion is not meant as a vigorous proof but a sketch of the physics involved.

Consider the amplitude for, say, quark-quark scattering which, as shown in Fig. 2.1a, proceeds in lowest order by single gluon exchange. For simplicity, take the case where $\theta = 90°$ so that there is only one invariant, $Q^2 = -t = s/2$. The "bare" diagram in Fig. 2.1a would have an amplitude

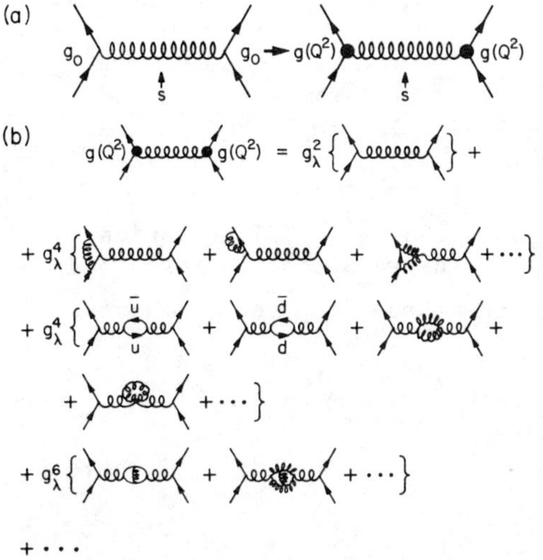

Fig. 2.1. Illustration of how a "bare" coupling g_o becomes a "running" coupling $g(Q^2)$ due to the interactions between quarks and gluons.

$$A_{qq}(Q^2, g_o^2) = g_o^2 f(Q^2), \quad (2.1)$$

where g_o^2 is the "bare" coupling. This amplitude is meaningless, however, since the first order corrections to the quark-quark-gluon vertex and to the gluon propagator (see Fig. 2.1b) are infinite. One must define a high Q^2 cut off, λ^2, so that the loop integrals converge. We are then left with an infinite series which for large λ looks something like

$$A_{qq}(Q^2, g_\lambda^2, \lambda^2) = f(Q^2)\{g_\lambda^2 + g_\lambda^4(c_1 + c_1' \log \frac{\lambda^2}{Q^2})$$

$$+ g_\lambda^6(c_2 + c_2' \log \frac{\lambda^2}{Q^2} + c_2'' \log \log \frac{\lambda^2}{Q^2}) \quad (2.2)$$

$$+ g_\lambda^8(\ldots.) + \ldots.\},$$

where g_λ is the coupling for the particular cut off λ and where the $\log\lambda^2$ terms come from single loop diagrams like

$$\underset{\text{large } \lambda^2}{\propto} \int^{\lambda^2} \frac{dq^2}{q^2} \propto \log\lambda^2, \quad (2.3)$$

and the log $\log\lambda^2$ terms come from two loop integrations like

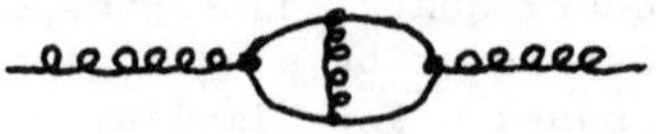

$$\underset{\text{large } \lambda^2}{\propto} \int^{\lambda^2} g^2 \frac{dq^2}{q^2} \qquad (2.4)$$

$$\propto \int^{\lambda^2} \frac{dq^2}{q^2 \log q^2} \propto \log \log \lambda^2 .$$

(In (2.4), one uses the one loop result that $g^2 \propto 1/\log q^2$ which reduces the degree of divergence.) All other contributions to (2.2) go to zero as $\lambda^2 \to \infty$. For example, the three loop terms behave as

$$\underset{\text{large } \lambda^2}{\propto} \int^{\lambda^2} g^4 \frac{dq^2}{q^2} \qquad (2.5)$$

$$\propto \int^{\lambda^2} \frac{dq^2}{q^2 \log^2 q^2} \sim \frac{1}{\log \lambda}$$

which converges. Notice also that eq. (2.2) has the property

$$A_{qq}(Q^2, g_\lambda^2, \lambda^2) = A_{qq}(Q^2/\lambda^2, g_\lambda^2), \qquad (2.6)$$

since all logs must have a dimensionless argument and Q^2 is the only other quantity which has the dimension of λ^2.

Since the theory is renormalizable, changing the cut off λ^2 cannot change the physical predictions. From (2.6), we see that changing λ^2 corresponds to changing the scale of Q^2, so if the answer is not to change, then the effective coupling $\bar{g}(Q^2)$ must change with Q^2 in just such a manner as to compensate for the change in λ^2 and leave the answer unchanged. This means that

$$A_{qq}(Q^2/\lambda^2, g_\lambda^2) = A_{qq}(Q^2, \bar{g}^2(Q^2)), \qquad (2.7)$$

and the dependence of the amplitude on the cut-off determines the dependence of $\bar{g}(Q^2)$ on Q^2. For example, if one neglects the log log terms (and constants), eq. (2.2) becomes (after summing)

$$A_{qq}(Q^2/\lambda^2, g_\lambda^2) = \frac{g_\lambda^2 f(Q^2)}{1 - a g_\lambda^2 \log(\lambda^2/Q^2)} \equiv \bar{g}^2(Q^2) f(Q^2), \tag{2.8}$$

where

$$1/\bar{g}^2(Q^2) = 1/g_\lambda^2 - a\log(\lambda^2/Q^2), \tag{2.9a}$$

and

$$a = (11 - \frac{2}{3} n_f)/16\pi^2, \tag{2.9b}$$

where n_f is the number of quark flavors (u, d, s, etc.). One generally chooses to remove all reference to the cut off λ^2. This is done by choosing a point of reference ("renormalization" point), say, $Q^2 = \mu^2$, at which the coupling is given by

$$1/g_\mu^2 = 1/g_\lambda^2 - a\log(\lambda^2/\mu^2). \tag{2.10}$$

From (2.9), one can now express the effective coupling in terms of the coupling at this renormalization point (and let $\lambda \to \infty$). We now have

$$1/\bar{g}^2(Q^2) = 1/g_\mu^2 - a\log(\mu^2/Q^2) \tag{2.11a}$$

or

$$\bar{g}^2(Q^2) = \frac{1}{1/g_\mu^2 + a\log(Q^2/\mu^2)}, \tag{2.11b}$$

which can be rewritten in the form

$$\alpha_s(Q^2) = \bar{g}^2(Q^2)/4\pi = \frac{4\pi}{(11 - \frac{2}{3} n_f)\log(Q^2/\Lambda^2)}, \tag{2.12}$$

where Λ is the one parameter that determines the size of $\alpha_s(Q^2)$ and thus the amount of "scale breaking." It is related to the renormalization point by

$$\Lambda^2 = \mu^2 \exp(-1/(a g_\mu^2)). \tag{2.13}$$

Including the log log terms in (2.2) yields

$$\alpha_s(Q^2) = \frac{4\pi}{(11 - \frac{2}{3} n_f)(\log(Q^2/\Lambda^2) + a'\log\log(Q^2/\Lambda^2))}, \quad (2.14)$$

where $a' = 3(306 - 38\, n_f)/(33 - 2\, n_f)^2$.

In the theory of QED, only the electron loops (vacuum polarization) in Fig. 2.2a contribute to the charge renormalization. In lowest order perturbation theory

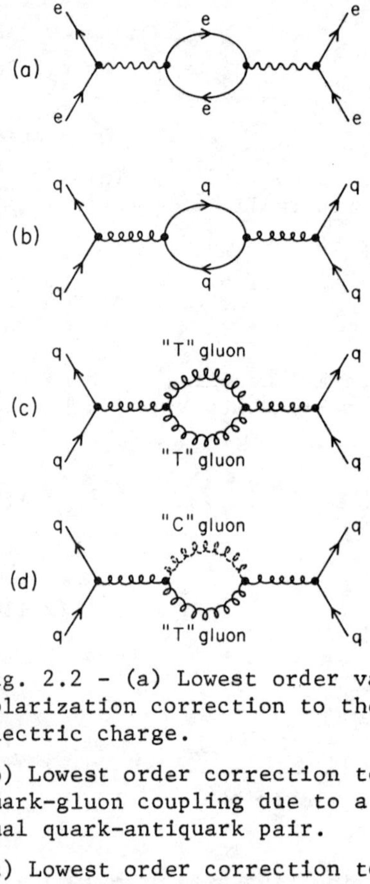

Fig. 2.2 - (a) Lowest order vacuum polarization correction to the electric charge.

(b) Lowest order correction to the quark-gluon coupling due to a virtual quark-antiquark pair.

(c) Lowest order correction to the quark-gluon coupling due to a virtual pair of transverse ("T") gluons in the coulomb gauge.

(d) Lowest order correction to the quark-gluon coupling due to a virtual pair of gluons, one transverse ("T") and one "coulomb" ("C") in the coulomb gauge.

$$\alpha_{QED}(Q^2) = \alpha_{QED}(m_e^2) \left[1 + \frac{\alpha_{QED}(m_e^2)}{3\pi} \log(Q^2/m_e^2)\right]. \quad (2.15a)$$

In higher orders, a whole series of the type $[\alpha_{QED}(m_e^2)\log m_e^2/Q^2]^N$ appears like in eq. (2.2). Summing the leading logarithms yields

$$\alpha_{QED}(Q^2) = \frac{1}{1/\alpha_o - (\frac{1}{3\pi})\log(Q^2/m_e^2)}. \quad (2.15b)$$

No matter how small α_o one has, one can always increase Q^2 to a point where $\alpha_{QED}(Q^2)$ becomes infinite. This means that perturbation theory breaks down at high Q^2 in QED. One needs to include higher and higher orders in α_{QED} as Q^2 increases. At low Q^2, on the other hand, $\alpha_{QED}(Q^2)$ is small ($\approx \alpha_o = 1/137$) and perturbation theory works well.

The physical reason for the increasing effective charge with increased Q^2 of the probing photon is clear. If Q^2 is small then the photon cannot resolve small distances and "sees" a "point" charge shielded by the vacuum polarization of the infinite sea of electron-positron pairs. As Q^2 increases, the photon "sees" a smaller and smaller spatial area and shielding effect is less.

In QCD the behavior of the effective coupling constant is strikingly different. The reason for this difference is the new feature of QCD, that the gluons carry charge (color) and interact with each other. The amount of the contributions of the various diagrams in Fig. 2.1 is gauge dependent. However, the situation is most clear in the coulomb gauge. In this gauge, the lowest order correction to $\alpha_s(Q^2)$ is given by

$$\alpha_s(Q^2) = \alpha_s(\mu^2)[1 + \tilde{a}\log Q^2/\mu^2], \tag{2.16a}$$

where

$$\tilde{a} = (\tfrac{2}{3}\, n_f + 5 - 16)/4\pi. \tag{2.16b}$$

The $+\frac{2}{3}\, n_f$ and the +5 come from the quark loop and the transverse gluon loop in Fig. 2.2 and are of the same sign as the QED case (2.15a). These contributions must be positive since the diagrams can be cut across the bubble and represent contributions to the physical rate for producing quark pairs or transverse gluon pairs which must be positive. The -16 in (2.16b) comes from the diagram with one transverse and one "coulomb" gluon in the bubble (Fig. 2.2d). This contribution need not be positive since the instantaneous "coulomb" gluon is not physical. Indeed, if $2/3\, n_f < 11$ then $\alpha_s(Q^2)$ in (2.16) decreases with increasing Q^2 and approaches zero at $Q^2 = \infty$ (asymptotic freedom) as shown in Fig. 2.4. This means that for QCD, perturbation theory works well at high Q^2 (short distances) but breaks down at small Q^2 (large distances) where $\alpha_s(Q^2)$ becomes large and hopefully (?) confines quarks within hadrons. It is interesting to notice that eq. (2.12) becomes infinite at $Q^2 = \Lambda^2$ which corresponds to a distance of about 0.5 Fermi for $\Lambda = 500$ MeV. (This is crude, however, since (2.12) is not applicable when $\alpha_s(Q^2)$ becomes large.)

A physical reason for the behavior of $\alpha_s(Q^2)$ is illustrated in Fig. 2.3. Quark-antiquark vacuum polarization shields the color charge as was the case in QED. However, now since the source can radiate charge (i.e., change from red to blue by emitting a

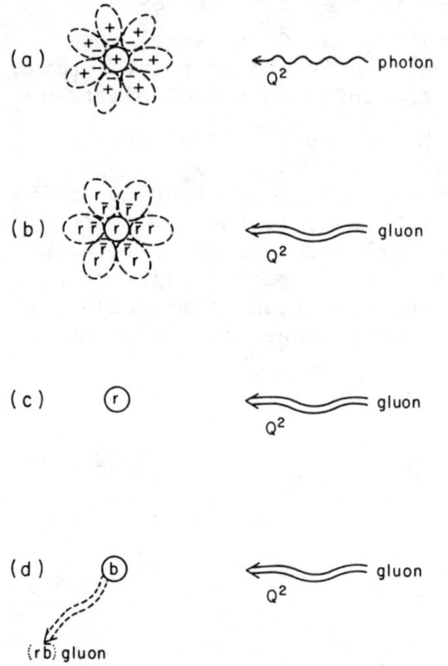

Fig. 2.3 - (a) Illustration of how vacuum polarization in QED will "shield" a bare positive charge when placed in a vacuum.

(b) The same shielding as in (a) but for a "red" charge in QCD.

(c) and (d) Show how a "red" charge can, in QCD, radiate away its red charge, r, and become a blue charge, b, via the emission of a virtual $r\bar{b}$ gluon.

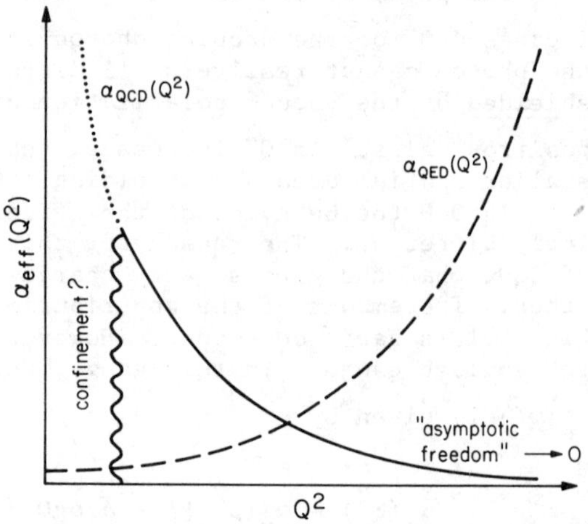

Fig. 2.4. Illustration of the behavior of the effective coupling calculated in perturbation theory in QED and QCD. In QED the effective coupling, $\alpha_{QED}(Q^2)$, is small at small Q^2, but becomes large at large Q^2 (small distances). In QCD, on the other hand, the effective coupling is large at small Q^2 (large distances) where confinement may occur, but decreases toward zero at Q^2 increases ("asymptotic freedom"). Perturbation theory should work well for QCD (QED) at large Q^2 (small Q^2).

red-blue-bar gluon), the charge is no longer located at a definite place in space. It is diffusely spread out due to gluon emission and absorption. As one increases the Q^2 of the incoming gluon probe, thereby looking at smaller and smaller spatial distances, it becomes less likely to find the "charge" (red in Fig. 2.3). This latter effect is stronger than the former and the effective charge thus appears weaker and weaker as Q^2 of the probe increases.

One should exercise care when using eq. (2.12) for if one has calculated only to order $\bar{g}^2(Q^2)$, then one should write

$$\alpha_s(Q^2) = \frac{4\pi}{(11 - \frac{2}{3} n_f)(\log Q^2/\Lambda^2 + c)}, \tag{2.17}$$

where c is a constant that may, in general, differ from process to process, but which is, in principle, determinable by calculating to order $\bar{g}^4(Q^2)$ since

$$\frac{1}{\log Q^2/\Lambda^2 + c} \underset{\text{large } Q^2}{\approx} \frac{1}{\log Q^2/\Lambda^2} - \frac{c}{(\log Q^2/\Lambda^2)^2}, \tag{2.18}$$

where the coefficient of c in the second term is order $\bar{g}^4(Q^2)$.

It did not matter that I chose the process, $qq \to qq$ in Fig. 2.1 to deduce the large Q^2 behavior of $\alpha_s(Q^2)$. I could have chosen quark + gluon → quark + gluon or gluon-gluon scattering. Although the individual correction diagrams would be different, since there is only one coupling, g_μ^2, in the theory, the high Q^2 result for $\alpha_s(Q^2)$ would be the same. (The $\log Q^2$ and $\log \log Q^2$ terms would be the same, however, the constant c in (2.17) would differ from process to process.) In addition, if the process involves more than one invariant (like s and t), then the above arguments hold when the variables become large but have some fixed ratio, say, $\xi = t/s$. In this case, the constant c in (2.17) would be a function of ξ.

The nature of the QCD coupling constant $\alpha_s(Q^2)$ takes a bit of getting used to. In QED it is easy to define what one means by the charge of an electron e, by the large distance behavior of the electric potential (Thomson limit). One cannot do this for QCD since the $Q^2 \to 0$ limit of $\alpha_s(Q^2)$ cannot be calculated by perturbation theory. Instead one must define some arbitrary point μ^2 at which the coupling is g_μ^2. It, however, does not matter which point μ^2 one chooses. If one chooses instead the point M^2 then the two couplings are related (to lowest order) by eq. (2.11)

$$1/g_\mu^2 - a\log\mu^2 = 1/g_M^2 - a\log M^2. \tag{2.19}$$

Thus the "real" parameter in the theory is not g_μ^2 or μ^2 but rather a mass scale Λ that is independent of μ^2 and given to this order by

$$\log\Lambda^2 \equiv -1/(a g_\mu^2) + \log\mu^2. \tag{2.20a}$$

Then for any arbitrary point μ^2 g_μ^2 is given by

$$g_\mu^2 = 1/(a\log\mu^2/\Lambda^2). \tag{2.20b}$$

The mass scale Λ, also defined by (2.13), must be determined experimentally.

III. CORRECTIONS TO R IN $e^+e^- \to$ HADRONS

The naive quark parton model result from the amplitude A_o in Fig. 3.1 is

$$R \equiv \sigma(e^+e^-\to\text{hadrons})/\sigma(e^+e^-\to\mu^+\mu^-) = 3 \sum_{q_i=u,d,\ldots}^{n_f} e_{q_i}^2\, \theta(s-4m_i^2), \tag{3.1}$$

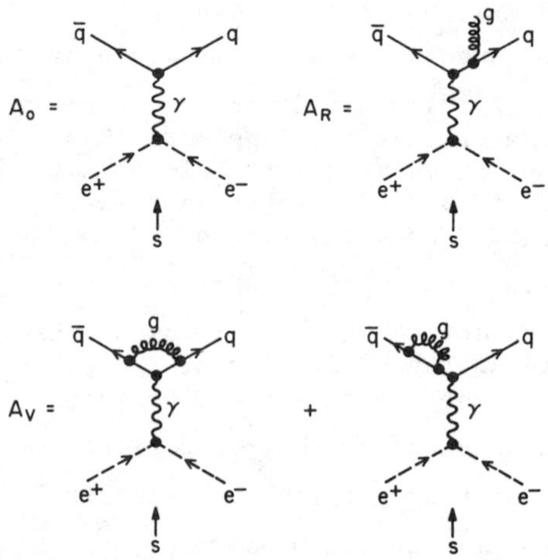

Fig. 3.1. The "leading order" amplitude, A_o, for the production of a quark and antiquark pair through the annihilation of an electron-positron pair together with the diagrams for real, A_R, and virtual, A_V, gluon corrections.

where e_{q_i} and m_i are the charges and the mass of the quarks q_i and where

$$\sigma(e^+e^-\to\mu^+\mu^-) = 4\pi\alpha^2/3s \tag{3.2}$$

with s the c.m. energy squared of the incoming e^+e^- pair.

Let us ignore for the moment the fact that $q^2 = s$ is timelike rather than spacelike and examine the lowest order corrections to R in (3.1) due to the real (A_R) and virtual (A_V) gluon emissions in Fig. 3.1b and 3.1c, respectively. The amplitude squared for the emission of a real gluon of 4-momentum q (see Fig. 3.1b) is

$$|A_R|^2 = \frac{16g_s^2}{s(p_2\cdot q)(p_1\cdot q)} \{(p_1\cdot k_1)^2 + (p_1\cdot k_2)^2 + (p_2\cdot k_1)^2 + (p_2\cdot k_2)^2\}, \tag{3.3}$$

where k_1 and k_2 are the momenta of the incoming electrons and where I have taken all particles as massless. This probability diverges when $p_2\cdot q$ or $p_1\cdot q$ are zero (p_1 and p_2 are the 4-momentum of the outgoing quarks). The quantity $p\cdot q$ can vanish in two ways. Let $\vec{p}$ be in the $\hat{z}$ direction and let ω, $q_\perp$, q_z be the energy, perpendicular, and $\hat{z}$-component of the emitted gluon,

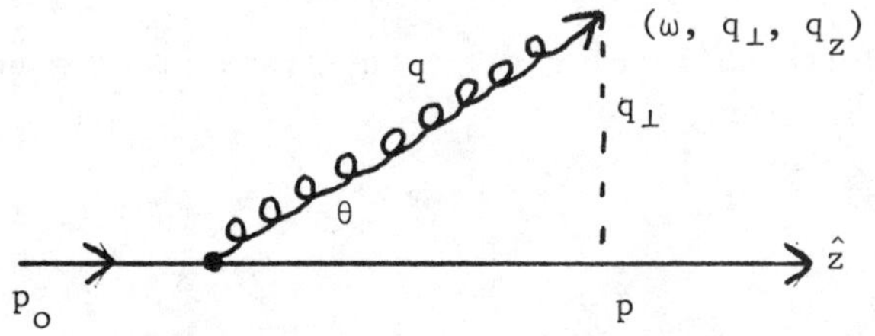

then

$$\frac{d(p\cdot q)}{p\cdot q} \propto \left(\frac{dq_\perp^2}{q_\perp^2}\right)\left(\frac{dq_z}{\omega}\right), \tag{3.4}$$

for $\vec{p}$ large and $q_\perp$ small. The first type divergence occurs when the emitted gluon is parallel to $\vec{p}$ (i.e., $q_\perp = 0$ or $\cos\theta = 1$). A massless parton can "decay" into two massless partons which causes p_o to be on-shell and hence the propagator in (3.3) diverges. A second type of divergence occurs when the energy of the gluon, $\omega = (q_\perp^2+q_z^2)^{1/2}$, goes to zero ("soft" divergence).

To get a finite answer to the integral of $|A_R|^2$ over all gluon momenta, we must introduce an infrared cut-off, m^2, whereupon eq. (3.3) becomes (symbolically)

$$\sigma_R = \int |A_R|^2 \, d(p_1 \cdot q) d(p_2 \cdot q)$$

$$= g^2 (c\log \frac{m^2}{Q^2} + c')(d\log \frac{m^2}{Q^2} + d') \qquad (3.5)$$

$$= g^2 (cd\log^2 \frac{m^2}{Q^2} + (c'd+cd')\log \frac{m^2}{Q^2} + c'd').$$

We get a $\log^2 \frac{m^2}{Q^2}$ divergence because of the two divergences in (3.4). In this example, however, both the $\log^2 \frac{m^2}{Q^2}$ and $\log \frac{m^2}{Q^2}$ divergences in (3.5) are canceled (to this order in g) by the virtual gluon correction in Fig. 3.1c. Namely,

$$\sigma_V = \int 2A_o A_V^* d(\text{loop})$$

$$= g^2 (-cd\log^2 \frac{m^2}{Q^2} - (c'd+cd')\log \frac{m^2}{Q^2} - c'd' + 1/4\pi^2), \qquad (3.6)$$

which need not be positive since it is an interference term between the born amplitude A_o and the virtual gluon vertex correction A_V. To order g^2, we have

$$R = \left(3 \sum_{i=1}^{n_f} e_i^2\right) (1 + \frac{1}{4\pi^2} g^2 + O(g^4)). \qquad (3.7)$$

In higher order, one has ultraviolet divergences that do not cancel and in particular one has

$$R = \left(3 \sum_{i=1}^{n_f} e_i^2\right) (1 + \frac{1}{4\pi^2} g_\lambda^2 + A g_\lambda^4 (\log \frac{\lambda^2}{Q^2} + c) + \ldots). \qquad (3.8)$$

We know from a more general theorem that R cannot depend on any cut-off m. In fact, the result for R must be infrared finite (i.e., no divergences) when perturbative corrections are summed. This more general theorem is due to Kinoshita[4], Lee and Nauenburg[5] (KLN theorem) and states that if one performs an incoherent sum over all possible incoming and outgoing quark and gluon states, then the result is infrared finite. The theorem applies in this case since the incoming state is e^+e^- and we are summing over all final quark and gluon states when we ask for the total cross section $\sigma(e^+e^- \to \text{hadrons})$.

In addition knowing that (3.8) does not depend on the ultraviolet cut-off λ allows us to determine automatically the coefficient of the $g_\lambda^4 \log(\lambda^2/Q^2)$ term. Suppose we change the cut-off from λ to μ then from (2.19), we have

$$\frac{1}{g_\lambda^2} - a\log\lambda^2 = \frac{1}{g_\mu^2} - a\log\mu^2 \tag{3.9}$$

or $g_\lambda^2 = g_\mu^2 + g_\mu^4 a\log\frac{\mu^2}{\lambda^2}$. Substituting into (3.8) yields

$$R = \left(3\sum_{i=1}^{n_f} e_i^2\right)\left(1 + \frac{1}{4\pi^2} g_\mu^2 + \frac{a}{4\pi^2} g_\mu^4 \log\frac{\mu^2}{\lambda^2} + A g_\mu^4 \log\frac{\lambda^2}{Q^2} + \ldots\right) \tag{3.10a}$$

implying that $A = a/4\pi^2$ and

$$R = \left(3\sum_{i=1}^{n_f} e_i^2\right)\left(1 + \frac{1}{4\pi^2} g_\mu^2 + a g_\mu^4 \log\frac{\mu^2}{Q^2} + \ldots\right)$$
$$= \left(3\sum_{i=1}^{n_f} e_i^2\right)\left(1 + \frac{1}{\pi}\,\frac{g_\mu^2/4\pi}{1 + g_\mu^2 a\log\frac{Q^2}{\mu^2}} + \ldots\right) \tag{3.10b}$$

$$= \left(3\sum_{i=1}^{n_f} e_i^2\right)\left(1 + \frac{1}{\pi}\alpha_s(Q^2) + \text{order}\,(\alpha_s^2)\right). \tag{3.10c}$$

where $\alpha_s(Q^2)$ is defined by (2.12) and does not depend on λ or μ only on the mass parameter Λ in (2.13). Since $\alpha_s(Q^2) \to 0$ as Q^2 increases,

R in (3.10c) approaches the naive parton model results as $Q^2 \to \infty$.

At Q = 7 GeV and with 4 quark flavors, one gets (to order α_s with $\Lambda = 0.4$ GeV)

$$R(Q = 7\text{ GeV}) = 10/3 + 0.28 + 1 \approx 4.6, \qquad (3.11)$$

which lies about 0.7 units below the data shown in Fig. 3.2. (The +1 in (3.11) is the contribution from a heavy lepton which has not been subtracted from the data.)

One should not be too concerned about the disagreement between (3.11) and the data in Fig. 3.2. Firstly, the data suffer from systematic uncertainty in overall normalization and can probably be trusted only to about 0.5 units. Secondly, and more important, the above analysis (although pedagogically useful) is too naive. In the continuation from spacelike to timelike q^2 at fixed $|q^2|$, the sums of leading logarithms $((g^2/4\pi)\log \frac{q^2}{\mu^2})^N$ become $(\alpha_s \log|\frac{q^2}{\mu^2}| + i\pi\alpha_s)^N$. For very large q^2, the complex part can be ignored. However, even at $|q^2| = 50\text{ GeV}^2$ $\log|\frac{q^2}{\mu^2}|$ is only 3.9 and the complex term $i\pi$ found in continuing from the spacelike to timelike region is certainly not negligible. This is discussed in great detail by Moorhouse, Pennington and Ross[6]. They avoid the problem of singularities in the timelike region by comparing the theory with "smeared" data $\tilde{R}(q^2)$ defined by

$$\tilde{R}(s,\Delta) = \frac{\Delta}{\pi}\int_{4m_\pi^2}^{\infty} \frac{ds' R(s')}{(s'-s)^2+\Delta^2}. \qquad (3.12)$$

Their results are shown in Fig. 3.3 and do not disagree with data.

Finally, it should be remarked that in the region, $\alpha_s(Q^2) \gtrsim 0.3$, nonperturbative effects may play a role and I don't believe we are in a position to give reliable estimates of corrections due to nonperturbative phenomenon.

IV. THE QUARK AND GLUON DISTRIBUTIONS

A. The Naive Parton Model Results

In the naive parton model, one defines parton distributions, $G_{h\to q}(x)$, as the number of quarks q with fraction of momentum between x and x + dx within a hadron of type h of high momentum. In particular, there are six functions necessary to describe the quark

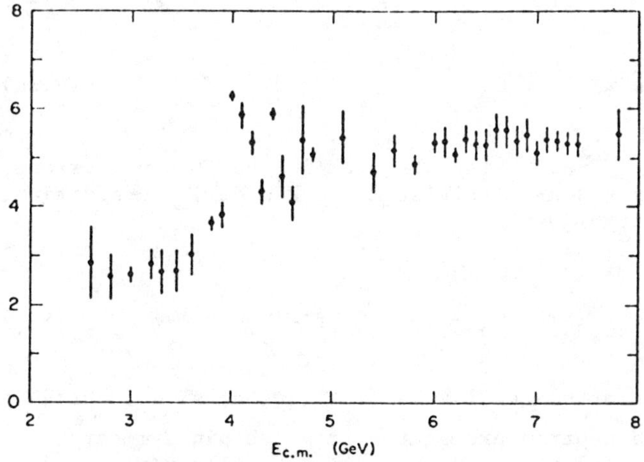

Fig. 3.2. Experimental values for $R = \sigma(e^+e^- \to \text{hadrons})/\sigma(e^+e^- \to \mu^+\mu^-)$ versus the c.m. energy. The data are from SLAC (Ref. 7) and have not been corrected for heavy lepton production.

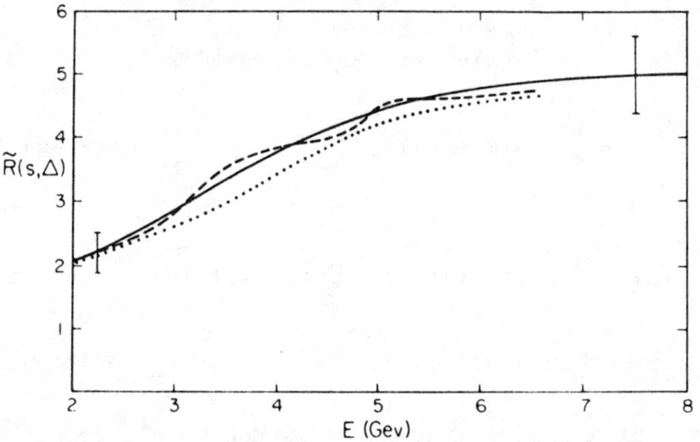

Fig. 3.3. Comparison of the theory (dashed curve) and data (solid curve with errors) for the "smeared" $\tilde{R}(s,\Delta)$ defined by eq. (3.12) from Moorhouse, Pennington and Ross[6].

distributions in a proton:

$$u(x) \equiv G_{p\to u}(x),$$

$$d(x) \equiv G_{p\to d}(x),$$

$$s(x) \equiv G_{p\to s}(x),$$

$$\bar{u}(x) \equiv G_{p\to \bar{u}}(x),$$

$$\bar{d}(x) \equiv G_{p\to \bar{d}}(x),$$

$$\bar{s}(x) \equiv G_{p\to \bar{s}}(x), \tag{4.1}$$

where u, d and s refer to up, down and strange quarks, respectively, and $\bar{u}$, $\bar{d}$ and $\bar{s}$ to their antiquarks. The distribution of gluons within a proton is defined by

$$g(x) \equiv G_{p\to g}(x),$$

where g stands for gluon. These distributions satisfy the following sum rules:

$$\int_0^1 [u(x) - \bar{u}(x)]dx = 2 \tag{4.2a}$$

$$\int_0^1 [d(x) - \bar{d}(x)]dx = 1 \tag{4.2b}$$

$$\int_0^1 [s(x) - \bar{s}(x)]dx = 0. \tag{4.2c}$$

That is the net number of each kind of quark is just the number one arrives at in the simple non-relativistic quark model. In addition, momentum conservation implies

$$\int_0^1 \left\{ \sum_{i=1}^{n_f} x(G_{p\to q_i}(x) + G_{p\to \bar{q}_i}(x)) + xg(x) \right\} dx = 1, \tag{4.3}$$

where n_f is the number of quark flavors (i.e., u, d, s, ..., etc.). The distributions in a neutron are gotten from isospin symmetry, which implies that $G_{n\to u}(x) = G_{p\to d}(x) = d(x)$, $G_{n\to d}(x) = u(x)$, $G_{n\to s}(x) = s(x)$, etc.

In the naive parton model, complete knowledge of the deep inelastic structure functions for electron, neutrino and antineutrino scattering off protons and neutrons is sufficient to obtain $u(x)$, $d(x)$, $\bar{u}(x)$, $\bar{d}(x)$, $g(x)$ and $s(x) + \bar{s}(x)$. For example, with the standard notation,

$$\nu W_2^{ep}(x) \equiv F_2^{ep}(x) = \frac{4}{9} x[u(x)+\bar{u}(x)] + \frac{1}{9} x[d(x)+\bar{d}(x)] + \frac{1}{9} x[s(x)+\bar{s}(x)], \tag{4.4a}$$

and

$$\nu W_2^{en}(x) \equiv F_2^{en}(x) = \frac{4}{9} x[d(x)+\bar{d}(x)] + \frac{1}{9} x[u(x)+\bar{u}(x)] + \frac{1}{9} x[s(x)+\bar{s}(x)], \tag{4.4b}$$

which are only functions of x and, in the naive parton model, do not depend separately on the energy loss of the leptons $\nu = E - E'$ or on the four-momentum transfer $q^2 = -Q^2$ [8]. This is because the basic interaction is assumed to be a photon of momentum q^2 interacting with a parton of momentum p ($p = \xi P$) producing a parton of momentum $p' = p + q$.

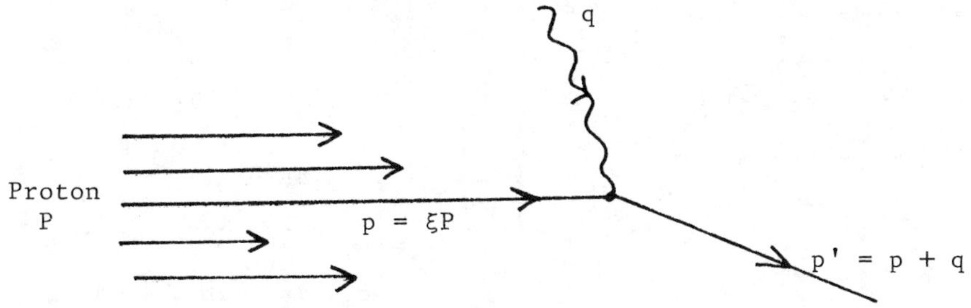

(4.5)

The condition that $(p')^2 = m^2$ implies

$$2\xi(P\cdot q) + q^2 + \xi^2 M^2 = m^2, \tag{4.6a}$$

which as $-q^2 \to \infty$ and $P\cdot q = M\nu \to \infty$ yields

$$\xi = \frac{2P\cdot q}{-q^2} = x. \tag{4.6b}$$

B. QCD "Scale Breaking": $G_i(x,Q^2)$

In QCD one must correct the naive parton model by including the possibility that the quark in Fig. 4.1a can radiate a gluon before or after the interaction with the virtual photon γ^* (see Fig. 4.1b). One might, at first sight, think that since the strong interaction coupling between the quark and gluon, $\alpha_s(Q^2)$, decreases with increasing Q^2, that one could go to a high enough Q^2 so that all corrections due to gluon emission are negligible and thereby regain the naive parton model at high Q^2. (This was the case for $\sigma(e^+e^-\to$hadrons) in Section III). This is not the case, however, for $\nu W_2(x,Q^2)$. Since transverse momentum is not bounded in QCD (as it was for the naive parton approach), the QCD predictions deviate more and more from the naive parton model as Q^2 increases.

If we define, for the purpose of discussion, a quark distribution $G_{p\to q}(x,\Delta)$ as the probability of finding a quark with fractional momentum x and transverse momentum less than some fixed Δ (let Δ be the usual naive parton model value of 300 - 500 MeV), then if the gluon radiation in Fig. 4.1b is soft, it is included in $G(x,\Delta)$. Hard gluon corrections must be included explicitly when calculating any specific process. For example, for $ep \to e + X$, one would write (symbolically)

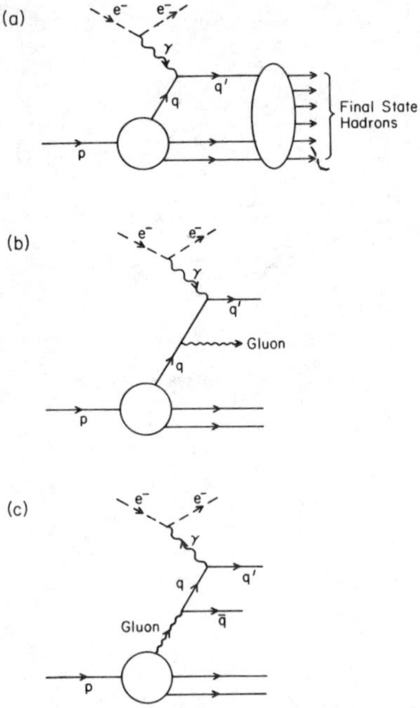

Fig. 4.1 - (a) Inelastic electron scattering in the naive parton model approximation.

(b) A typical gluon Bremsstrahlung correction to (a).

(c) A correction to (a) that involves the gluon distribution inside the proton.

$$\sigma(ep\to e+X) = \sum_i \int dy\, G_{p\to q_i}(y,\Delta) \{\hat{\sigma}_0(eq_i\to eq_i) + \hat{\sigma}_1(eq_i\to eq_i g,\Delta) + \hat{\sigma}_2(eq_i\to eq_i gg,\Delta) + \ldots\}, \tag{4.7}$$

where $\hat{\sigma}_0(eq_i\to eq_i)$ is just the usual elementary electron-quark cross section (Fig. 4.1a) and $\hat{\sigma}_1$ is the two-to-three subprocess eq $\to$ eq + gluon (Fig. 4.1b) where the gluon-quark invariant mass (or perpendicular momentum) is greater than Δ. The usual theory of Bremsstrahlung (see (3.4)) gives

$$\hat{\sigma}_1(eq\to eqg,\Delta) \alpha\ \bar{g}^2(Q^2)\log^2\frac{Q^2}{\Delta^2} \tag{4.8}$$

and since $\bar{g}^2(Q^2)$ only decreases like $1/\log Q^2$, the overall probability of gluon radiation outside Δ increases like $\log Q^2$. As Q^2 increases, it becomes more and more likely that the quarks radiate hard gluons and so higher and higher order corrections, $\hat{\sigma}_2(\Delta)$, $\hat{\sigma}_3(\Delta)$, ..., etc., must be included in (4.7). At any finite Q^2 eq. (4.7) contains only a finite number of terms since eventually it becomes impossible to emit another gluon and still be outside Δ of all the incoming and outgoing quarks. (The number of terms, of course, increases with Q^2.) It appears that the hope of using perturbation theory is lost, since higher order terms are proportional to $(\bar{g}^2(Q^2)\log^2\frac{Q^2}{\Delta^2})^N$ which is not small even though $\bar{g}^2(Q^2)$ may be.

The utility of perturbation theory can be restored in a clever manner[9]. Since the results of any calculation cannot depend on my choice of an arbitrary Δ in (4.7), I can let $\Delta^2 = \xi Q^2$. Now $\hat{\sigma}_1$

behaves like $\bar{g}^{-2}(Q^2)\log^2\xi$ and can be neglected at sufficiently high Q^2 whereupon (4.1) becomes

$$\sigma(ep\to e+X) = \sum_i \int dy G_{p\to q_i}(y,Q^2)\hat{\sigma}_o(eq_i\to eq_i), \tag{4.9}$$

where now the photon "sees" an "effective" parton distribution $G_{p\to q}(x,Q^2)$ that depends on Q^2. As Q^2 increases, more and more of the quarks' energy will radiate away and the photon will find less and less high x quarks in the proton.

Of course, I have not really accomplished anything since we still must calculate the Q^2 dependence of $G(x,Q^2)$. In addition, defining quark distributions $G(x,Q^2)$ is only useful if they are process independent. Do we have to perform a separate calculation of $G(x,Q^2)$ for every process or can we determine $G(x,Q^2)$ in one process (like $ep\to eX$) and use them elsewhere (like $pp\to \mu^+\mu^- + X$)? We will see in Section VI that to leading order, the $G_{p\to i}(x,Q^2)$ distributions are universal as in the naive parton model. We will not be able to calculate $G_{p\to i}(x,Q^2)$, but if they are given at some reference momentum, say Q_o^2, then the renormalization group (a consequence of the arbitrariness of Q_o^2) can be used to calculate them at any other Q^2 as long as Q_o^2 and Q^2 are large so $\alpha_s(Q^2)$ is small. I will discuss this more in Section VI. Let me now simply quote some results.

For reasons that will become clear, it is convenient to define the distribution for a quark minus antiquark (called non-singlet)

$$q^{NS}(x,\tau) \equiv q(x,\tau) - \bar{q}(x,\tau), \tag{4.10}$$

where $\tau = \log Q^2/Q_o^2$. Then QCD predicts that[10-13)]

$$M_n^{NS}(\tau) = M_n^{NS}(o)[\alpha(o)/\alpha(\tau)]^{A_n^{NS}/2\pi b} \tag{4.11a}$$

$$= M_n^{NS}(o)\left[\log \frac{Q^2}{\Lambda^2}\Big/\log \frac{Q_o^2}{\Lambda^2}\right]^{A_n^{NS}/2\pi b}, \tag{4.11b}$$

where $\quad b = (33-2n_f)/12\pi$ (4.11c)

and from (2.12) $\alpha(o)/\alpha(\tau) = 1 + b\alpha(o)\tau$ and where the moments of the distribution (4.6) are defined by

$$M_n^{NS}(\tau) = \int_0^1 x^{n-1} q^{NS}(x,\tau)dx. \quad (4.12)$$

The "anomalous dimensions" A_n^{NS} will be calculated from the theory in Section VI.

Equation (4.11) is the solution of the following differential equation

$$dM_n^{NS}(\tau)/d\tau = (\alpha(\tau)/2\pi)A_n^{NS}M_n^{NS}(\tau), \quad (4.13)$$

with the condition $M_n^{NS}(\tau=o) = M_n^{NS}(o)$. On the other hand, (4.13) is equivalent to the following convolution equation[13)]

$$\frac{dq^{NS}(x,\tau)}{d\tau} = \frac{\alpha(\tau)}{2\pi}\int_x^1 \frac{dy}{y} q^{NS}(y,\tau)P(\frac{x}{y}), \quad (4.14a)$$

provided

$$A_n^{NS} = \int_0^1 z^{n-1}P(z)dz. \quad (4.14b)$$

The change of $q^{NS}(x,\tau)$ with respect to $\log Q^2$ is then

$$\frac{dq^{NS}(x,\tau)}{d\tau} = \int_0^1 dy \int_0^1 dz\ \delta(zy-x)q^{NS}(x,\tau)[\frac{\alpha(\tau)}{2\pi} P(z)] \quad (4.15a)$$

or

$$q^{NS}(x,\tau) + dq^{NS}(x,\tau) = \quad (4.15b)$$

$$\int_0^1 dy \int_0^1 dz\ \delta(zy-x)q^{NS}(y,\tau)\{\delta(z-1) + \frac{\alpha(\tau)}{2\pi} P(z)d\tau\}.$$

Following Altarelli and Parisi[13)], the quantity

$$\tilde{P}_{qq} + d\tilde{P}_{qq} = \delta(z-1) + \frac{\alpha_s}{2\pi} P(z)d\tau, \quad (4.16)$$

is the probability density of finding, inside a quark, another quark with fraction z of the parent momentum. The change with τ of this probability produces the variation of the quark distribution function. The quantity $\alpha_s P(z)/2\pi$ is the variation per unit τ (of order α) of the probability density of finding a quark with fractional momentum z within a quark.

In general, we have

$$\frac{dq_i(x,\tau)}{d\tau} = \frac{\alpha(\tau)}{2\pi}\int_x^1 \frac{dy}{y}\,[q_i(y,\tau)P_{q\leftarrow q}(\frac{x}{y}) + g(y,\tau)P_{q\leftarrow g}(\frac{x}{y})], \tag{4.17a}$$

$$\frac{dg(x,\tau)}{d\tau} = \frac{\alpha(\tau)}{2\pi}\int_x^1 \frac{dy}{y}\,[\sum_{i=1}^{2n_f} q_i(y,\tau)P_{g\leftarrow q}(\frac{x}{y}) + g(y,\tau)P_{g\leftarrow g}(\frac{x}{y})], \tag{4.17b}$$

where $q_i(x,\tau)$ and $g(x,\tau)$ are the individual quark and gluon distributions, respectively, and where

$$\alpha_s P_{q\leftarrow q}(z)d\tau/2\pi =$$ 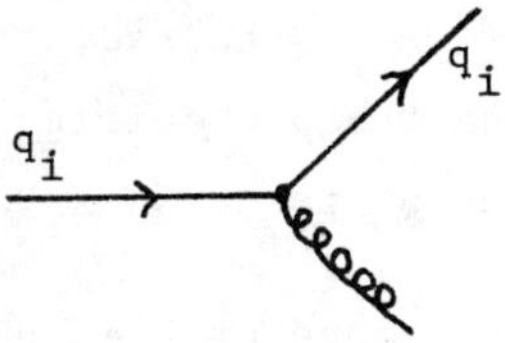

(4.18a)

is the probability of finding a quark carrying a fraction of momentum z of a parent quark that has radiated a gluon. Similarly,

$$\alpha_s P_{q\leftarrow g}(z)d\tau/2\pi =$$ 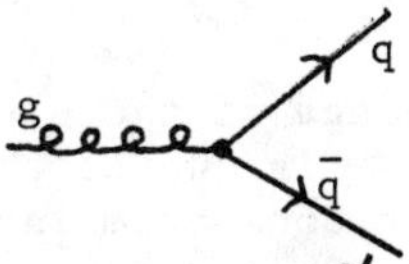

, (4.18b)

$$\alpha_s P_{g\leftarrow q}(z)d\tau/2\pi =$$ 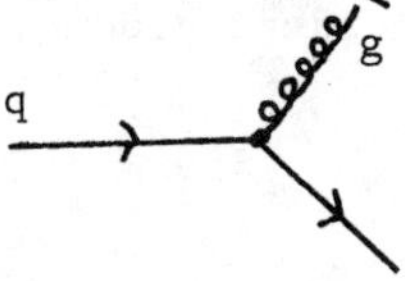

, (4.18c)

and

$$\alpha_s P_{g\leftarrow g}(z)d\tau/2\pi =$$ 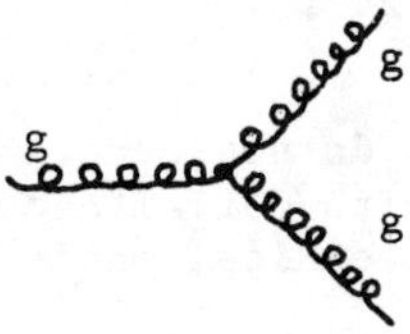

. (4.18d)

Equation (4.17) is illustrated in Fig. 4.2.

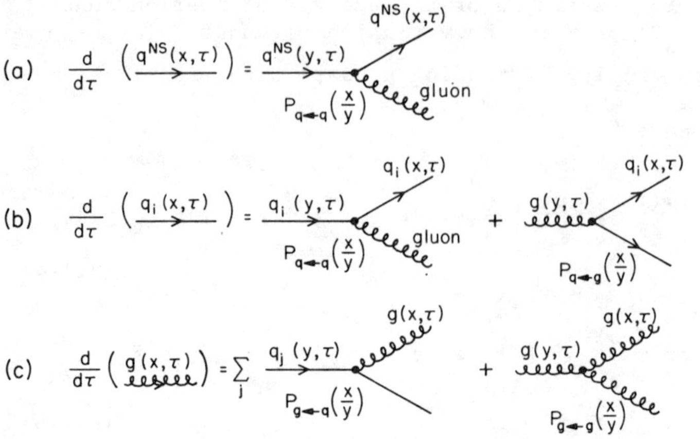

Fig. 4.2 - (a) Illustrates that the changes of the non-singlet quark distribution, $q^{NS}(x,\tau)$, w.r.t. $\tau = \log(Q^2/Q_0^2)$ is due to the radiation of a gluon with $P_{q\leftarrow q}(z)$ being the probability of finding a quark with momentum fraction z "within" a quark.

(b) Illustrates that the change of a quark distribution, $q_i(x,\tau)$, w.r.t. τ is due to the Bremsstrahlung radiation of a gluon, $P_{q\leftarrow q}(z)$, and the production of quark-antiquark pairs from a gluon, $P_{q\leftarrow g}(z)$.

(c) Illustrates that the change of the gluon distribution, $g(x,\tau)$, w.r.t. τ is due to the Bremsstrahlung radiation of gluons from incident quarks, $P_{g\leftarrow q}(z)$, and from incident gluons, $P_{g\leftarrow g}(z)$.

The quark distribution $q_i(x,\tau)$ changes with τ (or $\log Q^2$) due to the Bremsstrahlung of gluons $P_{q\to q}(z)$ and due to gluons that pair produce quarks of type q_i, $P_{g\to q}(z)$. The gluon distribution $g(x,\tau)$ changes with τ because of the Bremsstrahlung of gluons $P_{g\to g}(\tau)$ and from gluons that are produced from quark Bremsstrahlung, $P_{q\to g}(z)$. The equations are non-diagonal; both the quark and gluon distributions at $\tau + d\tau$ depend on the quark <u>and</u> gluon distributions at τ.

The four Feynman diagrams for P(z) in (4.18) diverge for massless partons, behaving as

$$d\tilde{P}(z) = \frac{\alpha_s}{2\pi} F_D(z) dk_\perp^2/k_\perp^2. \tag{4.19}$$

These "parallel" divergences do not cancel as in the $e^+e^- \to$ hadrons case since we are starting with definite quark (or gluon) states. The divergences are just a technical manifestation of the fact that we are trying to define a "quark" or a "gluon" distribution individually. If we had started with an initial proton (color singlet) and then summed over all diagrams, we would expect (by the KLN theorem)[4,5] to get a finite result. What is done here is to choose some regularization scheme and since one is (for the moment) only interested in the change of $\tilde{P}(z)$ with Q^2, the results will not depend on the precise nature of the regularization procedure. This will be discussed in more detail in Section VI. We will consider the quarks and gluons to be slightly off their mass shell with $p_q^2 < 0$ (spacelike). Then integrating (4.19) yields

$$\tilde{P}(z) = \frac{\alpha_s}{2\pi}\left(F_D(z)\log\left|\frac{Q^2}{p_q^2}\right| + F_c(z)\right), \tag{4.20}$$

where $F_D(z)$ and $F_c(z)$ are the coefficients of the divergent $\log(\frac{Q^2}{-p_q^2})$ and non-divergent part, respectively. From the definition (4.16)

$$P(z) = \frac{d\tilde{P}(z)}{d\tau} = \frac{d\tilde{P}(z)}{d\log Q^2} = \frac{\alpha_s}{2\pi} F_D(z). \tag{4.21}$$

That is, the P(z) functions that determine the Q^2 evolution of the parton distributions are determined by finding the coefficient of the divergent $\log(Q^2/-p_q^2)$ of the Feynman diagrams in (4.18). One finds that (see Section VI)

$$P_{q\leftarrow q}(z) = \left(\frac{4}{3}\right)\frac{1+z^2}{1-z} \qquad (z < 1) \tag{4.22a}$$

$$P_{q\leftarrow g}(z) = \frac{1}{2}\left(z^2 + (1-z)^2\right) \tag{4.22b}$$

$$P_{g\leftarrow q}(z) = \left(\frac{4}{3}\right)\frac{1+(1-z)^2}{z} \tag{4.22c}$$

$$P_{g\leftarrow g}(z) = 6\left[\frac{1-z}{z} + \frac{z}{1-z} + z(1-z)\right] \qquad (z < 1). \tag{4.22d}$$

The formulas for $P_{q\leftarrow q}(z)$ and $P_{g\leftarrow g}(z)$ diverge as $z \to 1$ from the "soft" divergence discussed in (3.4). These "soft" divergences are canceled by virtual gluon corrections as in the $e^+e^- \to$ hadrons. The final result is

$$P_{q\to q}(z) = (\tfrac{4}{3}) \left[\frac{1 + z^2}{(1 - z)_+} + \frac{3}{2}\,\delta(z-1)\right] \tag{4.23a}$$

$$P_{g\leftarrow g}(z) = 6\left[\frac{z}{(1-z)_+} + \frac{1 - z}{z} + z(1-z) + \frac{(11-n_f)}{12}\,\delta(z-1)\right], \tag{4.23b}$$

where $(1-z)_+^{-1}$ is defined by

$$\int_x^1 \frac{dz\,h(z)}{(1-z)_+} \equiv \int_x^1 \log(1-z)\,\frac{dh(z)}{dz}\,dz = h(z{=}1)\log(1-x)$$

$$+ \int_x^1 \frac{(h(z)-h(z{=}1))}{(1-z)}\,dz, \tag{4.24}$$

so that

$$\int_0^1 \frac{dz}{(1-z)_+} = 0. \tag{4.25}$$

Since the total number of quarks minus antiquarks is conserved, the probability of finding a quark in a quark, integrated over all z, must add up to one. It follows that (to order α)

$$\int_0^1 dz\,P_{q\leftarrow q}(z) = 0. \tag{4.26}$$

In addition, the P(z) functions obey the following sum rules

$$\int_0^1 dz\,z[P_{q\leftarrow q}(z) + P_{g\leftarrow q}(z)] = 0 \tag{4.27a}$$

$$\int_0^1 dz\,z[2n_f\,P_{q\leftarrow g}(z) + P_{g\leftarrow g}(z)] = 0. \tag{4.27b}$$

If we define a "singlet" distribution by

$$q^S(x,\tau) = \sum_{i=1}^{2n_f} q_i(x,\tau), \tag{4.28}$$

then we can write the following matrix equation

$$\frac{d\underset{\sim}{q}(x,\tau)}{d\tau} = \frac{\alpha(\tau)}{2\pi}\int_x^1 \frac{dy}{y}\, \underset{\sim}{P}\left(\frac{x}{y}\right)\underset{\sim}{q}(y,\tau) \tag{4.29}$$

where

$$\underset{\sim}{q}(x,\tau) = \begin{pmatrix} q^s(x,\tau) \\ g(x,\tau) \end{pmatrix} \tag{4.30a}$$

and

$$\underset{\sim}{P}(z) = \begin{pmatrix} P_{q\leftarrow q}(z) & 2n_f P_{q\leftarrow g}(z) \\ P_{g\leftarrow q}(z) & P_{g\leftarrow g}(z) \end{pmatrix}. \tag{4.30b}$$

The non-singlet is given by

$$\frac{dq_i^{NS}(x,\tau)}{d\tau} = \frac{\alpha(\tau)}{2\pi}\int_x^1 \frac{dy}{y}\, P_{q\leftarrow q}\left(\frac{x}{y}\right)q_i^{NS}(y,\tau), \tag{4.31a}$$

where

$$q_i^{NS}(x,\tau) = q_i(x,\tau) - \bar{q}_i(x,\tau). \tag{4.31b}$$

Defining

$$N_i(x,\tau) = \int_0^1 q_i^{NS}(x,\tau)dx, \tag{4.32}$$

as the total number of quarks of flavor i, then from (4.26), we see that

$$\frac{dN_i(x,\tau)}{d\tau} = 0. \tag{4.33}$$

This means that if $N_u(x,0) = 2$ for a proton as in (4.2a), it will remain 2 at any Q^2. In addition, (4.27a) and (4.27b) guarantee that

$$\frac{d}{d\tau}\int_0^1 dx\, x[\sum_{i=1}^{2n_f} q_i(x,\tau) + g(x,\tau)] = 0, \tag{4.34}$$

so that the total momentum of the proton (i.e., of all partons) is unchanged as Q^2 changes.

Transforming back to moments via (4.12) yields

$$\frac{dM_n^{NS}(\tau)}{d\tau} = \frac{\alpha(\tau)}{2\pi} A_n^{NS} M_n^{NS}(z), \tag{4.35a}$$

and

$$\frac{d\underset{\sim}{M}_n(\tau)}{d\tau} = \frac{\alpha(\tau)}{2\pi} \underset{\sim}{A}_n \underset{\sim}{M}_n(z), \tag{4.35b}$$

where

$$\underset{\sim}{M}_n(\tau) = \begin{pmatrix} M_n^s(\tau) \\ \\ M_n^g(\tau) \end{pmatrix}, \tag{4.36a}$$

and

$$\underset{\sim}{A} = \begin{pmatrix} A_n^{NS} & 2n_f A_n^{qg} \\ \\ A_n^{gq} & A_n^{gg} \end{pmatrix}, \tag{4.36b}$$

where M_n^{NS}, M_n^s and M_n^g are the moments of the non-singlet, singlet and gluon distributions, respectively. The "anomalous dimensions" A_n are given by

$$\underset{\sim}{A} = \int_0^1 dz\, z^{n-1} \underset{\sim}{P}(z), \tag{4.37}$$

and by (4.36). We are left with three independent Q^2 evolution equations for the moments. Namely,

$$M_n^{NS}(\tau) = M_n^{NS}(o) \left(\frac{\alpha(o)}{\alpha(\tau)}\right)^{A_n^{NS}/2\pi b} \tag{4.38a}$$

$$M_n^{+}(\tau) = M_n^{+}(o) \left(\frac{\alpha(o)}{\alpha(\tau)}\right)^{A_n^{+}/2\pi b} \tag{4.38b}$$

$$M_n^-(\tau) = M_n^-(o)\left(\frac{\alpha(o)}{\alpha(\tau)}\right)^{A_n^-/2\pi b} \quad , \tag{4.38c}$$

where $M_n^{\pm}(\tau)$ and $A_n^{\pm}$ are the eigenvectors and eigenvalues obtained upon diagonalizing eq. (4.36). From (4.22), (4.23) and (4.37), one obtains

$$A_n^{NS} = \left(\frac{4}{3}\right)\left[-\frac{1}{2} + \frac{1}{n(n+1)} - 2\sum_{j=2}^{n}\frac{1}{j}\right] \tag{4.39a}$$

$$A_n^{gq} = \left(\frac{4}{3}\right)\frac{2+n+n^2}{n(n^2-1)} \tag{4.39b}$$

$$A_n^{qg} = \left(\frac{1}{2}\right)\frac{2+n+n^2}{n(n+1)(n+2)} \tag{4.39c}$$

$$A_n^{gg} = 3\left[-\frac{1}{6} + \frac{2}{n(n-1)} + \frac{2}{(n+1)(n+2)} - 2\sum_{j=2}^{n}\frac{1}{j} - \frac{1}{9}n_f\right]. \tag{4.39d}$$

C. Analysis of Deep Inelastic Electron and Muon Scattering-Analysis of the Parton Distributions

Following the analysis of G. C. Fox[14)], the moments of the quark and gluon distributions

$$M_i(n,Q^2) = \int_0^1 x^n G_{p\to i}(x,Q^2)\,dx \tag{4.40a}$$

are given in terms of the moments at some reference momentum, Q_o^2, by

$$M_j(n,Q^2) = \sum_{i=1}^{9} M_i(n,Q_o^2)R_{ij}(n,Q^2,Q_o^2,\Lambda), \tag{4.40b}$$

where $R_{ij}(n,Q^2,Q_o^2,\Lambda^2)$ is the matrix constructed from eq. (4.36) and i corresponds to the constituent types ($u,d,s,c,\bar{u},\bar{d},\bar{s},\bar{c}$,glue). The matrix R_{ij} depends on $\alpha(Q^2)$ (i.e., on Λ^2 from (2.12)) and on the calculable anomalous dimensions A_n^{NS}, A_n^{qg}, A_n^{gq} and A_n^{gg} given by

(4.39). The resulting distributions at Q^2 are calculated in terms of those at Q_0^2 by diagonalizing (4.40b) and inverting (4.40a) by an inverse Mellin transform (eq. (13) of Ref. 14)).

Figure 4.3 shows the expected dependence of $\nu W_2(x,Q^2)$ resulting from an analysis of ep and μp data. The x dependence of the parton distributions at the reference momenta, $Q_0^2 = 4\ \text{GeV}^2$, was chosen to agree with experiment and Λ was varied to produce the observed amount of "scale breaking." The analysis of ep and μp data is sensitive to the gluon distribution only through diagrams like that of Fig. 4.1c. We have taken

$$xG_{p\to g}(x,Q_o^2) = (1+9x)(1-x)^4, \qquad (4.41)$$

however, the analysis of ep and μp is not sensitive to this precise choice. The resulting Q^2 dependence of $G_{p\to g}(x,Q^2) = g(x,Q^2)$ is shown in Fig. 4.3b. Figure 4.4 shows the behavior of

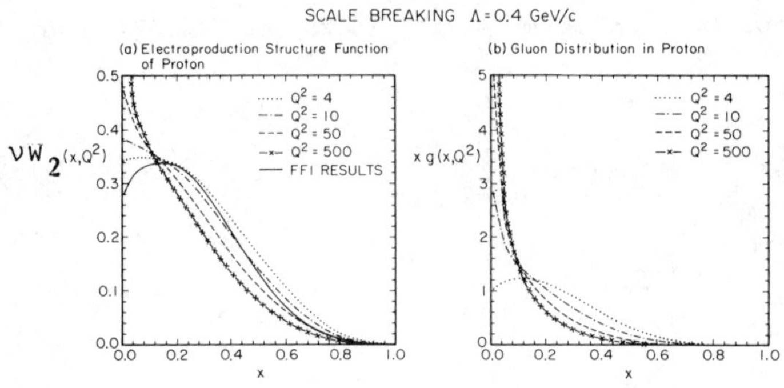

Fig. 4.3 - (a) Shows the predicted Q^2 dependence (scale breaking) of the electroproduction structure function for the proton, $\nu W_2(x,Q^2)$, arising from the constituent (quarks, antiquarks and gluons) distributions $G_i(x,Q^2)$ used in this analysis. The distributions at high Q^2 are calculated from the distributions at the reference momentum $Q_0^2 = 4\ \text{GeV}^2$ using a QCD moment analysis with $\Lambda = 0.4$ GeV/c. In asymptotically free theories, one expects a decrease in the number of high x constituents and an increase in the number of low x constituents as Q^2 increases. Also shown is the value of $\nu W_2(x)$ (independent of Q^2) used in the quark-quark "black-box" model of FFF[20].
(b) Shows the predicted Q^2 dependence of the distribution of gluons within the proton $xG_{p\to g}(x,Q^2)$ used in this analysis. The distribution at high Q^2 is calculated in terms of a distribution at the reference momentum $Q_0^2 = 4\ \text{GeV}^2$ chosen to be $xg(x,Q_o^2) = (1+9x)(1-x)^4$.

the quark and antiquark distributions. Both $\nu W_2(x,Q^2)$ and $xg(x,Q^2)$ exhibit the characteristic rise at small x and decrease at large **x** as Q^2 increases.

Figure 4.5 shows how the total

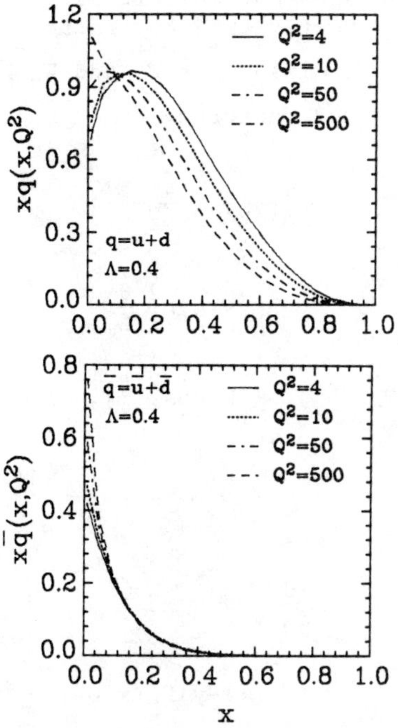

Fig. 4.4. Shows the predicted Q^2 dependence (scale breaking) of the quark ($q = u + d$) and antiquark ($\bar{q} = \bar{u} + \bar{d}$) distributions from the QCD moment analysis with $\Lambda = 0.4$ GeV/c.

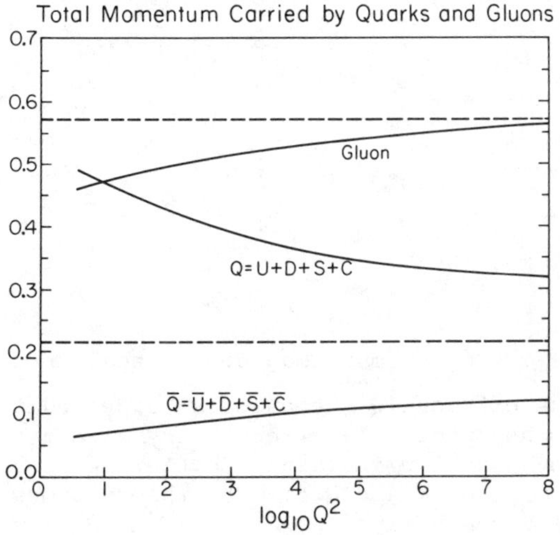

Fig. 4.5. Shows the predicted Q^2 dependence of the total momentum carried by quarks ($Q = U + D + S + C$), antiquarks ($\bar{Q} = \bar{U} + \bar{D} + \bar{S} + \bar{C}$) and gluons from a QCD analysis with $\Lambda = 0.4$ GeV/c. As Q^2 becomes large (very large!) $\bar{Q} \sim Q \sim 0.22$ and gluon ~ 0.57 for the number of quark flavors, n_f, equal to 4.

momentum carried by quarks, antiquarks and gluons within a proton is predicted to change with increasing Q^2. For $n_f = 4$, one expects the total momentum carried by quarks to approach that carried by antiquarks and to become 22%. The gluons carried (asymptotically) the remaining 57% of the proton momentum. However, as Fig. 4.5 indicates, the approach to asymptopia is quite gentle.

Fits to the existing ep and μp data are shown in Fig. 4.6 and Fig. 4.7. The "scale" breaking is precisely as expected from QCD and the size of the breaking indicates that Λ in (2.12) is in the range 0.4-0.6 GeV, where only data with Q > 1.5 GeV/c and hadron mass W > 2 GeV were considered.

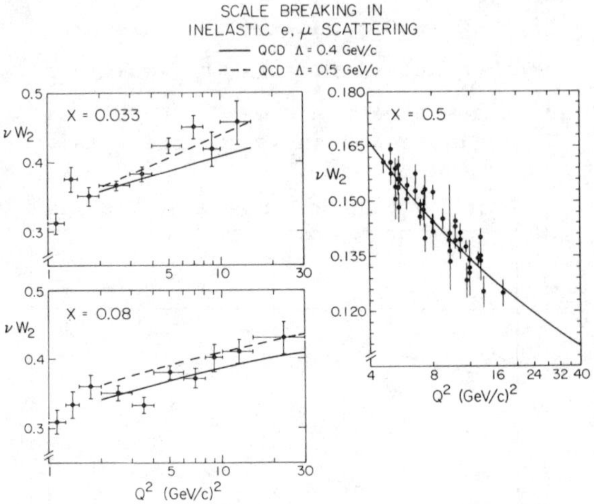

Fig. 4.6. Comparison of the scale breaking effects (Q^2 dependence) expected from an asymptotically free theory with data on ep and μp inelastic scattering at x = 0.033, 0.08 and 0.5. The theory comes from the analysis of Ref. 14 using Λ = 0.4 GeV/c (solid curve) and Λ = 0.5 GeV/c (dashed curve).

Experimentally one does not measure directly νW_2, but rather a mixture of W_1 and W_2. One measures

$$\frac{d^2\sigma}{d\Omega' dE'} = \frac{4\alpha^2 E'^2}{Q^4} [2W_1 \sin^2\frac{\theta}{2} + W_2 \cos^2\frac{\theta}{2}], \quad (4.42)$$

or

$$\frac{d\sigma}{dxdy} \propto [(1-y)F_2(x,Q^2) + \frac{1}{2} y^2 (2xF_1(x,Q^2)], \quad (4.43)$$

where the final lepton has energy E' and has scattered through an angle θ, $y = \nu/E$, and where

$$MW_1 = F_1, \quad (4.44a)$$

$$\nu W_2 = F_2. \quad (4.44b)$$

One needs to know the ratio of W_1 to W_2, or equivalently the value of $R = \sigma_L/\sigma_T$, to extract νW_2. The experimental determinations of R have large systematic errors and although they are consistent with QCD at low x, they are much larger than the QCD predictions at large x. I will discuss this problem in Section IV.F. The discrepancy between the theory and experimental value of R has the effect that νW_2 extracted using the theoretical R (as in Fox's analysis) varies with scattering angle θ. If the theoretical R is smaller than

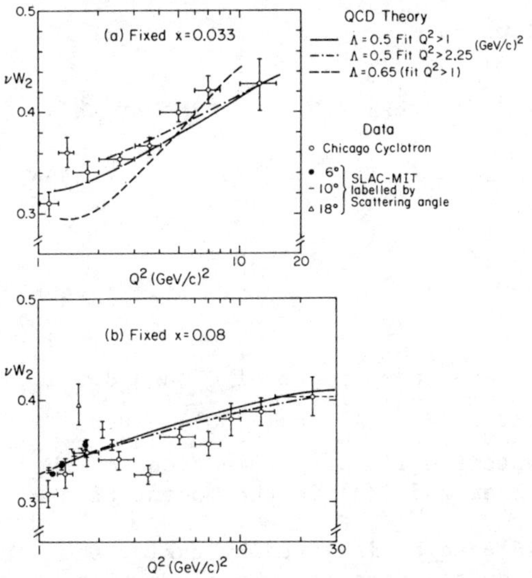

Fig. 4.7. $\nu W_2(x,Q^2)$ at fixed x = 0.033 and 0.08 for ep and μp inelastic scattering compared with the QCD predictions from Ref. 14.

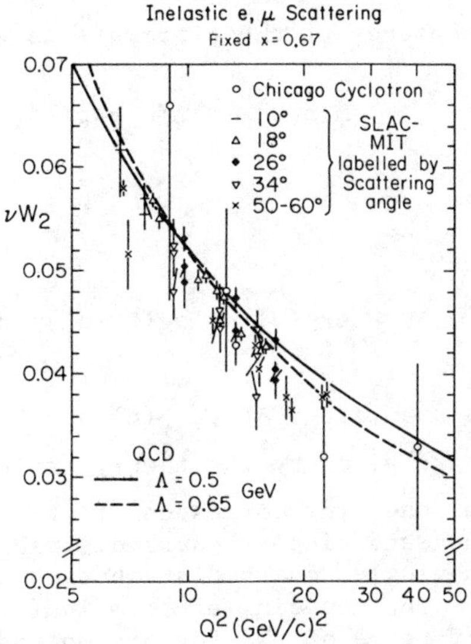

Fig. 4.8. $\nu W_2(x,Q^2)$ at fixed x = 0.67 for ep and μp inelastic scattering compared with the QCD predictions from Ref. 14. The experimental values of νW_2 have been extracted from the measured quantities assuming the theoretical form for $R = \sigma_L/\sigma_T$. The discrepancy between the theory and experiment in this figure is because R(theory) is smaller than R(expt) as can be seen in Fig. 4.11.

experiment, νW_2 will decrease with increasing scattering angle. This is seen clearly in Fig. 4.8 where the largest angle data (50, 60°) lie systematically below the best fit to the bulk of the data. It remains to be seen whether this is caused by systematic normalization errors in the experiment or if it is a true effect.

D. The Moment Method

Another method to test the theory is to form the moments

$$M_2(n,Q^2) = \int_0^1 x^{n-2}\nu W_2(x,Q^2)dx \tag{4.45}$$

directly from the data. As we have seen, in general, the Q^2

dependence of these moments is given by

$$M(n,Q^2) = A(n)\exp(-d_A w) + B(n)\exp(-d_B w) + C(n)\exp(-d_C w), \tag{4.46a}$$

where

$$\exp(w) = \log(Q^2/\Lambda^2)/\log(Q_0^2/\Lambda^2). \tag{4.46b}$$

The quantities A(n), B(n) and C(n) are independent of Q^2 and d_A, d_B, d_C are given by the theory (related to A_n^{NS}, $A_n^{\pm}$ in (4.38)). Here A(n) and B(n) correspond to two specific linear combinations of the gluon and singlet quark distributions and C(n) is the moment of the non-singlet quark distributions.

The non-singlet term in (4.46) can be isolated by taking the difference of νW_2 for proton and neutron targets. This has been done by Quirk[15)]. He finds $\Lambda = 0.68 \pm 0.1$ GeV/c by observing the Q^2 dependence of the $\exp(-d_C s)$ term. The results of a full analysis of both proton and deuteron data by Anderson, Matis and Myrianthopoulos[16)] are shown in Fig. 4.9. In both theory and experiment, the higher moments (large n) decrease faster with increasing Q^2 than does the n = 2 moment. The fact that a simultaneous fit is possible to three moments (n = 2, 4, 6) is a direct test of QCD. The Anderson, Matis and Myrianthopoulos moment analysis yields $\Lambda = 0.66 \pm 0.08$ GeV/c.

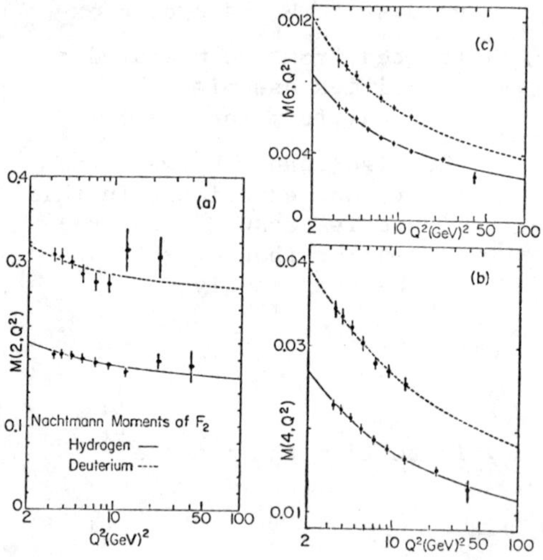

Fig. 4.9. Fits to the "Nachtmann" moments of νW_2 for ep and μp scattering from an analysis by Anderson and collaborators[16)] using $\Lambda = 0.66$ GeV/c.

It is important to realize that these moment analyses of the ep data include the elastic contributions ep → ep at x = 1 and that a lot of the Q^2 dependence of the moments comes from the decrease of the elastic term. On the other hand, the parton distribution method discussed in Section IV.C excluded the elastic term and even the resonance region (only W > 2 GeV was included). Clearly the difference in the $\Lambda \approx 0.5$ GeV/c determined in the latter analysis and the $\Lambda \simeq 0.7$ GeV/c determined by the direct moment analysis lies in these elastic or resonance $1/Q^2$ effects. The two methods agree at high Q^2.

E. Corrections of Order $1/Q^2$

One simple (almost kinematic) explanation of some of the observed scale breaking is that the structure functions W_1 and νW_2 might depend on only one variable but it is not x but a different variable $\tilde{x}$ that has the property that $\tilde{x}(x,Q^2) \to x$ as $Q^2 \to \infty$. The two usual choices for $\tilde{x}$ are

$$\tilde{x} = x' = \frac{x}{1 + M^2x/Q^2} \approx x - M^2x^2/Q^2 \tag{4.47a}$$

and

$$\tilde{x} = \xi = \frac{2x}{1 + \sqrt{1 + 4x^2M^2/Q^2}} \approx x - M^2x^3/Q^2, \tag{4.47b}$$

where I have shown the approximate form to order M^2/Q^2. The first form is the variable introduced by Bloom and Gilman[17], who pointed out that much of the (large x) scale violation could be removed by considering νW_2 as a function of x'. The ξ variable was introduced by Georgi and Politzer[11] to remove some of the (kinematic) M^2/Q^2 behavior and was derived from plausible but not watertight arguments based on the kinematic structure of Born terms. The parton model gives a useful mnemonic for the form of ξ in (4.47b). If a massless quark carries a fraction ξ of the proton momentum and is kicked onto its mass shell by the collision with a virtual photon of momentum $q^2 = -Q^2$

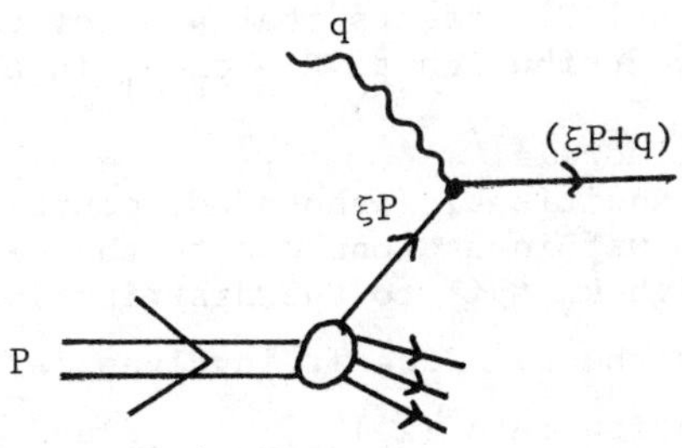

(4.48)

then

$$(\xi P+q)^2 = 0 = \xi^2 M^2 + 2\xi P\cdot q - Q^2 \qquad (4.49)$$

and ξ is the positive solution of this quadratic equation. In the analysis of the electroproduction data, Fox has used the ξ variable to remove some of the $1/Q^2$ effects.

One can also define moments of $F_2(x,Q^2) = \nu W_2(x,Q^2)$ in terms of the variable ξ. These Nachtmann moments[18)] are expected to be a better representation of the theory at low Q^2 and are defined by

$$M_2(n,Q^2) = \int_0^1 \frac{\xi^{n+1}}{x^3} F_2(x,Q^2) \qquad (4.50)$$

$$\frac{[(n^2+2n+3)+3(n+1)\sqrt{1+4M^2x^2/Q^2}+n(n+2)4M^2x/Q^2]dx}{(n+2)(n+3)}.$$

They approach the x-moments defined in (4.45) as Q^2 becomes large. These "Nachtmann" moments were used in the analysis of Quirk[15)] and Anderson et al.[16)] which is shown in Fig. 4.9.

There are other $1/Q^2$ effects that can not be estimated from perturbation theory. These effects are associated with the fact that partons are confined within hadrons. For example, quarks are confined in the transverse direction to within a hadron radius and, thereby from the uncertainty principle, they must have some transverse momentum. This transverse momentum, called primordial, is intrinsic to the basic parton wave function. It has developed over a long time scale and involves small Q^2 (large $\alpha_s(Q^2)$). It cannot be calculated by keeping a few terms in perturbation theory (one must know the wave function). At present, this primordial $k_\perp$ must be viewed as unknown but bounded (as in the naive parton model). The primordial $k_\perp$ can produce $1/Q^2$ effects that are not calculable. For example, it produces a contribution to $R = \sigma_L/\sigma_T$ in electroproduction of the form $4\langle k_\perp^2\rangle_{\text{primordial}}/Q^2$ [8)].

In QCD there is an additional (unbounded) contribution to the "effective" $k_\perp$ of quarks within hadrons due to the hard Bremsstrahlung of gluons. This high $k_\perp$ tail to the distribution can be calculated using perturbation theory since it involves large $k_\perp^2$ and thus small $\alpha(k_\perp^2)$. The separation into $(k_\perp)_{\text{primordial}}$ and

$(k_\perp)_{\text{perturbative}}$ is a bit artificial. There is only one function $f(k_\perp)$ that represents the effective transverse momentum of partons within hadrons. The large $k_\perp$ tail is calculated by perturbation theory (it behaves like $d^2k_\perp^2/k_\perp^2$) and the low $k_\perp$ part is not.

The results from the analysis of the quark and gluon distributions (4.40) agree with direct moment analyses of the data at large Q^2 but lie systematically below the data at low Q^2. (Remember that the direct moment analysis includes the elastic and quasi-elastic events while the distribution method does not.) This was noticed by De Rujula, Georgi and Politzer[11] who suggested that the effect is due to non-leading ("higher twist") corrections to QCD. These effects can be parameterized according to

$$\tilde{M}(n,Q^2) = M(n,Q^2)[1 + (n-2)\Delta/Q^2], \tag{4.51}$$

where $\tilde{M}(n,Q^2)$ are the experimentally determined moments defined by (4.50) and $M(n,Q^2)$ are the moments predicted from asymptotic QCD as in (4.40b). The value of Δ, which has dimensions $(\text{mass})^2$, is expected to be of order Λ^2 or $\langle k_\perp^2\rangle_{\text{primordial}}$. It represents final state interactions that, at low W, bind quarks into resonances. This additional multiplicative term in (4.51) decreases with increasing Q^2 and thus if one fits the moments with $\Delta = 0$, one gets a larger value of Λ than with $\Delta \neq 0$. The electroproduction analysis yields $\Delta \simeq 0.1$ to $0.2\ \text{GeV}^2$ for $0.5 < (n-2)/Q^2 < 1.5$.

F. Predictions for $R = \sigma_L/\sigma_T$ in ep Scattering

The naive parton model (with bounded transverse momentum) predicts that[8]

$$R = F_L/(2xF_1) \simeq 4(\langle k_\perp^2\rangle \pm \Delta^2)/Q^2, \tag{4.52}$$

where the longitudinal structure function F_L is given by $F_L = F_2 - 2xF_1$. In (4.52) $\langle k_\perp^2\rangle$ is the "primordial" transverse momentum of the quarks within hadrons and Δ represents corrections due to the bindings of the quarks.

In QCD, to order α_s, the longitudinal structure function is given by[19]

$$F_L(x,Q^2) = \frac{\alpha_s(Q^2)}{2\pi}\, x^2 \int_x^1 \frac{dy}{y^3} \tag{4.53}$$

$$\{\frac{8}{3} F_2(y,Q^2) + 2a_e y G_{p\to g}(y,Q^2)(1 - \frac{x}{y})\},$$

where $a_e = 20/9$ (for $n_f = 4$) and where the first term comes from the subprocess $q + \gamma^* \to q + g$ and the second term from $\gamma^* + g \to q + \bar{q}$. We will derive (4.53) in Section VI. Notice that since

$$F_2(x,Q^2) = \sum_{i=1}^{n_f} e_i^2 x(q_i(x,Q^2) + \bar{q}_i(x,Q^2)), \tag{4.54}$$

F_L is sensitive to both the quark and gluon distributions. To order α_s, the perturbative contribution to R is thus

$$R_{pert} = F_L/(2xF_1) = \alpha_s(Q^2) r(x,Q^2), \tag{4.55}$$

where the Q^2 dependence of $r(x,Q^2)$ comes from the Q^2 dependence of the quark and gluon distributions. Neglecting this latter Q^2 dependence yields

$$R_{pert} \propto \alpha_s(Q^2) \propto 1/\log(Q^2/\Lambda^2) \tag{4.56}$$

to compare with

$$R_{primordial} \propto 4\langle k_\perp^2\rangle_{primordial}/Q^2. \tag{4.57}$$

Thus if we are at sufficiently large Q^2 so that (4.57) can be neglected then we can predict precisely the behavior of R.

The function $r(x,Q^2)$ is large at small x and small at large x as can be seen in Fig. 4.10 where I show the results of R_{pert} calculated from Fox's analysis. Some corrections of order M^2/Q^2 have been included by using the ξ variable (4.47b). Figure 4.11 shows the predictions of R_{pert} with and without some target mass corrections from Buras, Floratos, Ross and Sachrajda[12]. The quality of the data is poor but there is some indication that the large x data are larger than the perturbative predictions. This may just be an indication for the presence of some $R_{primordial}$. For example, with

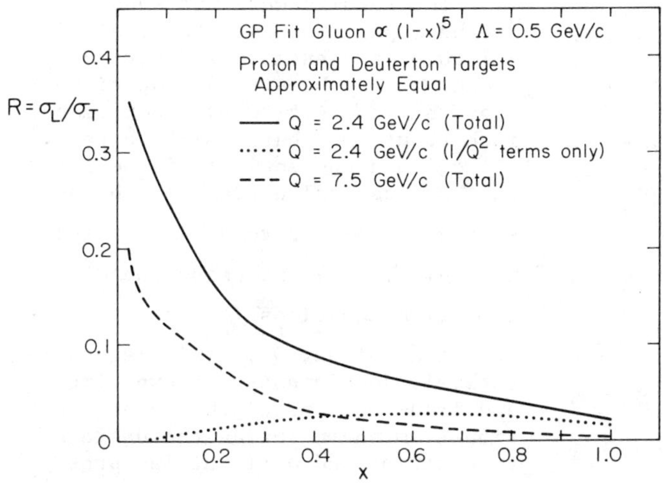

Fig. 4.10. Theoretical values of $R = \sigma_L/\sigma_T$ from the analysis of Ref. 14 with $\Lambda = 0.5$ GeV/c. The dotted curve shows the importance of $1/Q^2$ contribution to R.

$\langle k_\perp^2 \rangle_{primordial} = 0.25\ GeV^2$ (and $\Delta^2 \simeq \langle k_\perp^2 \rangle_{primordial}$) then $R_{primordial} \simeq 0.25$ at $Q^2 = 8\ GeV^2$ which is considerably larger than the perturbative contribution.

G. Conclusions from Electron and Muon Inelastic Scattering

We have seen that QCD provides a good description of the scale breaking in the ep and μp data with Λ in (2.12) lying between 0.4 - 0.7 GeV/c. The precise value of Λ depends on how one handles the $1/Q^2$ corrections to the theory. Both the distribution method (Section C) and the direct moment method (Section D) require $1/Q^2$ corrections at low Q^2. Most likely, these corrections are larger in the latter, which explains the different Λ's determined by the two methods.

The value of $R = \sigma_L/\sigma_T$ is large in both theory and experiment at small x. There is some indication, although the data suffer from possible systematic errors, that R is large ($\simeq 0.25$) at large x as well. The perturbative QCD contribution to R is small at large x so the theory may need corrections. However, the corrections may just be $1/Q^2$ terms (perhaps coming from the primordial non-perturbative $\langle k_\perp^2 \rangle$ of partons within hadrons).

V. ANALYSIS OF CHARGED CURRENT NEUTRINO AND ANTINEUTRINO PROCESSES

A. Comparison with Results from the Electroproduction Analysis

The analysis of electroproduction off proton and deuteron targets provided information concerning the distribution of $u + \bar{u}$ and

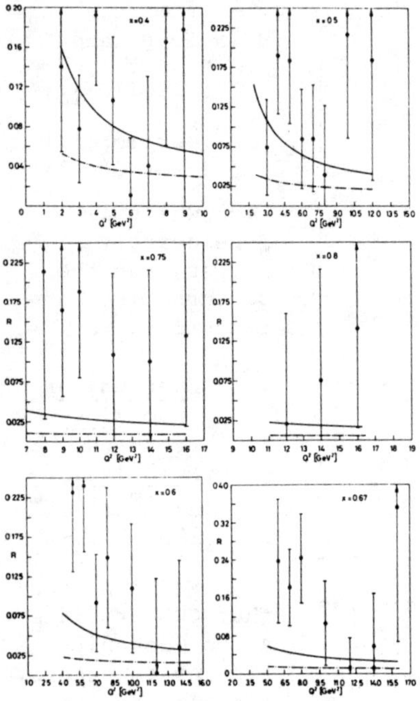

Fig. 4.11. Comparison of theoretical values of $R = \sigma_L/\sigma_T$ (solid curves) with data (from an analysis by A. J. Buras et al.[12]). The dashed curves are the predictions before M^2/Q^2 effects have been included.

$d + \bar{d}$ quarks as a function of x and Q^2. The results were not sensitive to the separation into the individual quark and antiquark distributions nor to the strange contribution, $s + \bar{s}$. Also we did not learn much about the charm quark content, $c + \bar{c}$, in a nucleon. It is believed that one can calculate $c + \bar{c}$ by assuming it to be zero at low Q^2 (say, $Q_o^2 = 4(\text{GeV}/c)^2$) and using the QCD evolution equations (4.17) to generate it at higher Q^2. This is qualitatively consistent with experiment and leads to a negligible contribution from charm inside the nucleon to neutrino and antineutrino processes[21,22]. We will now investigate whether what we have learned about the quark distributions and about the mass scale Λ is consistent with ν and $\bar{\nu}$ data.

The neutrino and antineutrino differential cross sections can be written in terms of the structure functions F_1, F_2 and F_3 as follows:

$$\frac{d\sigma^{\nu,\bar{\nu}}}{dxdy} = \frac{G^2ME}{\pi}\,[(1-y)F_2(x,Q^2) + y^2xF_1(x,Q^2) \pm y(1-\tfrac{1}{2}y)xF_3(x,Q^2)], \tag{5.1}$$

where G is the weak interaction coupling and where $x = Q^2/2M\nu$, $y = \nu/E$ and $\nu = E - E'$ (E and E' are the incident neutrino and final muon energy, respectively, and M is the proton mass). In the naive parton model and to leading order in QCD, F_2 and F_3 can be written in terms of the quark and antiquark distributions. Namely,

$$F_2(x,Q^2) = 2x[q(x,Q^2) + \bar{q}(x,Q^2)] \tag{5.2a}$$

$$xF_3(x,Q^2) = 2x[q(x,Q^2) - \bar{q}(x,Q^2)] \tag{5.2b}$$

or defining $F_A(x,Q^2)$ and $F_Q(x,Q^2)$ by

$$F_A(x,Q^2) = \frac{1}{2}\,(F_2(x,Q^2) - xF_3(x,Q^2)) \tag{5.2c}$$

$$F_Q(x,Q^2) = \frac{1}{2}\,(F_2(x,Q^2) + xF_3(x,Q^2)) \tag{5.2d}$$

gives

$$F_A(x,Q^2) = 2x\bar{q}(x,Q^2) \tag{5.2e}$$

$$F_Q(x,Q^2) = 2xq(x,Q^2). \tag{5.2f}$$

Substituting back into (5.1) yields

$$\frac{d^2\sigma^\nu}{dxdy} \simeq \frac{2G^2ME}{\pi}\,\{xq(x,Q^2) + x\bar{q}(x,Q^2)(1-y)^2 - \frac{y^2Rx}{2}\,(q(x,Q^2) + \bar{q}(x,Q^2))\} \tag{5.3a}$$

$$\frac{d^2\sigma^{\bar{\nu}}}{dxdy} \simeq \frac{2G^2ME}{\pi}\,\{xq(x,Q^2)(1-y)^2 + x\bar{q}(x,Q^2) - \frac{y^2Rx}{2}\,(q(x,Q^2) + \bar{q}(x,Q^2))\}, \tag{5.3b}$$

where $R = F_L/2xF_1$ with $F_L = F_2 - 2xF_1$ and where the quark distributions $q(x,Q^2)$ and antiquark distributions $\bar{q}(x,Q^2)$ to be used are as follows:

Reaction	Final State	Quark Distributions
$\nu p \to \mu^- + X$	Non-charm	$\bar{u}$, $d\cos^2\theta_c + s\sin^2\theta_c$
	Charm	$d\sin^2\theta_c + s\cos^2\theta_c$
$\bar{\nu}p \to \mu^+ + X$	Non-charm	u, $\bar{d}\cos^2\theta_c + \bar{s}\sin^2\theta_c$
	Charm	$\bar{d}\sin^2\theta_c + \bar{s}\cos^2\theta_c$,

(5.4)

where θ_c is the Cabibbo angle. The charm content in the nucleon has been neglected.

The contribution to the total rate from $R = (F_2 - 2xF_1)/F_2 = \sigma_L/\sigma_T$ in (5.3) is small and will be discussed in Section V.D. The QCD perturbative contribution to R (to order $\alpha_s(Q^2)$) is only about 5% integrated over x and y for antineutrino scattering and about 2.5% for the neutrino case. In addition, the effect is, as we shall see, concentrated at small x as for electroproduction.

In comparing with neutrino and antineutrino processes, one must separate the quark and antiquark distributions. We have taken[21-25]

$$x\bar{u}(x,Q_o^2) = 0.23(1-x)^8, \tag{5.5a}$$

$$x\bar{d}(x,Q_o^2) = 0.23(1-x)^7, \tag{5.5b}$$

at the reference momentum $Q_o^2 = 4\ (\mathrm{GeV}/c)^2$. We choose $\bar{u}(x) \neq \bar{d}(x)$ (broken SU(3)) because theoretically one does not expect SU(3) to be perfectly satisfied for these distributions[26]. Present experiments, however, are sensitive only to $\bar{u} + \bar{d}$ and the choices in (5.5) yield a ratio

$$\int_0^1 x(\bar{u}+\bar{d})dx \Big/ \int_0^1 x(u+d)dx \simeq 0.11 \tag{5.6}$$

at $Q^2 = 4\ (\mathrm{GeV}/c)^2$ which is consistent with neutrino experiments. This ratio, however, increases with increasing Q^2 (see Fig. 4.5). In the SU(3) limit, the amount of strange quarks, s, would equal the amount of u-bar (or d-bar) quarks. It is, however, expected that since strange quarks are more massive, that they occur at less than the SU(3) value.

Possibly the best way to study the strange content within nucleons is to examine the dilepton rate in neutrino and antineutrino charged current scattering. These events are assumed to come from charm decay, where the charm is produced predominantly from the strange antiquark in $\bar{\nu}$-nucleon collisions and about equally from the $\bar{s}$ and cabibbo suppressed $\bar{d}$ in ν-nucleon collisions. Figure 5.1 shows the dilepton cross section together with the theoretical predictions from Fox's analysis[21], where a 10% branching ratio for charm into muons is assumed. Predictions are given for both an SU(3) symmetric strange sea and for $s(x,Q_o^2) = \bar{s}(x,Q_o^2) = \frac{1}{2}\bar{u}(x,Q_o^2)$. The theory is not in particularly good agreement with the data especially at the lower energies. This might be due to experimental problems like decay background or acceptance corrections that are worse at low energy. On the other hand, at low Q^2, the theory is sensitive to kinematic ambiguities of order M_c^2/Q^2 (where M_c is the charm quark mass). Fox has used the normal scaling variable

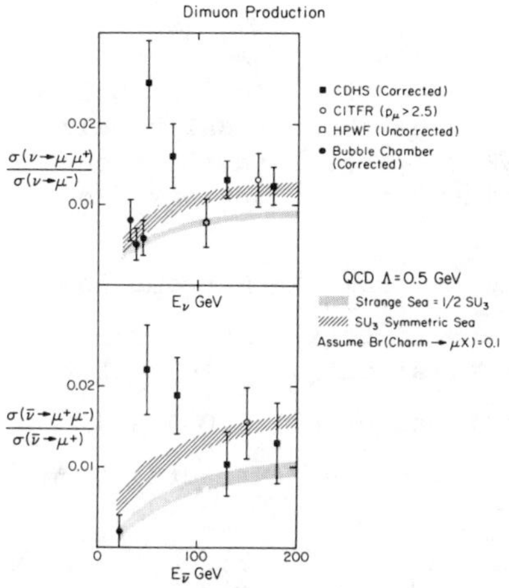

Fig. 5.1. Predictions from QCD for the rate of dimuon compared to single muon production in neutrino and antineutrino scattering taken from Ref. 21. The theory is shown for two choices for the strange quark distribution. The shaded regions indicate uncertainties due to changes in the antiquark distributions.

$$\xi_c = (Q^2+M_c^2)/2M\nu, \qquad (5.7)$$

for charm quark production which may not include all the M_c^2/Q^2 corrections.

As discussed in Section IV.C and shown in Fig. 4.5, the asymptotically free theory of QCD predicts that as Q^2 increases the number of quarks becomes equal to the number of antiquarks and the amount of momentum carried by gluons increases to 76% for $n_f = 3$ or 57% for $n_f = 4$. Thus we expect

$$\int_0^1 F_2(x,Q^2{=}\infty)dx = 0.24 \qquad n_f = 3$$
$$\qquad\qquad\qquad\qquad\quad 0.43 \qquad n_f = 4 \tag{5.8a}$$

and

$$\int_0^1 xF_3(x,Q^2{=}\infty)dx = 0. \tag{5.8b}$$

Data on these integrals are shown as a function of Q^2 in Fig. 5.2 together with the theoretical predictions. We see that both the theory and experiment are a long way from $Q^2 = \infty$.

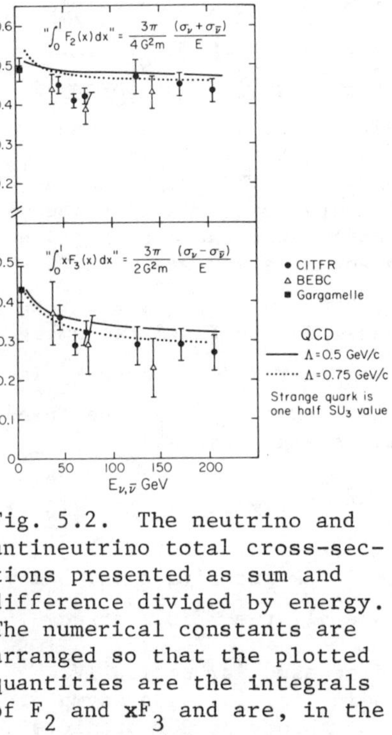

Fig. 5.2. The neutrino and antineutrino total cross-sections presented as sum and difference divided by energy. The numerical constants are arranged so that the plotted quantities are the integrals of F_2 and xF_3 and are, in the naive quark model, the total momentum carried by quarks plus antiquarks and quarks minus antiquarks, respectively. The QCD predictions are shown with the strange quark distribution at one half the SU_3 value.

The data from the Gargamelle and BEBC experiments[27)] have provided the first detailed look at the structure functions $F_2(x,Q^2)$ and $xF_3(x,Q^2)$ which they extracted from

$$\frac{d\sigma^{\nu,\bar\nu}}{dxdy} = \frac{G^2ME}{\pi}\{F_2(x,Q^2)(1-y+\frac{y^2}{2}) \pm xF_3(x,Q^2)(y-\frac{y^2}{2})\} \tag{5.9}$$

by adding and subtracting the neutrino and antineutrino cross sections (and assuming $F_1 = 2xF_2$). In the naive parton model, F_2 and xF_3 are proportional to $x(q+\bar q)$ and $x(q-\bar q)$ (see 5.2)) and are functions only of x. In QCD, we expect both F_2 and xF_3 to be functions of x and Q^2 but not to depend on the incident energy E. This is important since for fixed x, the full range of $Q^2 = 2MExy$ is gotten by combining data at different incident energies. However, a non-zero value of $R = \sigma_L/\sigma_T$

does introduce an explicit energy dependence of F_2 and xF_3 (if defined according to (5.9)). For example, F_2 extracted from (5.9) by adding ν and $\bar{\nu}$ differential cross sections is underestimated by a factor of R ($\approx$ 20% at low x) at energies where x and Q^2 convert to y = 1. Apart from this slight energy dependence, we can compare directly the theory and experimental determinations of F_2 and xF_3 as a function of x and Q^2. This is shown in Fig. 5.3 and Fig. 5.4, where the theory is calculated at energies where $y \leq 0.5$ so that the effect of R is small. The agreement between the theory and experiment is quite good. Remember that the normalization and shape of the curves come from the electroproduction analysis. The ν and $\bar{\nu}$ data have not been used in the determination of the structure functions or in the determination of Λ. It is interesting to note that the Gargamelle data, which gave some of the early support for the quark parton picture, have $\langle Q^2 \rangle = 1\ \mathrm{GeV}^2$ where even a believer in QCD would expect sizable $1/Q^2$ corrections. Nevertheless, the data join on quite smoothly to the larger Q^2 BEBC data.

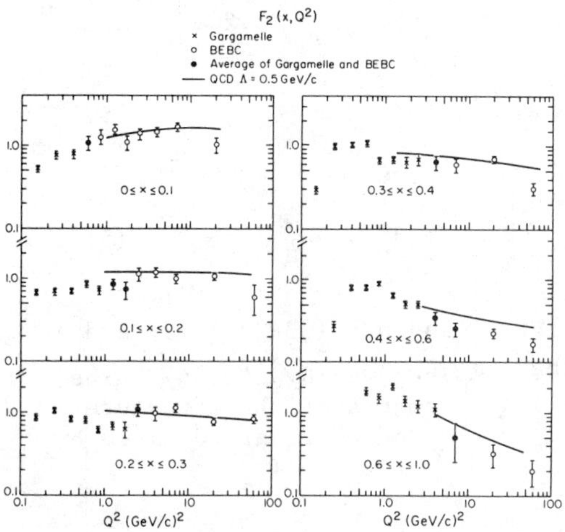

Fig. 5.3. The values of $F_2(x,Q^2)$ from the BEBC Gargamelle collaboration[27]. They are compared with the QCD calculations with Λ = 0.5 GeV/c and the strange quark at one half its SU_3 value. The effect of using an SU_3 symmetric sea is to raise the prediction by 15% in the lowest x bin 0 to 0.1 with smaller effects at higher x. Further, the nonzero value of $R = \sigma_L/\sigma_T$ (at small x) in the theory makes "$F_2(x,Q^2)$" a function of energy. This is again about a 20% effect with the lower energy, high y, point falling below the higher energy value at the same Q^2 = 2MExy. These two small effects are not shown on figure. (This figure is taken from Ref. 21.)

B. The Moment Method - Direct Tests of QCD

In electroproduction, one isolated the non-singlet distributions (4.10) by taking the difference of proton and neutron target data. In neutrino scattering, xF_3 is proportional to $q-\bar{q}$ which is directly the non-singlet combination. The non-singlet distributions (or moments)

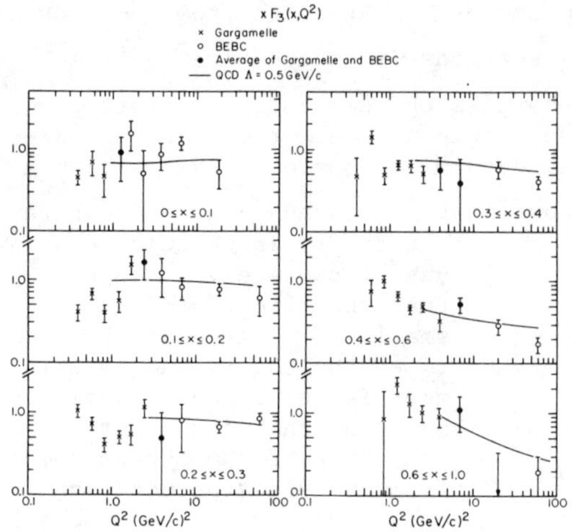

Fig. 5.4. Same as Fig. 7 except for $xF_3(x,Q^2)$ instead of $F_2(x,Q^2)$. The results are not sensitive to either R or the strange sea.

do not mix with gluon distributions (gluons contribute equally to q and $\bar{q}$) and their Q^2 evolution is given simply by (4.38a). Namely,

$$M_3(n,Q^2) = M_3(n,Q_o^2)$$

$$(\log Q^2/\Lambda^2/\log Q_o^2/\Lambda^2)^{-d_n}, \qquad (5.10a)$$

where

$$d_n = -A_n^{NS}/2\pi b \qquad (5.10b)$$

and $b = (33-2n_f)/12\pi$ and A_n^{NS} is given by (4.39a). The moments of the structure function F_3 are defined by

$$M_3(n,Q^2) = \int_0^1 x^{n-1} F_3(x,Q^2 dx, \qquad (5.11)$$

where $x = Q^2/2M\nu$ and are determined directly from the ν and $\bar{\nu}$ data as shown in Fig. 5.5a for n = 2 to 5. The moments fall off at large Q^2 at a rate that increases with increasing n.

Part of the decrease with increasing Q^2 of the x-moments defined by (5.11) is, at small Q^2, due to target mass correction terms of order M^2/Q^2 as discussed in Section IV.E. Some of the trivial kinematic M^2/Q^2 effects can be removed by defining moments in terms of the ξ variable defined in (4.47b). These are called Nachtmann moments and for xF_3 are given by

$$M_3(n,Q^2) = \int_0^1 \frac{\xi^{n+1}}{x^3} xF_3(x,Q^2) \frac{[1 + (n+1)\sqrt{1+4M^2x^2/Q^2}]}{(n+2)} dx, \qquad (5.12)$$

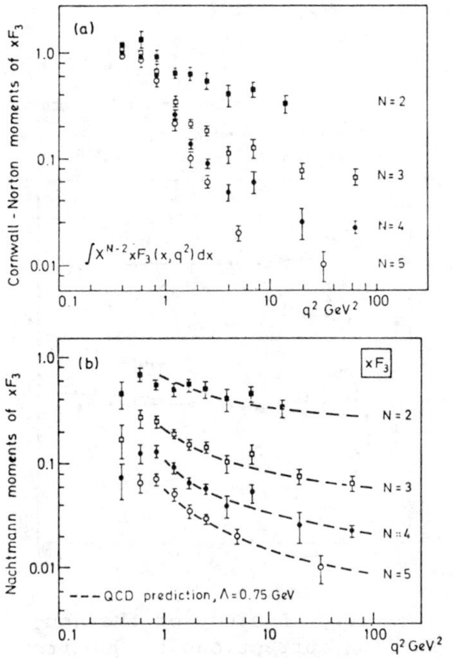

Fig. 5.5. Data from the BEBC collaboration (Ref. 27) on the ordinary (Cornwall-Norton) moments and Nachtmann moments of $xF_3(x,Q^2)$ versus Q^2.

which approaches the x-moments (5.11) as Q^2 becomes large. The Nachtmann moments of xF_3 for the neutrino data are shown in Fig. 5.5b and as can be seen, there is little difference between (5.11) and (5.12) for $Q^2 \geqslant 4\ (\mathrm{GeV/c})^2$. Figure 5.5b also shows the QCD predictions of eq. (5.10) with $\Lambda = 0.75$ GeV/c.

Taking the log of both sides of eq. (5.10) gives

$$\log M_3(n,Q^2)/d_n = C(n) - \log\log Q^2/\Lambda^2 \tag{5.13}$$

where $C(n)$ is independent of Q^2. This implies that a plot of $\log M_3(n,Q^2)$ versus $\log M_3(n',Q^2)$ is a straight line,

$$\log M_3(n,Q^2)/d_n - \log M_3(n',Q^2)/d_{n'} = C(n) - C(n'), \tag{5.14}$$

with a slope given by $d_n/d_{n'}$. In addition, from (5.10b) we see that the slope is independent of Λ and independent of the number of quark flavors, n_f.

Figure 5.6 shows log-log plots of some of the moments for $Q^2 > 1\ (\mathrm{GeV/c})^2$ from the BEBC analysis[27]. The anomalous dimensions d_n predicted by QCD appear to be confirmed at about the 10% level. This is the most significant test of QCD to date. Remember the naive parton model would predict that all the data along the lines in Fig. 5.6 should lie at one point (i.e., no Q^2 dependence).

The skeptic may, however, worry that the data in Fig. 5.6 go down to $Q^2 = 1\ (\mathrm{GeV/c})^2$ where the quantitative validity of the asymptotic predictions of QCD are not so obvious. There are undoubtedly $1/Q^2$ corrections like the ones discussed in Section IV.C. In addition, higher order QCD processes may play a role at this low Q^2.

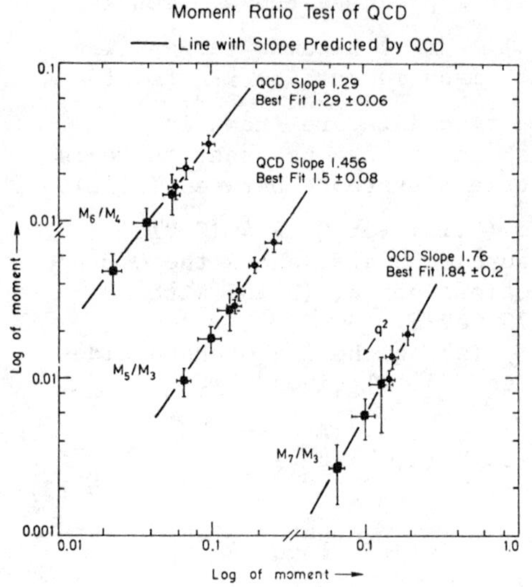

Fig. 5.6. Plot of $\log M_3(n,Q^2)$ versus $\log M_3(n',Q^2)$, where $M_3(n,Q^2)$ are moments of xF_3 from the BEBC analysis of neutrino and antineutrino interactions[27]. According to QCD, the data should lie along straight lines with a slope given by the theory. The data are plotted for $Q^2 > 1$ $(GeV/c)^2$ with the square points having $Q^2 \geq 4$ $(GeV/c)^2$.

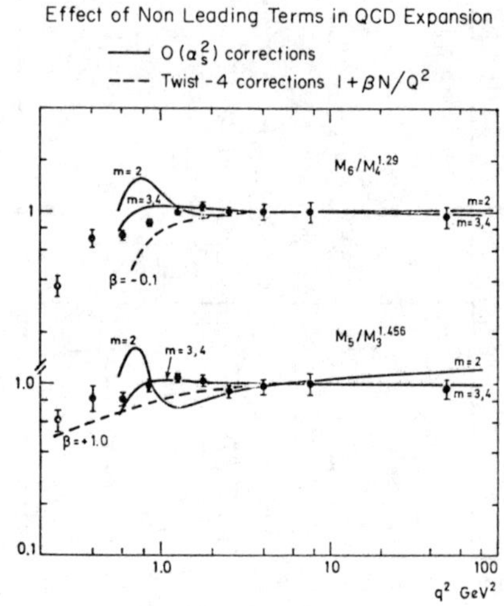

Fig. 5.7. A study of the non-leading corrections to QCD from Ref. 27. According to the theory, the quantity $M_3(n,Q^2)/[M_3(n-2,Q^2)]^{d_n/d_{n-2}}$ should be independent of Q^2 at large Q^2. Shown are typical twist-4 (i.e., M^2/Q^2) [dashes] $O(\alpha_s^2)$ [solid] corrections. The latter are shown for either m = 2, 3 or 4 flavors.

The low Q^2 problem is seen clearly in Fig. 5.7 where the BEBC group have plotted

$$M_3(n',Q^2)/[M_3(n,Q^2)]^{d_{n'}/d_n} \tag{5.15}$$

as a function of Q^2. According to (5.10), this quantity should be independent of Q^2 if the asymptotic QCD formulation is valid. Shown on the figure are the higher order g_s^4 corrections as calculated in Ref. 12. These corrections are surprisingly small and predict little deviation from asymptotic QCD for $Q^2 \gtrsim 1$ $(GeV/c)^2$. Also shown are typical multiplicative corrections of the form shown in (4.51). The data do not require any such corrections (i.e., $\Delta = 0$) but are consistent with a range $-0.1 \leq \Delta \leq 0.5$ GeV^2. This includes the value of 0.1 - 0.2 GeV^2 found in the corresponding analysis of the electroproduction moments.

C. Determination of Λ

Equation (5.10) can be written

$$[M_3(n,Q^2)]^{-1/d_n} = C(n)(\log Q^2 - \log\Lambda^2) \tag{5.16}$$

which means that Λ can be determined directly from the intercept at $\log Q^2 = 0$ of a linear fit of the form $\log Q^2 - \log\Lambda^2$ to M_3^{-1/d_n}. The results are shown in Fig. 5.8 where the BEBC group has fitted the n = 3, 5 and 7 moments for $Q^2 > 1$ (GeV/c)^2. This results in

neutrino moments:

$$\Lambda = 0.74 \pm 0.05 \text{ GeV/c} \tag{5.17a}$$

and can be compared with the results from electroproduction which yielded:

electroproduction moments:

$$\Lambda = 0.66 \pm 0.08 \text{ GeV/c}, \tag{5.17b}$$

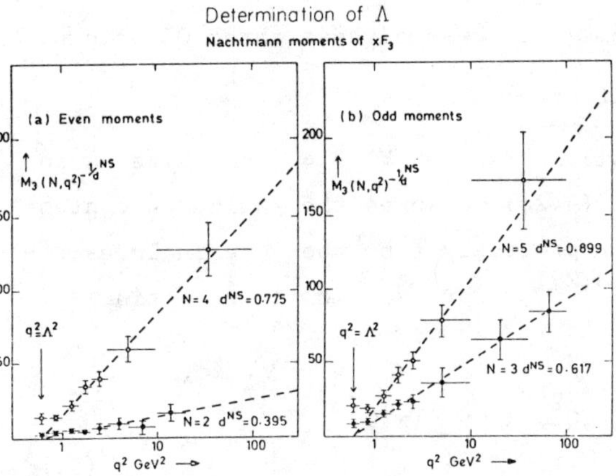

Fig. 5.8. The determination of Λ from the Q^2 dependence of $M_3(n,Q^2)$ for n = 2 to 5 from Ref. 27.

whereas the analysis of the electroproduction data in terms of the quark and gluon distributions gave

$$\text{structure function analysis:} \quad \Lambda = 0.4 - 0.5 \text{ GeV/c}. \tag{5.17c}$$

As discussed in Section IV, the moment analyses give larger values of Λ (i.e., more Q^2 dependence) than the analysis in terms of the quark and gluon distributions. Supposedly, the moment method needs corrections of order $1/Q^2$ (remember in constructing the moments, one has included the elastic and quasi-elastic events). For example, if one attempts to remove some of the non-leading $1/Q^2$ effects by using eq. (4.51) with $\Delta = 0.15$, then one finds from a neutrino and anti-neutrino moment analysis that $\Lambda = 0.52$ GeV in agreement with (5.17c).

We must conclude that the value of Λ is uncertain at present because most of the data are at low or moderate Q^2 values and we don't know precisely how to handle non-leading corrections to QCD. In my opinion, if one is in a region of Q^2 where the Nachtmann moments given by (5.12) differ from the ordinary moments (5.11) then the analysis is uncertain and should not be completely trusted. If, in the BEBC analysis, we restrict ourselves to $Q^2 > 4(\mathrm{GeV}/c)^2$, as we did for the structure function analysis of the electroproduction data, then we are left with only the highest three Q^2 data points in Figs. 5.6, 5.7 and 5.8. These points receive no contributions from elastic events. They are consistent with the quark and gluon distribution analysis of Section IV with $\Lambda = 0.5$ GeV/c, but one cannot prove that QCD is correct from just these highest three Q^2 points.

D. Some Order $\alpha_s(Q^2)$ Effects

As I will derive in detail in Section VI.B.4, the naive parton model result that $F_Q(x,Q^2)$ in (5.2e) measures the antiquark content within the proton is no longer precisely true when one includes effects of order $\alpha_s(Q^2)$ in QCD[28,29]. For example, if we define

$$\frac{F_2^{\nu N}(x,Q^2)}{x} = 2(G_{N\to d}(x,Q^2) + G_{N\to s}(x,Q^2)) + 2(G_{N\to \bar{u}}(x,Q^2) + G_{N\to \bar{c}}(x,Q^2)) \tag{5.18}$$

then to order $\alpha_s(Q^2)$

$$\frac{F_A^{\nu N}(x,Q^2)}{x} = \int_x^{1.0} \frac{dy}{y} \{(G_{N\to\bar{u}}(y,Q^2) + G_{N\to\bar{c}}(y,Q^2))$$

$$(2\delta(z-1) + \alpha_s(Q^2)\Delta f_{3,q}(z)) \qquad (5.19)$$

$$- (G_{N\to d}(y,Q^2) + G_{N\to s}(y,Q^2))\alpha_s(Q^2)\Delta f_{3,q}(z)\}$$

where $z = x/y$ and $\Delta f_{3,q}(z)$ is a function that will be calculated in Section VI.B.4 and where all M^2/Q^2 effects have been neglected. Equation (5.19) shows that to order $\alpha_s(Q^2)$, $F_A(x,Q^2)$ receives contributions from the quarks as well as antiquarks within the proton. The effects of this are not large, except at large x, as can be seen in Figs. 5.9 and 5.10. In these figures, I compare the leading order $F_A(x,Q^2) \propto G_{p\to\bar{q}}(x,Q^2)$ with the order $\alpha_s(Q^2)$ $F_A(x,Q^2)$ given by (5.19).

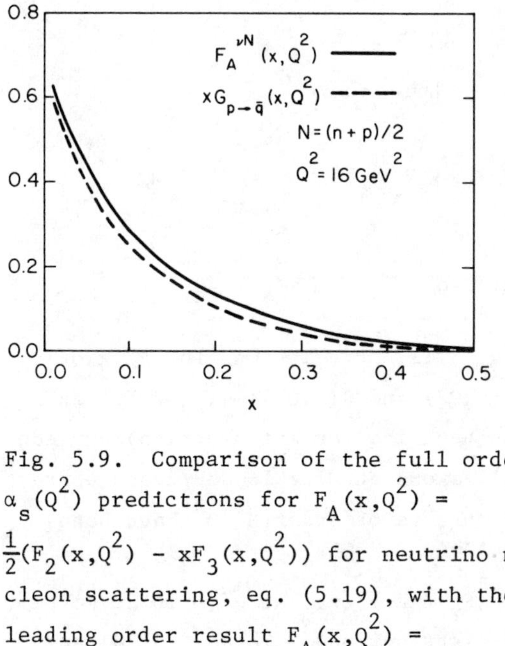

Fig. 5.9. Comparison of the full order $\alpha_s(Q^2)$ predictions for $F_A(x,Q^2) = \frac{1}{2}(F_2(x,Q^2) - xF_3(x,Q^2))$ for neutrino nucleon scattering, eq. (5.19), with the leading order result $F_A(x,Q^2) = xG_{p\to\bar{q}}(x,Q^2) = 2x(G_{N\to\bar{u}}(x,Q^2) + G_{N\to\bar{c}}(x,Q^2))$. The comparison is made at Q^2 = 16 GeV/c using Λ = 0.4 GeV/c.

The longitudinal structure function $F_L = F_2 - 2xF_1$ for neutrino scattering is given by perturbation theory to order $\alpha_s(Q^2)$, by

$$F_L^{\nu N}(x,Q^2) = \frac{\alpha_s(Q^2)}{2\pi} x^2 \int \frac{dy}{y^3} \{\frac{8}{3} F_2^{\nu N}(y,Q^2) + 2a_\nu yG_{N\to g}(y,Q^2)(1 - \frac{x}{y})\}, \qquad (5.20)$$

which is the same as the

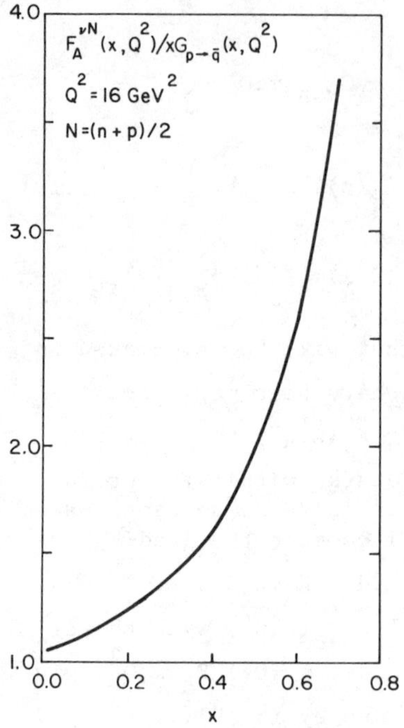

Fig. 5.10. Ratio of the full order $\alpha_s(Q^2)$ prediction for $F_A(x,Q^2) = \frac{1}{2}(F_2(x,Q^2) - xF_3(x,Q^2))$ for neutrino nucleon scattering to the leading order result $F_A(x,Q^2) = xG_{p\to\bar{q}}(x,Q^2) = 2x(G_{N\to\bar{u}}(x,Q^2) + G_{N\to\bar{c}}(x,Q^2))$. As in Fig. 5.10, the comparison is made at $Q^2 = 16$ GeV/c using $\Lambda = 0.4$ GeV/c.

electroproduction case (4.53) except now $a_\nu = 8$. Here gluons within the proton make an important contribution. Equation (5.20) yields an $F_L(x,Q^2)$ that is small at large x but is sizable at low x. Figure 5.11 shows

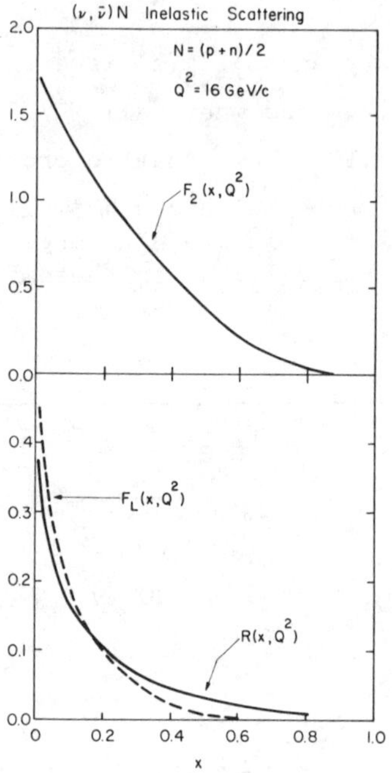

Fig. 5.11. Predictions for $F_2(x,Q^2)$, $F_L(x,Q^2)$ and $R(x,Q^2) = (F_2-2xF_1)/2xF_1$ for neutrino (or antineutrino) nucleon collisions at $Q^2 = 16$ GeV/c and where no effects of order M^2/Q^2 have been included.

$F_2(x,Q^2)$, $F_L(x,Q^2)$ and $R(x,Q^2) = (F_2-2xF_1)/2xF_1$ at $Q^2 = 16$ GeV/c resulting from our quark and gluon distributions. In eq. (5.20) and Fig. 5.11, I have included only the QCD perturbative contribution to R. As in the electroproduction case, one may have an additional primordial ("non-perturbative") contribution which cannot, at present, be calculated but which behaves like $1/Q^2$. Here, however, $1/Q^2$

effects may not be as serious since in (5.3), R is weighted by y^2. To see R, one needs $y \approx 1$ but then $Q^2 = 2MExy$ is large so $1/Q^2$ is small. I will discuss R more in Section VI.B.4.

The data from the BEBC analysis[27)] yield $R = 0.11 \pm 0.12$ integrated over x for $Q^2 > 1.0\ \text{GeV}^2$ while the CDHS group[30)] claims $R \leq 0.05$ with $\langle Q^2 \rangle = 22\ \text{GeV}^2$. This latter result is a bit worrisome since the QCD perturbative contribution to R is about 0.10 at this Q^2 (see Section VI.B.4). If QCD is responsible for the scale breaking of the structure functions, then at high Q^2 (after all $1/Q^2$ effects have died out), one must see the R given by (5.20). R cannot be zero.

VI. QCD PERTURBATION THEORY

A. General Formalism

I would like now to discuss the question of how quark and gluon distributions determined from one process, say, deep inelastic electron scattering compare to those determined from another process like neutrino scattering or large mass muon pair production. To leading order (and in the naive parton model), distributions determined in various processes are the same. In QCD there are corrections of order $\alpha_s(Q^2)$ that may be important and should be examined in detail. In Section VI.B, I will work out some specific examples. However, first let me write down the general QCD formalism which can later be compared to specific perturbation calculations. I will not derive this general formalism but refer the reader to papers by Georgi and Politzer[1,10,11)] and Gross and Wilczek[2)].

We will be interested in the three structure functions

$$\tilde{F}_1(x,Q^2) \equiv 2F_1(x,Q^2) \tag{6.1a}$$

$$\tilde{F}_2(x,Q^2) \equiv F_2(x,Q^2)/x \tag{6.1b}$$

$$\tilde{F}_3(x,Q^2) \equiv F_3(x,Q^2) \tag{6.1c}$$

where F_1, F_2, F_3 are defined in (4.43) and (5.1). The moments of these distributions are defined by

$$M_i(n,Q^2) = \int_0^1 dx\, x^{n-1}\tilde{F}_i(x,Q^2). \tag{6.2}$$

For simplicity for the moment, let me consider only the non-singlet distribution (see (4.10) and (5.10)) and write the moments as

$$M^{NS}(n,Q^2) = \int_0^1 dx x^{n-1}\tilde{F}_{NS}(x,Q^2)$$
$$\equiv C_n^{NS}(Q^2/\mu^2,g^2)q_o^{(n)}, \tag{6.3}$$

where g^2 is the strong interaction coupling and μ^2 is the (arbitrary) renormalization point (as in Section II) and where the presently uncalculuable non-perturbative part of the non-singlet moments has been absorbed into the (unknown) moments $q_o^{(n)}$. Perturbation theory will be applied to the moments $C_n^{NS}(Q^2/\mu^2,g^2)$ which is a function of Q^2/μ^2 and the coupling. Since the theory is renormalizable, C_n^{NS} cannot depend on the choice of μ^2. The equation (analogous to (2.19)) which expresses this fact is the renormalization group equation

$$(\mu \frac{\partial}{\partial\mu} + \beta(g)\frac{\partial}{\partial g} - \gamma_n(g))C_n^{NS}(Q^2/\mu^2,g^2) = 0, \tag{6.4}$$

which has the general solution

$$C_n^{NS}(Q^2/\mu^2,g^2) = C_n^{NS}(1,\bar{g}^2)\text{EXP}\{-\int_{\bar{g}(\mu^2)}^{\bar{g}(Q^2)} \frac{\gamma_n(t)}{\beta(t)}dt\}, \tag{6.5}$$

where $\beta(g)$ governs how the effective strong interaction coupling $\bar{g}^2(Q^2)$ depends on $\tau = \log Q^2/\mu^2$. Namely,

$$\frac{d\bar{g}^2}{d\tau} = \bar{g}\beta(\bar{g}) \tag{6.6}$$

with $\bar{g}(\tau=0) = g$.

From (6.3) and (6.5), it is easy to see that the change of M_n^{NS} w.r.t. $\alpha_s(Q^2) = \bar{g}^2(Q^2)/4\pi$ is given by

$$\frac{dM^{NS}(n,Q^2)}{d\alpha_s(Q^2)} = (-\frac{2\pi\gamma_n(\bar{g})}{\bar{g}\beta(\bar{g})} + \frac{1}{C_n^{NS}(1,\bar{g}^2)}\frac{d\bar{C}_n^{NS}(1,\bar{g}^2)}{d\alpha_s(Q^2)})M^{NS}(n,Q^2). \tag{6.7}$$

The functions $\gamma_n(\bar{g}^2)$, $\bar{g}\beta(\bar{g})$ and $C_n^{NS}(1,\bar{g}^2)$ can all be expanded in powers of $\alpha_s(Q^2)/4\pi$ yielding

$$\gamma_n(\bar{g}) = \gamma_o^{(n)} \frac{\alpha_s(Q^2)}{4\pi} + \gamma_1^{(n)} \left(\frac{\alpha_s(Q^2)}{4\pi}\right)^2 + \ldots \tag{6.8a}$$

$$\bar{g}\beta(\bar{g}) = -\beta_o\left(\frac{\alpha_s(Q^2)}{4\pi}\right)^2 - \beta_1\left(\frac{\alpha_s(Q^2)}{4\pi}\right)^3 + \ldots \tag{6.8b}$$

$$C_n^{NS}(1,\bar{g}^2) = 1 + B_n^{NS} \frac{\alpha_s(Q^2)}{4\pi} + \ldots \tag{6.8c}$$

In Section I, we discussed the effective strong interaction coupling $\bar{g}^2(Q^2)$ and although I did not present the results in the form of (6.8b), we already know β_o and β_1 from eq. (2.12) and (2.14). The strong interaction coupling calculated to leading order from (6.6) and (6.8b) is

$$\alpha_o(Q^2) = \frac{4\pi}{\beta_o \log Q^2/\Lambda^2} \tag{6.9}$$

and comparing with (2.12), we see that

$$\beta_o = (33-2n_f)/3. \tag{6.10}$$

In the next order (i.e., α_o^2), we have

$$\alpha(Q^2) = \alpha_o(Q^2)\left(1 - \frac{\beta_1}{\beta_o} \frac{\alpha_o(Q^2)}{4\pi} \log \log Q^2/\Lambda^2\right) \tag{6.11}$$

and from (2.14), we see that

$$\beta_1 = (306-38n_f)/3. \tag{6.12}$$

Now going back to (6.7) and inserting (6.8) we have, to lowest order

$$\alpha_s(Q^2)\,\frac{dM^{NS}(n,Q^2)}{d\alpha_s(Q^2)} = \left(\frac{\gamma_o^{(n)}}{2\beta_o}\right)M^{NS}(n,Q^2) \tag{6.13}$$

which implies

$$M^{NS}(n,Q^2) = q_n^{NS}(\alpha_s(Q^2))^{dn}, \tag{6.14a}$$

where $q_n^{NS} = q_o^{(n)}$ are unknown. Comparing two different Q^2 values yields

$$M^{NS}(n,Q^2) = M^{NS}(n,Q_o^2)[\alpha(Q^2)/\alpha(Q_o^2)]^{dn} \tag{6.14b}$$

$$= M^{NS}(n,Q_o^2)[\log(Q^2/\Lambda^2)/\log(Q_o^2/\Lambda^2)]^{-dn},$$

with

$$d_n = \frac{\gamma_o^{(n)}}{2\beta_o} = -\frac{A_n^{NS}}{2\pi b}, \tag{6.14c}$$

where A_n^{NS} and b have been defined earlier by eq. (4.11c) and eqs. (4.11b) and (4.39a), respectively. As discussed in Section IV, eq. (6.14) can be written in terms of a convolution as

$$\frac{dq^{NS}(x,\tau)}{d\tau} = \frac{\alpha_s(Q^2)}{2\pi}\int_x^1 \frac{dy}{y}\, q^{NS}(y,\tau)P_{q\leftarrow q}\left(\frac{x}{y}\right) \tag{6.15a}$$

provided

$$A_n^{NS} = \int_o^1 z^{n-1}P_{q\leftarrow q}(z)dz \tag{6.15b}$$

and where the quark distribution, $q^{NS}(x,\tau)$, is defined by the inverse of eq. (4.12).

Notice that

$$[\alpha_s(Q^2)/\alpha_s(Q_o^2)]^{dn} = [1-b\alpha_s(Q^2)\log(Q^2/Q_o^2)]^{dn}$$

$$\approx 1 + \frac{\alpha_s(Q^2)}{2\pi}A_n^{NS}\log(Q^2/Q_o^2) + \ldots. \tag{6.16}$$

is a power series in $\alpha_s(Q^2)\log Q^2 \sim O(1)$. Thus by the use of the renormalization group equation, we have succeeded in summing all terms of the form $(\alpha_s(Q^2)\log Q^2)^N$. If we instead had simply calculated using ordinary perturbation theory only to order $\alpha_s(Q^2)\log Q^2$, we would have arrived at

$$q^{NS}(x,\tau) = \int_x^1 \frac{dy}{y}\, q_o^{NS}(y)\{\delta(z-1) + \frac{\alpha_s(Q^2)}{2\pi} P_{q\leftarrow q}(z)\tau\} \quad (6.17a)$$

where $\tau = \log(Q^2/Q_o^2)$ or in terms of moments

$$M^{NS}(n,Q^2) = q_o^{(n)}\ (1 + \frac{\alpha_s(Q^2)}{2\pi} A_n^{NS}\log(Q^2/Q_o^2)). \quad (6.17b)$$

Actually, this is not quite correct. As we shall see in Section VI.B, ordinary perturbative calculations for massless quarks and gluons diverge logarithmically. To calculate with perturbation theory in QCD, one must adopt some procedure for handling these infrared divergences. Politzer has suggested a scheme for removing (or "regularizing") these divergences that is similar to renormalization and I shall follow his prescription. We shall let the incoming quark and gluons for any basic subprocess be slightly off mass shell and space-like, for example, $p_q^2 = -m_q^2$. The perturbative results then are finite and to order $\alpha_s(Q^2)\log Q^2$ will have a form similar to (6.17b). Namely,

$$M^{NS}(n,Q^2) = \tilde{q}_o^{(n)}(1 + \frac{\alpha(Q^2)}{2\pi} A_n^{NS}\log(Q^2/m_q^2)). \quad (6.18)$$

Since we expect that the physical observable $M^{NS}(n,Q^2)$ in (6.18) to be finite in the limit $m_q \to 0$, it must be that we have made an "artificial" divergence in the way we have done the perturbation calculation. That is, we have divided $M^{NS}(n,Q^2)$ into two pieces $\tilde{q}_o^{(n)}$ and $A_n^{NS}\log(Q^2/m_q^2)$ both of which diverge as $m_q \to 0$ but whose product is finite. After all, the $\tilde{q}_o^{(n)}$ are unknown and arbitrary. According to Politzer, we should now "renormalize" $\tilde{q}_o^{(n)}$ in (6.18). We imagine that it has an expansion of the form

$$\tilde{q}_o^{(n)} = q_o^{(n)}(1 + \frac{\alpha_s}{2\pi} A_n^{NS} \log(m_q^2/Q_o^2) + \ldots) \tag{6.19}$$

where $q_o^{(n)}$ are finite in the $m_q^2 \to 0$ limit. Equation (6.18) now becomes (to order α_s) equivalent to eq. (6.17b). The unwanted $\log(m_q^2)$ divergence has been absorbed into the arbitrary and unknown $\tilde{q}_o^{(n)}$. This seems a reasonable thing to do particularly since we will only be interested in the change of the distribution $q^{NS}(x,Q^2)$ relative to what it is at Q_o^2. The functions $P_{q\leftarrow q}(z)$ in (6.15a) will be calculated from ordinary perturbation theory by picking out the coefficient of the $\log(Q^2/m_q^2)$ term in eq. (6.17a) in a calculation to order $\alpha(Q^2)\log Q^2$. Equation (6.15) will then be used to sum all the orders of $(\alpha_s(Q^2)\log Q^2)^N$.

Including the next order in (6.7) and (6.8) results in the differential equation

$$\alpha_s(Q^2)\,\frac{dM^{NS}(n,Q^2)}{d\alpha_s(Q^2)} = \left\{\frac{\gamma_o^{(n)}}{2\beta_o} + (h_n+B_n^{NS})\alpha_s(Q^2)/4\pi\right\} M^{NS}(n,Q^2), \tag{6.20}$$

where

$$h_n = \frac{\gamma_1^{(n)}}{2\beta_o} - \frac{\beta_1\gamma_o^{(n)}}{2\beta_o^2}. \tag{6.21}$$

The solution of which is

$$M^{NS}(n,Q^2) = q_n^{NS}[\alpha_s(Q^2)/\alpha_s(Q_o^2]^{dn} \tag{6.22a}$$

$$\mathrm{Exp}[(h_n+B_n)\alpha_s(Q^2)/4\pi]$$

or

$$M^{NS}(n,Q^2) = M^{NS}(n,Q_o^2)[\alpha_s(Q^2)/\alpha_s(Q_o^2)]^{dn}$$

$$\text{Exp}[(h_n+B_n)(\alpha_s(Q^2)-\alpha_s(Q_o^2))/4\pi]$$

$$\approx M^{NS}(n,Q_o^2)[\alpha_s(Q^2)/\alpha_s(Q_o^2)]^{dn} \tag{6.22b}$$

$$[1+(h_n+B_n)(\alpha_s(Q^2)-\alpha_s(Q_o^2))/4\pi].$$

Expanding (6.22b) in powers of $\alpha_o(Q^2)$ yields (to order $\alpha_o^2(Q^2)$)

$$M^{NS}(n,Q^2) = M^{NS}(n,Q_o^2)[\alpha_o(Q^2)/\alpha_o(Q_o^2)]^{dn}$$

$$[1+(h_n+B_n+\ell_n)(\alpha_o(Q^2)-\alpha_o(Q_o^2)/4\pi] \tag{6.22c}$$

with

$$\ell_n = -\frac{\beta_1\gamma_o^{(n)}}{2\beta_o^2} \log \log Q^2/\Lambda^2. \tag{6.23}$$

I should point out that the individual terms h_n and B_n in (6.22) are, in general, not unique[28,29]. They will depend on the particular "regularization" scheme one has used. The sum, $h_n + B_n$, however, will be unique.

Since $\alpha_s(Q^2) - \alpha_s(Q_o^2)$ is of order $\alpha_o^2(Q^2)$, the corrections of (6.22) due to h_n and B_n are of order $\alpha_o^2(Q^2)$. Suppose, however, we are interested in comparing distributions determined in process A, $M_A^{NS}(n,Q^2)$, with those determined in process B, $M_B^{NS}(n,Q^2)$, then from (6.22b) we have (to order $\alpha_s(Q^2)$)

$$M_B^{NS}(n,Q^2) - M_A^{NS}(n,Q^2) = M_A^{NS}(n,Q^2)\Delta B_n^{NS}\alpha_s(Q^2)/4\pi, \tag{6.24a}$$

where

$$\Delta B_n^{NS} \equiv B_n^{NS,B} - B_n^{NS,A}. \tag{6.24b}$$

Thus the corrections due to B_n are of order $\alpha_s(Q^2)$ when comparing one process with another. Notice that since $\gamma_o^{(n)}$, $\gamma_1^{(n)}$, β_o and β_1 in (6.18a) are process independent, h_n cancels out when comparing one process with another. As before, eq. (6.24) can be written in terms of a convolution yielding

$$q_B^{NS}(x,\tau) = \int_x^1 \frac{dy}{y}\, q_A^{NS}(y,\tau)\{\delta(z-1)+\alpha_s(Q^2)\Delta f_q(z)\}, \tag{6.25a}$$

where $z = \frac{x}{y}$ and

$$\Delta B_n^{NS} = 4\pi \int_0^1 z^{n-1}\Delta f_q(z)dz. \tag{6.25b}$$

To order $\alpha_o(Q^2)$, both $q_B^{NS}(x,\tau)$ and $q_A^{NS}(x,z)$ satisfy (6.15).

Thus if we are interested in calculating to full order $\alpha_o(Q^2)$, we must pick a "reference reaction" A (we will follow Ref. 28 and choose F_2 in ep scattering) to define the quark distributions. They will evolve with Q^2 according to (6.15). The quark distributions in other processes (like F_1 in ep scattering or F_3 in νp scattering, etc.) can be calculated from (6.25) if we know the functions $\Delta f_q(z)$ (or equivalently ΔB_n).

To calculate $\Delta f(z)$, we notice that if we calculate using ordinary perturbation theory and keep terms of order $\alpha_s(Q^2)\log Q^2$ <u>and</u> terms of order $\alpha_s(Q^2)$ then we arrive at

$$q^{NS}(x,Q^2) = \int_x^1 \frac{dy}{y}\, \tilde{q}_o^{NS}(y) \tag{6.26a}$$

$$[\delta(z-1) + \frac{\alpha_s}{2\pi} P_{q\leftarrow q}(z)\log(Q^2/m_q^2)+\alpha_s\Delta f(z)]$$

or

$$M^{NS}(n,Q^2) = \tilde{q}_o^{(n)}[1 + \frac{\alpha_s}{2\pi} A_n^{NS}\log(Q^2/m_q^2) + \frac{\alpha_s(Q^2)}{4\pi} \Delta B_n], \quad (6.26b)$$

where again we have regularized by taking the incoming partons off-shell. After absorbing the $\log(m_q^2)$ divergence into the $\tilde{q}_o^{(n)}$ by the use of (6.19), we arrive at

$$q^{NS}(x,Q^2) = \int_x^{1.0} \frac{dy}{y} q_o^{NS}(y)[\delta(z-1) + \frac{\alpha_s(Q^2)}{2\pi} P_{q\leftarrow q}(z)\tau$$

$$+ \alpha_s(Q^2)\Delta f(z)]. \quad (6.26c)$$

In the more general case (like the singlet distributions) where there is mixing between initial quark, $G^{(o)}_{p\to q}(y)$, and gluon, $G^{(o)}_{p\to g}(y)$, distributions, one has

$$G_{p\to q}(x,Q^2) = \int_x^1 \frac{dy}{y} \{G^{(o)}_{p\to q}(y)$$

$$[\delta(z-1) + \frac{\alpha_s}{2\pi} P_{q\leftarrow q}(z)\tau + \alpha_s(Q^2)\Delta f_q(z)]$$

$$+ G^{(o)}_{p\to g}(y)[\frac{\alpha_s}{2\pi} P_{g\leftarrow q}(z)\tau + \alpha_s(Q^2)\Delta f_g(z)]\}, \quad (6.26d)$$

with $\tau = \log Q^2/Q_o^2$. From here we can read off not only the $P_{q\leftarrow q}(z)$ and $P_{g\leftarrow q}(z)$ functions that generate the Q^2 evolution of $G_{p\to q}(x,Q^2)$ according to (4.29) but also the terms $\Delta f_q(z)$, $\Delta f_g(z)$ that are proportional to $\alpha_s(Q^2)$.

In general, in any problem there will be a sum of terms of the form $[\alpha_s(Q^2)\log Q^2]^N$ that are all of leading order 1 and are summed by the use of eq. (4.29). In addition, there are terms proportional to $[\alpha_s(Q^2)]^N$. These are not summed but since $\alpha_s(Q^2)$ is small, hopefully one only needs to calculate to order $\alpha_o(Q^2)$ (or maybe $\alpha_o^2(Q^2)$) to get an accurate result. Let us now examine some specific examples.

B. Some Specific Calculations to Full Order $\alpha_s(Q^2)$

1. Gluon Contributions to Deep Inelastic Electron Scattering

The best way to understand perturbative QCD is to make some specific calculations. In this section, I will make some order $\alpha_s(Q^2)$ calculations that we can compare to the general formalism I wrote down in Section VI.A. We shall begin by calculating by the use of ordinary perturbation theory the full order $\alpha_s(Q^2)$ gluon contributions to the structure functions $\tilde{F}_1$ and $\tilde{F}_2$ in deep inelastic electron proton scattering (DIS). Gluons contribute to $\tilde{F}_1$ and $\tilde{F}_2$ (to order α_s) due to the subprocess $\gamma^* + g \to q + \bar{q}$ shown in Fig. 6.1. The differential cross section for this process is given by

$$\frac{d\hat{\sigma}^{\Sigma}_{DIS}}{d\hat{t}} = \frac{\pi\alpha\alpha_s e_q^2}{(\hat{s}+Q^2)^2}\left(\frac{1}{8}\right)\left|A^{\Sigma}_{DIS}(\hat{s},\hat{t})\right|^2$$
$$= \frac{\pi\alpha\alpha_s e_q^2}{(\hat{s}+Q^2)^2}\left\{\frac{\hat{u}}{\hat{t}} + \frac{\hat{t}}{\hat{u}} + \frac{2Q^2}{\hat{t}\hat{u}}(\hat{t}+\hat{u}+Q^2)\right\}, \tag{6.27}$$

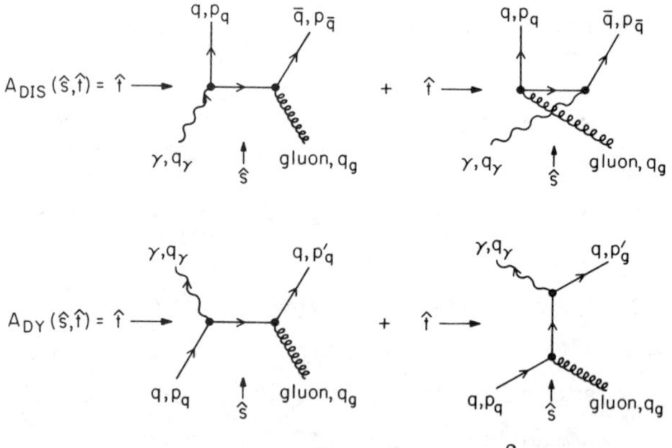

Fig. 6.1. Diagrams of order $\alpha_s(Q^2)$ that produce corrections to deep inelastic scattering (DIS) and to the "Drell-Yan" production of large mass muon pairs (DY) that are proportional to the probability of finding gluons with the initial hadron.

where $\hat{s}$, $\hat{t}$ and $\hat{u}$ are the usual s, t and u invariants but for the subprocess $\gamma^* + g \to q + \bar{q}$ and e_q^2 is the electric charge of the quark q. In addition, the virtual photon momentum is given by $q_\gamma^2 = -Q^2$ and I have taken the gluon and quark to have zero mass. The superscript Σ is to signify that in calculating (6.27), I have used $\Sigma\varepsilon_\mu\varepsilon^*_\nu = \delta_{\mu\nu}$ where ε_μ is the photon polarization.

This means that σ^{Σ}_{DIS} is related to a particular combination of $\tilde{F}_1$ and $\tilde{F}_2$ which I will write down shortly.

A problem now arises when we try to compute the total $\gamma^* + g$ cross section by integrating (6.27). The integral

$$\hat{\sigma}^{\Sigma}_{DIS}(\hat{s}) = \int_{\hat{t}_{max}}^{\hat{t}_{min}} d\hat{t}\, \frac{d\hat{\sigma}^{\Sigma}_{DIS}}{d\hat{t}}(\hat{s},\hat{t}) \tag{6.28}$$

diverges logarithmically like $\log(\hat{t}_{min}/\hat{t}_{max})$ since for massless quarks and gluons $\hat{t}_{min} = 0$. As discussed in Section VI.A, to calculate with perturbation theory in QCD, one must find some procedure for removing (or "regularizing") the infrared ($\hat{t} \to 0$, $\hat{u} \to 0$) divergences which occur. We will let the incoming gluon be slightly off-shell and spacelike (i.e., $q_g^2 = -m_g^2$ in Fig. 6.1). In this case

$$|A^{\Sigma}_{DIS}(\hat{s},\hat{t})|^2 = 8\{\frac{\hat{u}}{\hat{t}} + \frac{\hat{t}}{\hat{u}} + \frac{2Q^2}{\hat{t}\hat{u}}(\hat{t}+\hat{u}+Q^2) + \frac{2m_g^2}{\hat{t}\hat{u}}(\hat{t}+\hat{u}+m_g^2)$$
$$- Q^2 m_g^2 (\frac{1}{\hat{u}^2} + \frac{1}{\hat{t}^2} - \frac{4}{\hat{t}\hat{u}}) \tag{6.29}$$

and $\hat{t}_{min}$ and $\hat{t}_{max}$ in (6.28) become (approximately)

$$\hat{t}_{min} = \hat{u}_{min} = -m_g^2 Q^2/(\hat{s}+Q^2)$$
$$\hat{t}_{max} = \hat{u}_{max} = -(\hat{s}+Q^2). \tag{6.30}$$

The differential cross section can now be integrated to give

$$\hat{\sigma}^{\Sigma}_{DIS}(\hat{s}) = \frac{2\pi\alpha\alpha_s e_q^2}{(\hat{s}+Q^2)^2}\{-(\frac{2Q^2\hat{s}-(\hat{s}+Q^2)^2}{(\hat{s}+Q^2)})\log(\frac{(\hat{s}+Q^2)}{Q^2 m_g^2}) - 2(\hat{s}+Q^2)\} \tag{6.31}$$

where all terms that vanish in the limit $m_g^2 \to 0$ have been dropped.

In particular

$$\int_{\hat{t}_{max}}^{\hat{t}_{min}} d\hat{t} = (\hat{s}+Q^2) \tag{6.32a}$$

$$\int_{\hat{t}_{max}}^{\hat{t}_{min}} \frac{d\hat{t}}{\hat{t}} = \log\left(\frac{\hat{t}_{min}}{\hat{t}_{max}}\right) = -\log\left[\frac{(\hat{s}+Q^2)^2}{Q^2 m_g^2}\right] \tag{6.32b}$$

$$m_g^2 \int_{\hat{t}_{max}}^{\hat{t}_{min}} \frac{d\hat{t}}{\hat{t}^2} = -\frac{(\hat{s}+Q^2)}{Q^2}. \tag{6.32c}$$

We have thus succeeded in calculating $\hat{\sigma}^{\Sigma}_{DIS}(\hat{s})$. It contains two terms; one that diverges like $\log(m_g^2)$ as $m_g^2 \to 0$ and one that is finite in this limit (called the "constant" piece).

This subprocess must now be "embedded" in the desired observed process $\gamma^* + p \to X$. To do this, it is convenient to define

$$z \equiv \frac{Q^2}{2q_g \cdot q_\gamma} = \frac{x}{y} \tag{6.33a}$$

where

$$x = \frac{Q^2}{2P \cdot q_\gamma} \tag{6.33b}$$

and y is the fraction of the proton momentum, P, carried by the gluon (i.e., $q_g = yP$) as illustrated in Fig. 6.2. In terms of z and Q^2, the total $\gamma^* + g$ cross section becomes

$$\hat{\sigma}^{\Sigma}_{DIS}(Q^2) = \frac{2\pi\alpha\alpha_s e_q^2}{Q^2}\, z\{(z^2+(1-z)^2)\log\left(\frac{Q^2}{m_g^2 z^2}\right) - 2\} \tag{6.34}$$

and if we now use the fact that the total γ^*p cross section can be related in general to a structure function by

$$\frac{F_\Sigma(x,Q^2)}{x} = \int_x^{1.0} \frac{dy}{y}\, G_{p\to g}(y)\left(\frac{Q^2}{4\pi^2\alpha z}\right)\hat{\sigma}_\Sigma(Q^2) \tag{6.35}$$

we arrive at

$$\frac{F_\Sigma(x,Q^2)}{x} = 2e_q^2 \int_x^1 \frac{dy}{y} G_{p\to g}^{(o)}(y)$$

$$\{\frac{\alpha_s}{2\pi} P_{q\leftarrow g}(z)\log\frac{Q^2}{m_g^2} + \alpha_s f_{\Sigma,g}(z)\} \tag{6.36}$$

where I have included just one quark flavor and

$$P_{q\leftarrow g}(z) = \frac{1}{2}(z^2+(1-z)^2) \tag{6.37}$$

is the same function that I quoted in (4.22) and

$$\alpha_s f_{\Sigma,g}(z) = -\frac{\alpha_s}{2\pi}\frac{1}{2}$$

$$\{(z^2+(1-z)^2)(2\log z)+2\}. \tag{6.38}$$

As mentioned earlier, the structure function $F_\Sigma(x,Q^2)$ is related to $F_1(x,Q^2)$ and $F_2(x,Q^2)$. In particular

$$F_\Sigma(x,Q^2) = F_2(x,Q^2) - \frac{3}{2} F_L(x,Q^2), \tag{6.39a}$$

where the longitudinal structure function is given by

$$F_L(x,Q^2) = F_2(x,Q^2) - 2xF_1(x,Q^2). \tag{6.39b}$$

The perturbative calculation for F_L proceeds as before except now

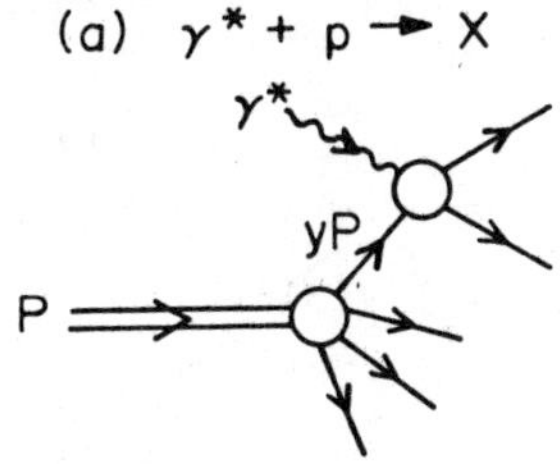

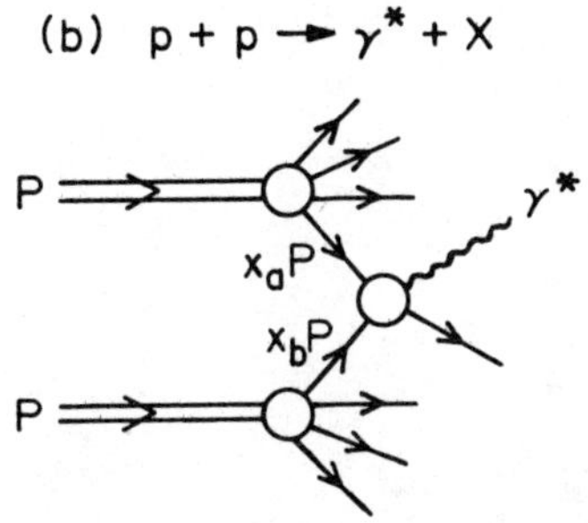

Fig. 6.2 - (a) Illustrates how the subprocess $\gamma^* + a \to b + c$ where a, b and c are constituents (quarks or gluons) is "embedded" within the experimentally measured process $\gamma^* + p \to X$, where p is a proton and y is the fraction of momentum of the proton carried by the constituent a.

(b) Illustrates how the constituent subprocess $a + b \to \gamma^* + c$ is "embedded" with the observed process $pp \to \gamma^* + X$. The virtual photon γ^* then fragments into a muon pair. The quantities x_a and x_b are the fractional momenta carried by constituents (quarks or gluons) a and b, respectively.

$$\frac{d\hat{\sigma}_L^{DIS}}{d\hat{t}} = \frac{\pi\alpha\alpha_s e_q^2}{(\hat{s}+Q^2)}\hat{s} \tag{6.40}$$

and there are no divergent pieces. Integrating (6.40) yields

$$\hat{\sigma}_L^{DIS} = \frac{\pi\alpha\alpha_s e_q^2}{(\hat{s}+Q^2)}\hat{s} = \pi\alpha_s\alpha(1-z). \tag{6.41}$$

By the use of

$$\frac{F_L(x,Q^2)}{x} = \int_x^{1.0} \frac{dy}{y} G_{p\to g}(y)\left(\frac{2z}{\pi^2\alpha}\right)\hat{\sigma}_L \tag{6.42}$$

we arrive at (again for one quark flavor)

$$\frac{F_L(x,Q^2)}{x} = 2e_q^2 \int_x^{1.0} \frac{dy}{y} G_{p\to g}(y)\alpha_s f_{L,g}(z) \tag{6.43}$$

with

$$\alpha_s f_{L,g}(z) = \frac{\alpha_s}{2\pi} 2z(1-z). \tag{6.44}$$

Combining (6.36), (6.39) and (6.43) yields

$$\tilde{F}_1(x,Q^2) = 2F_1(x,Q^2) = 2e_q^2 \int_x^1 \frac{dy}{y} G_{p\to g}(y) \left\{\frac{\alpha_s}{2\pi} P_{q\leftarrow g}(z)\log\frac{Q^2}{m_g^2} + \alpha_s f_{1,g}(z)\right\} \tag{6.45a}$$

$$\tilde{F}_2(x,Q^2) = \frac{F_2(x,Q^2)}{x} = 2e_q^2 \int_x^1 \frac{dy}{y} G_{p\to g}(y) \left\{\frac{\alpha_s}{2\pi} P_{q\leftarrow g}(z)\log\frac{Q^2}{m_g^2} + \alpha_s f_{2,g}(z)\right\} \tag{6.45b}$$

with $z = \frac{x}{y}$ and $P_{q\leftarrow g}(z)$ given by eq. (6.37) and where

$$\alpha_s f_{2,g}(z) = \frac{\alpha_s}{2\pi}\left(\frac{1}{2}\right)\left[(z^2+(1-z)^2)(-2\log z)-2+6z-6z^2\right] \tag{6.46a}$$

$$\alpha_s(f_{2,g}(z)-f_{1,g}(z)) = \frac{\alpha_s}{2\pi}\,2z(1-z). \tag{6.46b}$$

2. Contributions to Deep Inelastic Scattering from $\gamma^* + q \to q + g$

In this case, we regulate by keeping the incoming quark off-shell and spacelike ($p_q^2 = -m_q^2$). The differential cross section for the subprocess $\gamma^* + q \to q + g$ shown in Fig. 6.3 is given by

$$\frac{d\sigma^{\Sigma}_{DIS}}{d\hat{t}}(\hat{s},\hat{t}) = \frac{\pi\alpha\alpha_s e_q^2}{(\hat{s}+Q^2)^2}\left(\frac{8}{3}\right)\left\{-\frac{\hat{t}}{\hat{s}}-\frac{\hat{s}}{\hat{t}}+\frac{2Q^2(\hat{s}+\hat{t}+Q^2)}{\hat{s}\hat{t}}+\frac{m_q^2Q^2}{\hat{t}^2}\right\}, \tag{6.47a}$$

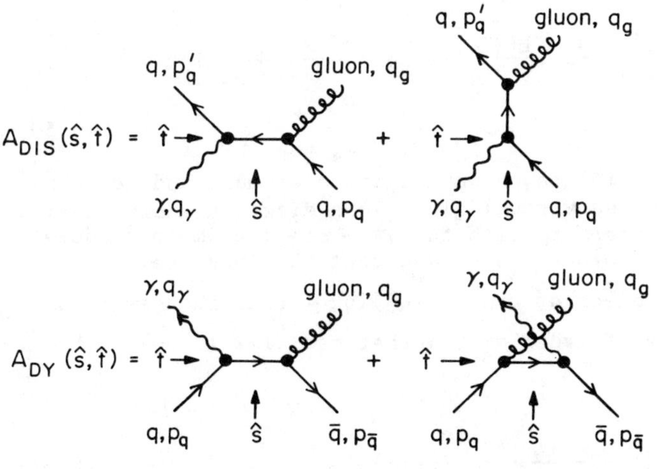

Fig. 6.3. Diagrams of order $\alpha_s(Q^2)$ that produce corrections to deep inelastic scattering (DIS) and to the "Drell-Yan" production of large mass muon pairs (DY) which are due to real gluon emission.

where all terms that contribute nothing to the total cross section, σ^{Σ}_{DIS}, in the limit $m_q \to 0$ have been dropped[31]. Integrating over $\hat{t}$ as in eq. (6.28) yields

$$\hat{\sigma}^{\Sigma}_{DIS}(Q^2) = \frac{\pi\alpha\alpha_s e_q^2}{Q^2}\left(\frac{8}{3}\right)z\left\{\frac{(1+z^2)}{1-z}\log\left(\frac{Q^2}{m_g^2 z^2}\right) - \frac{3}{2}\frac{1}{1-z} + 1\right\} \tag{6.47b}$$

where z is given by (6.33a) and $\hat{t}_{min}$ and $\hat{t}_{max}$ are the same as (6.30) except m_g^2 is replaced by m_q^2. Equation (6.35) now implies that

$$\frac{F_\Sigma(x,Q^2)}{x} = e_q^2 \int_x^1 \frac{dy}{y} G^{(o)}_{p\to q}(y) \tag{6.48}$$

$$\{\delta(z-1) + \frac{\alpha_s}{2\pi} P_{q\leftarrow q}(z)\log(\frac{Q^2}{m_q^2}) + \alpha_s f_{\Sigma,q}(z)\}$$

where the $\delta(z-1)$ term comes from the leading $\gamma^* + q \to q$ term and where

$$P_{q\leftarrow q}(z) = (\frac{4}{3})(\frac{1+z^2}{1-z}) \qquad (z < 1) \tag{6.49}$$

and

$$\alpha_s f_{\Sigma,q}(z) = \frac{\alpha_s}{2\pi}(\frac{4}{3})\{(\frac{1+z^2}{1-z})(-2\log z) - \frac{3}{2}\frac{1}{1-z} + 1\} \qquad (z < 1). \tag{6.50}$$

These results are not complete, however, since we must add to them the contributions from the virtual gluon loop diagrams shown in Fig. 6.4. These diagrams interfere with the $\gamma^* + q \to q$ term to produce $\delta(z-1)$ contributions of order α_s. These contributions are of two types. A "divergent" piece of order $\alpha(Q^2)\log Q^2$ that changes (6.49) and a "constant" piece of order $\alpha_s(Q^2)$ that modifies (6.50). The final results are[32)]

$$P_{q\leftarrow q}(z) = (\frac{4}{3})\{\frac{1+z^2}{(1-z)_+} + \frac{3}{2}\delta(z-1)\}, \tag{6.51}$$

and

$$\alpha_s f_{\Sigma,q}(z) = \frac{\alpha_s}{2\pi}(\frac{4}{3})\{\frac{1+z^2}{1-z}(-2\log z) - \frac{3}{2}\frac{1}{(1-z)_+} + 1$$

$$- \frac{2}{3}\pi^2\delta(z-1)\}, \tag{6.52}$$

where $1/(1-z)_+$ is defined by eq. (4.24).

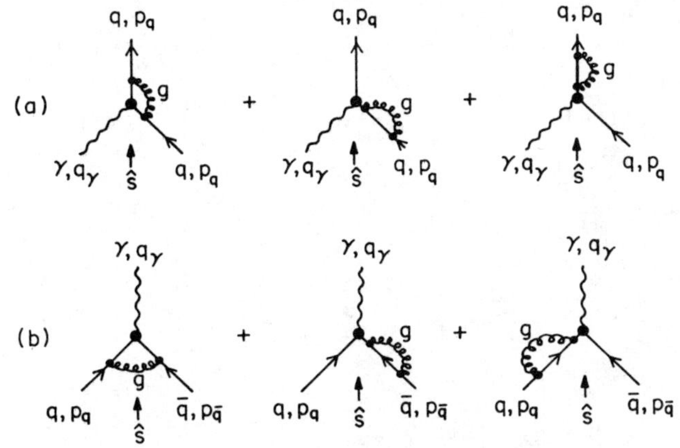

Fig. 6.4. Virtual gluon corrections to the deep inelastic scattering subprocess $\gamma^* + q \to q$ (a) and to the Drell-Yan subprocess $q + \bar{q} \to \gamma^*$ (b). These diagrams interfere with the order zero "Born diagrams" producing effects of order $\alpha_s(Q^2)$.

The longitudinal differential cross section is

$$\frac{d\hat{\sigma}_L}{d\hat{t}} = \frac{\pi\alpha\alpha_s}{(\hat{s}+Q^2)^2}\, e_q^2 \left(\frac{4}{3}\right) \{Q^2+\hat{t}+\hat{s}\} \qquad (6.53)$$

which implies

$$\hat{\sigma}_L = \pi\alpha\alpha_s \left(\frac{4}{3}\right) e_q^2 \qquad (6.54)$$

whereupon (6.42) yields (for one quark flavor)

$$\frac{F_L(x,Q^2)}{x} = e_q^2 \int_x^1 \frac{dy}{y}\, G^{(o)}_{p\to q}(y)\, \alpha_s(Q^2)\, f_{L,q}(z) \qquad (6.55)$$

with

$$\alpha_s f_{L,q}(z) = \frac{\alpha_s}{2\pi} \left(\frac{4}{3}\right) 2z. \qquad (6.56)$$

Finally, using (6.39), we arrive at

$$\tilde{F}_1(x,Q^2) = 2F_1(x,Q^2) = e_q^2 \int_x^1 \frac{dy}{y}\, G_{p\to q}(y) \qquad (6.57a)$$

$$\left\{\delta(z-1) + \frac{\alpha_s}{2\pi} P_{q\leftarrow q}(z) \log \frac{Q^2}{m_q^2} + \alpha_s f_{1,q}(z)\right\}$$

$$\tilde{F}_2(x,Q^2) = \frac{F_2(x,Q^2)}{x} = e_q^2 \int_x^1 \frac{dy}{y} G_{p\to q}(y) \tag{6.57b}$$

$$\{\delta(z-1) + \frac{\alpha_s}{2\pi} P_{q\leftarrow q}(z) \log \frac{Q^2}{m_q^2} + \alpha_s f_{2,q}(z)\}$$

where $P_{q\leftarrow q}(z)$ is given by (6.51) and

$$\alpha_s f_{2,q}(z) = \frac{\alpha_s}{2\pi} (\tfrac{4}{3}) [- \frac{2(1+z^2)}{1-z} \log z - \frac{3}{2} \frac{1}{(1-z)_+} \tag{6.58a}$$

$$+ 1 + 3z - \frac{2\pi^2}{3} \delta(z-1)]$$

$$\alpha_s [f_{2,q}(z) - f_{1,q}(z)] = \frac{\alpha_s}{2\pi} (\tfrac{4}{3}) 2z. \tag{6.58b}$$

3. Final Order $\alpha_s(Q^2)$ Results for Deep Inelastic Electron Proton Scattering

If we now define quark distributions $G^{(1)}(x,Q^2)$ and $G^{(2)}(x,Q^2)$ by

$$\tilde{F}_1(x,Q^2) = 2F_1(x,Q^2) \equiv \sum_{i=1}^{n_f} e_{q_i}^2 (G^{(1)}_{p\to q_i}(x,Q^2) + G^{(1)}_{p\to \bar{q}_i}(x,Q^2)), \tag{6.59a}$$

and

$$\tilde{F}_2(x,Q^2) = \frac{F_2(x,Q^2)}{x} \equiv \sum_{i=1}^{n_f} e_{q_i}^2 (G^{(2)}_{p\to q_i}(x,Q^2) + G^{(2)}_{p\to \bar{q}_i}(x,Q^2)), \tag{6.59b}$$

then by combining eqs. (6.45) and (6.57), we arrive at

$$G^{(1)}_{p\to q}(x,Q^2) = \int_x^1 \frac{dy}{y} \{G^{(o)}_{p\to q}(y)[\delta(z-1) + \frac{\alpha_s}{2\pi} P_{q\leftarrow q}(z)\log \frac{Q^2}{m_q^2}$$

$$+ \alpha_s f_{1,q}(z)] + G^{(o)}_{p\to g}(y)[\frac{\alpha_s}{2\pi} P_{q\leftarrow g}(z)\log \frac{Q^2}{m_q^2}$$

$$+ \alpha_s f_{1,g}(z)]\}, \tag{6.60a}$$

and

$$G^{(2)}_{p\to q}(x,Q^2) = \int_x^1 \frac{dy}{y} \{G^{(o)}_{p\to q}(y)[\delta(z-1) + \frac{\alpha_s}{2\pi} P_{q\leftarrow q}(z)\log \frac{Q^2}{m_q^2}$$

$$+ \alpha_s f_{2,q}(z)] + G^{(o)}_{p\to g}(y)[\frac{\alpha_s}{2\pi} P_{q\leftarrow g}(z)\log \frac{Q^2}{m_g^2}$$

$$+ \alpha_s f_{2,g}(z)]\}, \tag{6.60b}$$

where $P_{q\leftarrow q}(z)$ and $P_{q\leftarrow g}(z)$ are given by (6.51) and (6.37), respectively, and the f-functions are given by (6.46) and (6.58). As explained in Section VI.A, we can absorb the $\log(m^2)$ divergences into the unknown distributions $G^{(o)}(y)$ and use the "renormalization group" improved equations in (4.29) to sum all the leading $\alpha(Q^2)\log Q^2$ terms. Equations (6.60) then become

$$G^{(i)}_{p\to q}(x,Q^2) = \int_x^1 \frac{dy}{y} \{G_{p\to q}(y,Q^2)[\delta(z-1)+\alpha_s(Q^2)f_{i,q}(z)]$$

$$+ G_{p\to g}(y,Q^2)\alpha_s(Q^2)f_{i,g}(z)\} \tag{6.61}$$

for i = 1,2 and where the Q^2 dependences of $G^{(i)}_{p\to q}(x,Q^2)$ and $G^{(i)}_{p\to g}(x,Q^2)$ are governed to order $\alpha_s(Q^2)$ by eq. (4.29). We are not quite finished since as mentioned in Section VI.A the functions $f_{i,q}(z)$ and $f_{i,g}(z)$ in (6.61) are not unique. They depend on the particular "regularization" scheme we have employed. Only the $\Delta f_q(z)$ are unique. We will thus define the quark distributions determined by a measurement of $F_2(x,Q^2)$ in deep inelastic electron scattering to be the "reference" distributions. That is, we will take

$$\frac{F_2(x,Q^2)}{x} \equiv \sum_{i=1}^{n_f} e_{q_i}^2 (G_{p\to q_i}(x,Q^2)+G_{p\to \bar{q}_i}(x,Q^2)) \tag{6.62}$$

to be the definition of $G_{p\to q_i}(x,Q^2)$ to full order $\alpha_s(Q^2)$. With this definition, we have to order $\alpha_s(Q^2)$

$$G^{(2)}_{p\to q_i}(x,Q^2) = G_{p\to q_i}(x,Q^2) \tag{6.63a}$$

$$G^{(1)}_{p\to q_i}(x,Q^2) = \int_x^1 \frac{dy}{y} \{G_{p\to q_i}(x,Q^2)[\delta(z-1)+\alpha_s(Q^2)\Delta f_{1,q}(z)] + G_{p\to g}(x,Q^2)\alpha_s(Q^2)\Delta f_{1,g}(z)\} \tag{6.63b}$$

with

$$\alpha_s \Delta f_{1,q}(z) \equiv \alpha_s(f_{1,q}(z)-f_{2,q}(z)) = -\frac{\alpha_s}{2\pi}\left(\frac{4}{3}\right)2z \tag{6.63c}$$

$$\alpha_s \Delta f_{1,g}(z) = \alpha_s(f_{1,g}(z) - f_{2,g}(z)) = -\frac{\alpha_s}{2\pi}\, 2z(1-z) \tag{6.63d}$$

and where $G^{(i)}_{p\to q_i}(x,Q^2)$ satisfy eq. (4.29). The $\Delta f(z)$ functions in (6.63) should now be unique.

Having developed all the machinery of eqs. (6.63), it is an easy matter to write down the expression for

$$R = \sigma_L/\sigma_T = F_L/2xF_1. \tag{6.64}$$

From (6.39b) and (6.63) we have

$$\frac{F_L(x,Q^2)}{x} = -\alpha_s(Q^2)\int_x^1 \frac{dy}{y}\left\{\sum_{i=1}^{n_f} e_{q_i}^2 (G_{p\to q_i}(x,Q^2)\right.$$

$$+ G_{p\to\bar{q}_i}(x,Q^2))\Delta f_{1,q}(z) \tag{6.65}$$

$$\left. + \left(\sum_{i=1}^{2n_f} e_{q_i}^2\right) G_{p\to g}(x,Q^2)\Delta f_{1,g}(z)\right\}$$

or after substituting in for $\Delta f(z)$ and using (6.62) becomes

$$F_L(x,Q^2) = \frac{\alpha_s(Q^2)}{2\pi} x^2 \int_x^1 \frac{dy}{y^3} \tag{6.66}$$

$$\{\frac{8}{3} F_2(y,Q^2)+2a_e yG_{p\to g}(y,Q^2)(1 - \frac{x}{y})\},$$

where

$$a_e = \left(\sum_{i=1}^{2n_f} e_i^2\right) = \frac{20}{9} \qquad (\text{for } n_f = 4), \tag{6.67}$$

which is the formula used in Section IV.F. The quantity R is now given by

$$R = F_L/(F_L+F_2) = R_2/(1+R_2) \tag{6.68}$$

where

$$R_2 = F_L/F_2 \tag{6.69}$$

and F_2 is given by (6.62).

A particularly easy quantity to calculate is

$$\bar{R}_2(Q^2) = \bar{F}_L(Q^2)/\bar{F}_2(Q^2) \tag{6.70a}$$

with

$$\bar{F}_i(Q^2) = \int_0^1 F_i(x,Q^2)dx, \tag{6.70b}$$

for i = 2,L. Using (6.63c) and (6.63d), we see that

$$\alpha_s \int_0^1 z\Delta f_{1,q}(z)dz = -\frac{\alpha_s}{2\pi}\left(\frac{8}{9}\right) = -0.141\ \alpha_s \tag{6.71a}$$

$$\alpha_s \int_0^1 z\Delta f_{1,g}(z)dz = -\frac{\alpha_s}{2\pi}\left(\frac{1}{6}\right) = -0.027\ \alpha_s \tag{6.71b}$$

then from (6.66), we have

$$\bar{F}_L(Q^2) = \frac{\alpha_s}{2\pi}\left\{\left(\frac{8}{9}\right)\bar{F}_2(Q^2) + \left(\frac{1}{6}\right) a_e G(Q^2)\right\}, \tag{6.72}$$

where $G(Q^2)$ is the total momentum carried by gluons:

$$G(Q^2) = \int_0^1 xG_{p\to g}(x,Q^2)dx. \tag{6.73}$$

At Q^2 = 16 GeV our distributions give

$$\bar{F}_2(Q^2=16\ \text{GeV}) = 0.164 \tag{6.74a}$$

$$G(Q^2=16\ \text{GeV}) = 0.514 \tag{6.74b}$$

so

$$\bar{R}_2(Q^2=16\ \text{GeV}) = 0.107 \tag{6.74c}$$

$$\bar{R}(Q^2=16\ \text{GeV}) = 0.12. \tag{6.74d}$$

At this Q^2, about half of $\bar{F}_L(Q^2)$ is due to the gluon term $G(Q^2)$ and about half due to the quark term $\bar{F}_2(Q^2)$. These results, of course, contain none of the M^2/Q^2 corrections of the type discussed in Section IV.F.

4. Some Order $\alpha_s(Q^2)$ Results for Deep Inelastic Neutrino and Antineutrino Nucleon Scattering

The results for $G^{(1)}_{p\to q}(x,Q^2)$ and $G^{(2)}_{p\to q}(x,Q^2)$ given by (6.63) are the same for neutrino scattering. Here, however, we have one further structure function $F_3(x,Q^2)$. The quark probabilities for

$F_3(x,Q^2)$ are

$$G^{(3)}_{p\to q_i}(x,Q^2) - G^{(3)}_{p\to \bar{q}_i}(x,Q^2) = \int_x^{1.0} \frac{dy}{y}$$
$$(G_{p\to q_i}(y,Q^2)-G_{p\to\bar{q}_i}(y,Q^2))[\delta(z-1)+\alpha_s(Q^2)\Delta f_{3,q}(z)], \tag{6.75a}$$

where again the reference distribution is $G^{(2)}_{p\to q}(x,Q^2)$ in (6.62). The function $\Delta f_{3,q}(z)$ is calculated as in the previous section with the result[28)]

$$\alpha_s\Delta f_{3,q}(z) = \alpha_s(f_{3,q}(z)-f_{2,q}(z)) = \frac{\alpha_s}{2\pi}(-\frac{4}{3})(1+z). \tag{6.75b}$$

The subprocess $\gamma^* + g \to q + \bar{q}$ does not contribute to (6.75a).

Given eqs. (6.63) and (6.75), it is an easy matter to calculate neutrino and antineutrino observables to order $\alpha_s(Q^2)$. As discussed in Section V to leading order (and in the naive parton model), the structure functions

$$F_Q(x,Q^2) = \frac{1}{2}(F_2(x,Q^2)+xF_3(x,Q^2)) \tag{6.76a}$$

$$F_A(x,Q^2) = \frac{1}{2}(F_2(x,Q^2)-xF_3(x,Q^2)) \tag{6.76b}$$

measure the quark and antiquark distributions, respectively, within hadrons. Equation (6.75) shows that to order α_s, both F_Q and F_A receive contributions from quarks and antiquarks. In particular, for νN scattering

$$\frac{F_Q^{\nu N}(x,Q^2)}{x} = \int_x^{1.0}\frac{dy}{y}\{(G_{N\to d}(y,Q^2)+G_{N\to s}(y,Q^2))(2\delta(z-1)$$
$$+\alpha_s(Q^2)\Delta f_{3,q}(z))-(G_{N\to\bar{u}}(y,Q^2) \tag{6.77a}$$
$$+G_{N\to\bar{c}}(y,Q^2))\alpha_s(Q^2)\Delta f_{3,q}(z)\},$$

$$\frac{F_A^{\nu N}(x,Q^2)}{x} = \int_x^{1.0} \frac{dy}{y} \{(G_{N\to\bar{u}}(y,Q^2)+G_{N\to\bar{c}}(y,Q^2))(2\delta(z-1)$$

$$+ \alpha_s(Q^2)\Delta f_{3,q}(z))-(G_{N\to d}(y,Q^2) \tag{6.77b}$$

$$+ G_{N\to s}(y,Q^2))\alpha_s(Q^2)\Delta f_{3,q}(z),$$

and for $\bar{\nu}N$ scattering, we have

$$\frac{F_Q^{\bar{\nu}N}(x,Q^2)}{x} = \int_x^{1.0} \frac{dy}{y} \{(G_{N\to u}(y,Q^2)+G_{N\to c}(y,Q^2))(2\delta(z-1)$$

$$+ \alpha_s(Q^2)\Delta f_{3,q}(z))-(G_{N\to\bar{d}}(y,Q^2)$$

$$+ G_{N\to\bar{s}}(y,Q^2))\alpha_s(Q^2)\Delta f_{3,q}(z)\} \tag{6.78a}$$

$$\frac{F_A^{\bar{\nu}N}(x,Q^2)}{x} = \int_x^{1.0} \frac{dy}{y} \{(G_{N\to\bar{u}}(y,Q^2)+G_{N\to\bar{s}}(y,Q^2))(2\delta(z-1)$$

$$+ \alpha_s(Q^2)\Delta f_{3,q}(z))-(G_{N\to u}(y,Q^2)$$

$$+ G_{N\to c}(y,Q^2))\alpha_s(Q^2)\Delta f_{3,q}(z)\}. \tag{6.78b}$$

For the longitudinal structure function defined by (6.39b), we have for νN scattering

$$\frac{F_L^{\nu N}(x,Q^2)}{x} = 2\int_x^{1.0} \frac{dy}{y} \{(G_{N\to d}(y,Q^2)+G_{N\to s}(y,Q^2)+G_{N\to\bar{u}}(y,Q^2)$$

$$+ G_{N\to\bar{s}}(y,Q^2))(2\delta(z-1)+\alpha_s(Q^2)\Delta f_{1,q}(z))$$

$$+ 4G_{p\to g}(y,Q^2)\alpha_s(Q^2)\Delta f_{1,g}(z)\} \tag{6.79}$$

and similarly $\bar{\nu}N$ scattering. If we use

$$\frac{F_2^{\nu N}(x,Q^2)}{x} = 2(G_{N\to d}(x,Q^2)+G_{N\to s}(x,Q^2)+G_{N\to\bar{u}}(x,Q^2)+G_{N\to s}(x,Q^2)) \tag{6.80}$$

and substitute in for $\Delta f_{1,q}(z)$ and $\Delta f_{1,g}(z)$ from (6.63c,d), we arrive at

$$F_L^{(\nu,\bar{\nu})N}(x,Q^2) = \frac{\alpha_s(Q^2)}{2\pi}\, x^2 \int_x^1 \frac{dy}{y^3} \tag{6.81a}$$

$$\{\frac{8}{3}\, F_2^{(\nu,\bar{\nu})N}(y,Q^2)+2a_\nu(1-\frac{x}{y})yG_{N\to g}(y,Q^2)\},$$

which is the same as (6.66) except now

$$a_\nu = 8. \tag{6.81b}$$

Integrating F_A, F_Q and F_2 over x yields

$$\bar{F}_Q^{\nu N}(Q^2) = (U+D+2S)(1-0.0885\ \alpha_s(Q^2))$$

$$+ (\bar{U}+\bar{D}+2\bar{C})(0.0885\ \alpha_s(Q^2)) \tag{6.82a}$$

$$\bar{F}_Q^{\bar{\nu}N}(Q^2) = (U+D+2C)(1-0.0885\ \alpha_s(Q^2))$$

$$+ (\bar{U}+\bar{D}+2\bar{S})(0.0885\ \alpha_s(Q^2)) \tag{6.82b}$$

$$\bar{F}_A^{\nu N}(Q^2) = (\bar{U}+\bar{D}+2\bar{C})(1-0.0885\ \alpha_s(Q^2))$$

$$+ (U+D+2S)(0.0885\ \alpha_s(Q^2)) \tag{6.82c}$$

$$\bar{F}_A^{\bar{\nu}N}(Q^2) = (\bar{U}+\bar{D}+2\bar{S})(1-0.0885\ \alpha_s(Q^2))$$

$$+ (U+D+2C)(0.0885\ \alpha_s(Q^2)) \tag{6.82d}$$

$$\bar{F}_L^{\nu N}(Q^2) = 0.141\ \alpha_s(Q^2)(U+D+2S+\bar{U}+\bar{D}+2\bar{C}) + 0.212\ \alpha_s(Q^2)G \tag{6.82e}$$

$$\bar{F}_L^{\bar{\nu} N}(Q^2) = 0.141\ \alpha_s(Q^2)(U+D+2C+\bar{U}+\bar{D}+2\bar{S}) + 0.212\ \alpha_s(Q^2)G, \tag{6.82f}$$

where $N = \frac{1}{2}(n+p)$ and where I have used (6.71a,b) together with

$$\alpha_s \int_0^1 z\Delta f_{3,q}(z)dz = -\frac{10}{9}\frac{\alpha_s}{2\pi} = -0.177\ \alpha_s \tag{6.83}$$

and where U, D, S, etc., are the total fraction of momentum carried by u, d, s, quarks, respectively,

$$U(Q^2) = \int_0^1 xG_{p\to u}(x,Q^2)dx \tag{6.84}$$

and G is the fraction of momentum carried by gluons as in (6.73). At $Q^2 = 16\ \text{GeV}^2$, our distributions give

$$\begin{array}{llll} U = 0.267 & D = 0.149 & S = \bar{S} = 0.016 & \bar{U} = 0.027 \\ \bar{D} = 0.030 & G = 0.477 & C = \bar{C} = 0.006 & \end{array} \tag{6.85}$$

which gives for an isoscalar target ($N = \frac{1}{2}(p-n)$)

$$\begin{aligned} \bar{F}_2^{\nu N} &= \bar{F}_2^{\bar{\nu} N} = 0.527 \\ \bar{F}_L^{\nu N} &= \bar{F}_L^{\bar{\nu} N} = 0.057 \\ \bar{F}_A^{\nu N} &= 0.081 \\ \bar{F}_A^{\bar{\nu} N} &= 0.101 \\ \bar{F}_Q^{\nu N} &= 0.447 \\ \bar{F}_Q^{\bar{\nu} N} &= 0.426, \end{aligned} \tag{6.86a}$$

where $\alpha_s(Q^2=16\text{ GeV}) = 0.327$ (i.e., $\Lambda = 0.4$ GeV/c).

Thus at this Q^2

$$\bar{R} = (\bar{F}_2 - 2x\bar{F}_1)/2x\bar{F}_1 = 0.123 \tag{6.86b}$$

for νN and $\bar{\nu}N$ scattering. It is also interesting to note that

$$\bar{F}_A^{\nu N}/(\bar{U}+\bar{D}+2\bar{C}) = 1.16 \tag{6.86c}$$

at this Q^2. In the naive (leading order) parton model $\bar{F}_A^{\nu N} = \bar{U} + \bar{D} + 2\bar{C}$.

5. Large Mass Muon Pair Production in Proton-Proton Collisions

As a next example of QCD perturbation theory let us calculate the order $\alpha_s(Q^2)$ corrections to the production of large mass muon pairs in proton-proton collisions ("Drell-Yan" process). Starting with the "Compton" term $g + q \to \gamma^* + q$ shown in Fig. 6.1 and taking the gluon off-mass shell $q_g^2 = -m_g^2$, we have

$$\frac{d\hat{\sigma}_{DY}^{c}}{d\hat{t}}(\hat{s},\hat{t}) = \frac{\pi\alpha\alpha_s e_q^2}{\hat{s}^2}\left(\frac{1}{3}\right)\left\{-\frac{\hat{t}}{\hat{s}} - \frac{\hat{s}}{\hat{t}} + \frac{2M^2}{\hat{s}\hat{t}}(\hat{s}+\hat{t}-M^2) - \frac{M^2 m_g^2}{\hat{t}^2}\right\}, \tag{6.87}$$

where M^2 is the mass of the virtual photon and all terms that give no contribution to the total rate

$$\hat{\sigma}_{DY}(\hat{s}) = \int_{\hat{t}_{max}}^{\hat{t}_{min}} d\hat{t}\,\frac{d\hat{\sigma}_{DY}}{d\hat{t}}(\hat{s},\hat{t}) \tag{6.88}$$

in the limit $m_g \to 0$ have been dropped. In this case

$$\hat{t}_{min} = -\frac{M^2 m_g^2}{\hat{s}} \tag{6.89a}$$

$$\hat{t}_{max} = M^2 - \hat{s}. \tag{6.89b}$$

Integration yields

$$\hat{\sigma}^{c}_{DY}(\hat{s}) = \frac{\pi\alpha\alpha_s e_q^2}{\hat{s}^2}\left(\frac{1}{3}\right)\{(\hat{s} + \frac{2M^4}{\hat{s}} - 2M^2)\log\left(\frac{\hat{s}(\hat{s}-m^2)}{M^2 m_g^2}\right)$$

$$+ \frac{1}{2}\frac{(\hat{s}-m^2)^2}{\hat{s}} + \frac{2M^2}{\hat{s}}(\hat{s}-m^2) - 1\}, \qquad (6.90)$$

where I have used

$$\int_{t_{max}}^{t_{min}} d\hat{t} = \hat{s} - M^2, \quad \int_{t_{max}}^{t_{min}} t dt = -\frac{1}{2}(\hat{s}-m^2)^2$$

$$\int_{t_{max}}^{t_{min}} \frac{d\hat{t}}{\hat{t}} = \log\frac{t_{min}}{t_{max}} = -\log\left(\frac{(\hat{s}-M^2)\hat{s}}{M^2 m_g^2}\right) \qquad (6.91)$$

$$m_g^2 \int_{t_{max}}^{t_{min}} \frac{dt}{t^2} = -\frac{\hat{s}}{M^2}.$$

To embed this subprocess inside the desired $p + p \to \mu^+\mu^- + X$ reaction, it is convenient to define

$$\hat{\tau} = M^2/\hat{s} \qquad (6.92a)$$

which is related to the observed $\tau = M^2/s$ by

$$\hat{\tau} = \tau/x_a x_b, \qquad (6.92b)$$

$$\text{since } \hat{s} = x_a x_b s \qquad (6.92c)$$

where, as illustrated in Fig. 6.2, x_a and x_b are the fraction of momentum of the initial protons carried by the gluon and quark, respectively. In terms of $\hat{\tau}$, (6.90) becomes

$$\hat{\sigma}^{c}_{DY}(\hat{s}) = \frac{\pi\alpha\alpha_s e_q^2}{\hat{s}} \left(\frac{1}{3}\right) \tag{6.93}$$

$$\{(\hat{\tau}+(1-\hat{\tau})^2)\log\left(\frac{M^2(1-\hat{\tau})}{m_g^2\hat{\tau}^2}\right) - \frac{1}{2} + \hat{\tau} - \frac{3}{2}\hat{\tau}^2\}.$$

If we now use the relationship

$$\frac{d\sigma_{DY}}{dM^2}(s,M^2;pp\to\mu^+\mu^-+X) = \frac{\alpha}{3\pi M^2}\int_{\tau}^{1.0} dx_a \int_{\tau/x_a}^{1.0} dx_b \tag{6.94}$$

$$G_{p\to a}(x_a)G_{p\to b}(x_b)\hat{\sigma}_{DY}(\hat{s};a+b\to\gamma^*+c),$$

where the $(\alpha/3\pi M^2)$ factor comes from integrating over the muon pair angular distribution, then we arrive at (for one quark flavor)

$$\frac{d\sigma^{c}_{DY}}{dM^2}(s,M^2) = \left(\frac{4\pi}{9}\right)\frac{\alpha^2 e_q^2}{sM^2}\int_{\tau}^{1.0}\frac{dx_a}{x_a}\int_{\tau/x_b}^{1.0}\frac{dx_b}{x_b}(G^{(o)}_{p\to q}(x_a)G^{(o)}_{p\to g}(x_b)$$

$$+ (x_a\leftrightarrow x_b))\left(\frac{\alpha_s}{2\pi}P_{q\leftarrow g}(\hat{\tau})\log\frac{M^2}{m_g^2} + \alpha_s f_g^{DY}(\hat{\tau})\right), \tag{6.95}$$

with

$$P_{q\leftarrow g}(\hat{\tau}) = \frac{1}{2}(\hat{\tau}+(1-\hat{\tau})^2) \tag{6.96a}$$

which is identical to (6.37) and

$$\alpha_s f_g^{DY}(\hat{\tau}) = \frac{\alpha_s}{2\pi}\left(\frac{1}{2}\right)\{(\hat{\tau}^2+(1-\hat{\tau})^2)[-2\log\hat{\tau}+\log(1-\hat{\tau})]$$

$$- \frac{1}{2} + \hat{\tau} - \frac{3}{2}\hat{\tau}^2\}. \tag{6.96b}$$

The annihilation term in Fig. 6.3 yields the differential cross section

$$\frac{d\hat{\sigma}^A_{DY}}{d\hat{t}}(\hat{s},\hat{t}) = \frac{\pi\alpha\alpha_s e_q^2}{\hat{s}^2}\left(\frac{8}{9}\right)\left(\frac{\hat{u}}{\hat{t}} + \frac{\hat{t}}{\hat{u}} + \frac{2M^4}{t^u} - \frac{2M^2}{t} - \frac{2M^2}{u}\right.$$

$$\left. - \frac{m_q^2 M^2}{\hat{u}^2} - \frac{m_{\bar{q}}^2 M^2}{\hat{t}^2}\right), \qquad (6.97)$$

where the incoming quark and antiquark are off-shell with $p_q^2 = -m_q^2$ and $p_{\bar{q}}^2 = -m_{\bar{q}}^2$, respectively. In addition, all terms in (6.97) that don't contribute to the total integrated rate, $\hat{\sigma}^A_{DY}$, in the limit $m_q \to 0$, $m_{\bar{q}} \to 0$ have been dropped. Integrating over $\hat{t}$ and using $\hat{t}_{min} = -m_{\bar{q}}M^2/\hat{s}$, $\hat{u}_{min} = -m_q M^2/\hat{s}$ yields

$$\hat{\sigma}^A_{DY}(\hat{s}) = \frac{\pi\alpha\alpha_s}{\hat{s}}\left(\frac{8}{9}\right)\left\{\left(\frac{1+\hat{\tau}^2}{1-\hat{\tau}}\right)\log\left(\frac{M^2(1-\hat{\tau})}{m_{\bar{q}}^2\hat{\tau}^2}\right)\right.$$

$$\left. + \left(\frac{1+\hat{\tau}^2}{1-\hat{\tau}}\right)\log\left(\frac{M^2(1-\hat{\tau})}{m_q^2\hat{\tau}^2}\right) + 2\hat{\tau} - 4\right\}, \qquad (6.98)$$

where $\hat{\tau}$ is given by (6.92b). Using (6.94) now results in (for one quark flavor)

$$\frac{d\sigma^A}{dM^2}(s,M^2) = \left(\frac{4\pi}{9}\right)\frac{\alpha^2 e_q^2}{sM^2}\int_\tau^{1.0}\frac{dx_a}{x_a}\int_{\tau/x_b}^{1.0}\frac{dx_b}{x_b}$$

$$\left(G^{(o)}_{p\to q}(x_a)G^{(o)}_{p\to\bar{q}}(x_b)+G^{(o)}_{p\to q}(x_b)G^{(o)}_{p\to\bar{q}}(x_a)\right)$$

$$\left\{\delta(\hat{\tau}-1)+\frac{\alpha_s}{2\pi}P_{q\leftarrow q}(\hat{\tau})\right.$$

$$\left.\left(\log\frac{Q^2}{m_q^2} + \log\frac{Q^2}{m_{\bar{q}}^2}\right)+2\alpha_s f_q^{DY}(\hat{\tau})\right\} \qquad (6.99)$$

where the $\delta(\hat{\tau}-1)$ term comes from the leading $q + \bar{q} \to \gamma^*$ subprocess and where

$$P_{q\leftarrow q}(\hat{\tau}) = \frac{4}{3}\left(\frac{1+\hat{\tau}^2}{1-\hat{\tau}}\right) \qquad (\hat{\tau} < 1) \tag{6.100a}$$

and

$$\alpha_s f_q^{DY}(\hat{\tau}) = \frac{\alpha_s}{2\pi}\left(\frac{4}{3}\right)\left\{\left(\frac{1+\hat{\tau}^2}{1-\hat{\tau}}\right)(\log(1-\hat{\tau})-2\log\hat{\tau}) + \hat{\tau} - 2\right\}.$$

$$(\hat{\tau} < 1) \tag{6.100b}$$

As in the electroproduction case, we must add to (6.100a) and (6.100b) order $\alpha_s(Q^2)\ \delta(\hat{\tau}-1)$ contributions that arise from the virtual gluon corrections in Fig. 6.4 interfering with $q + \bar{q} \rightarrow \gamma^*$. These terms in (6.100) then become[33)]

$$P_{q\leftarrow q}(\hat{\tau}) = \frac{4}{3}\left[\frac{1+\hat{\tau}^2}{(1-\hat{\tau})_+} + \frac{3}{2}\,\delta(\hat{\tau}-1)\right] \tag{6.101a}$$

$$\alpha_s f_q^{DY}(\hat{\tau}) = \frac{\alpha_s}{2\pi}\left(\frac{4}{3}\right)\left\{(1+\hat{\tau}^2)\left(\frac{\log(1-\hat{\tau})}{1-\hat{\tau}}\right)_+ - \frac{2(1+\hat{\tau}^2)}{1-\hat{\tau}}\log\hat{\tau}+\hat{\tau}-2+\left(\frac{11}{4}\right)\delta(\hat{\tau}-1)\right\} \tag{6.101b}$$

where $1/(1-\hat{\tau})_+$ is defined as before by eq. (4.24) and $\left(\frac{\log(1-\hat{\tau})}{1-\tau}\right)_+$ is defined by

$$\int_x^1 h(z)\left(\frac{\log(1-z)}{1-z}\right)_+ dz \equiv \int_x^1 \frac{1}{2}\log^2(1-z)\,\frac{dh(z)}{dz}\,dz \tag{6.102}$$

$$= \frac{1}{2}\,h(z=1)\log^2(1-x)+\int_x^1 \frac{(h(z)-h(z=1))}{(1-z)}\log(1-z)dz.$$

Combining (6.99) and (6.95) yields

$$\frac{d\sigma_{DY}}{dM^2}(s,M^2) = (\frac{4\pi}{9})\frac{\alpha^2 e_q^2}{sM^2}\int_{\tau}^{1.0}\frac{dx_a}{x_a}\int_{\tau/x_b}^{1.0}\frac{dx_b}{x_b}$$

$$\{(G^{(o)}_{p\to q}(x_a)G^{(o)}_{p\to\bar{q}}(x_a)+(x_a\leftrightarrow x_b))$$

$$(\delta(\hat{\tau}-1)+\frac{\alpha_s}{2\pi}P_{q\leftarrow q}(\hat{\tau})\log(\frac{M^4}{m_q^2 m_{\bar{q}}^2})+2\alpha_s f_q^{DY}(\hat{\tau}))$$

$$+\ (G^{(o)}_{p\to q}(x_a)G^{(o)}_{p\to g}(x_b)+(x_a\leftrightarrow x_b))$$

$$(\frac{\alpha_s}{2\pi}P_{q\leftarrow g}(\hat{\tau})\log(\frac{M^2}{m_g^2})+\alpha_s f_g^{DY}(\hat{\tau}))\}. \quad (6.103)$$

If we now <u>define</u> "Drell-Yan" quark distributions by

$$\frac{d\sigma_{DY}}{dM^2}(s,M^2) = \frac{4\pi}{9}\frac{\alpha^2}{sM^2}\int_{\tau}^{1.0}\frac{dx_a}{x_a}\sum_{i=1}^{n_f}e_q^2$$

$$[G^{DY}_{p\to q_i}(x_a,M^2)G^{DY}_{p\to\bar{q}_i}(x_b,M^2) \quad (6.104)$$

$$+\ G^{DY}_{p\to q_i}(x_b,M^2)G^{DY}_{p\to\bar{q}_i}(x_a,M^2)],$$

which is the usual formula where $x_a x_b = \tau$, then

$$G^{DY}_{p\to q_i}(x,M^2) = \int_x^{1.0}\frac{dy}{y}\{G^{(o)}_{p\to q_i}(y)[\delta(z-1)+\frac{\alpha_s}{2\pi}P_{q\leftarrow q}(z)\log(\frac{M^2}{m_q^2})$$

$$+\ \alpha_s f_q^{DY}(z)]+G^{(o)}_{p\to g}(y)[\frac{\alpha_s}{2\pi}P_{q\leftarrow g}(z)\log(\frac{M^2}{m_g^2})$$

$$+\ \alpha_s f_g^{DY}(z))\} \quad (6.105a)$$

and

$$G^{DY}_{p\to\bar{q}_i}(x,M^2) = \int_x^{1.0} \frac{dy}{y} \{G^{(o)}_{p\to\bar{q}_i}(y)[\delta(z-1)+ \frac{\alpha_s}{2\pi} P_{q\leftarrow q}(z)\log(\frac{M^2}{m^2_{\bar{q}}})$$

$$+ \alpha_s f^{DY}_q(z)]+G^{(o)}_{p\to g}(y)[\frac{\alpha_s}{2\pi} P_{q\leftarrow g}(z)\log(\frac{M^2}{m^2_g})$$

$$+ \alpha_s f^{DY}_g(z)]\}, \tag{6.105b}$$

where $z = \frac{x}{y}$. One can easily verify this by substituting into (6.103) and keeping terms only to order α_s. We now play the same game of absorbing the $\log(m^2)$ divergences into the distributions $G^{(o)}(y)$ and using (4.29) to sum all the $\alpha_s(M^2)\log M^2$ terms, we arrive at

$$G^{DY}_{p\to q}(x,M^2) = \int_x^{1.0} \frac{dy}{y} \{G_{p\to q}(y,M^2)[\delta(z-1)+\alpha_s(M^2)\Delta f^{DY}_q(z)]$$

$$+ G_{p\to g}(y,M^2)\alpha_s(M^2)\Delta f^{DY}_g(z)\} \tag{6.106a}$$

and similarly for $G^{DY}_{p\to\bar{q}}(x,M^2)$, and where[34)]

$$\alpha_s\Delta f^{DY}_q(z) = \alpha_s(f^{DY}_q(z)-f_{2,q}(z)) = \frac{\alpha_s}{2\pi}(\frac{2}{3})[2(1+z^2)(\frac{\log(1-z)}{1-z})_+$$

$$+ \frac{3}{(1-z)_+} -6-4z+(\frac{4\pi^2}{3} + \frac{11}{2})\delta(z-1)], \tag{6.106b}$$

$$\alpha_s\Delta f^{DY}_g(z) = \alpha_s(f^{DY}_g(z)-f_{2,g}(z))$$

$$= \frac{\alpha_s}{2\pi}(\frac{1}{2})[(z^2+(1-z)^2)\log(1-z)+ \frac{1}{2}(9z^2-10z+3)]. \tag{6.106c}$$

In (6.106a), $G_{p\to q}(x,M^2)$ and $G_{p\to g}(x,M^2)$ are the quark distributions defined by F_2 in ep scattering in (6.62) with Q^2 replaced by M^2. Their evolution with M^2 is governed by the "renormalization" eqs. (4.29). Substituting (6.106a) back into (6.104) gives

$$\frac{d\sigma_{DY}}{dM^2}(s,M^2) = \frac{4\pi}{9}\frac{\alpha^2}{sM^2}\int_{\tau}^{1.0}\frac{dx_a}{x_a}\sum_{i=1}^{n_f} e_{q_i}^2$$

$$[G_{p\to q_i}(x_a,M^2)G_{p\to\bar{q}_i}(x_b,M^2)+(x_a\leftrightarrow x_b]$$

$$+\frac{4\pi\alpha^2}{9sM^2}\int_{\tau}^{1.0}\frac{dx_a}{x_a}\int_{\tau/x_a}^{1.0}\frac{dx_b}{x_b}\sum_{i=1}^{n_f} e_q^2$$

$$[G_{p\to q_i}(x_a,M^2)G_{p\to\bar{q}_i}(x_b,M^2)+(x_a\leftrightarrow x_b)]2\alpha_s\Delta f_q^{DY}(\hat{\tau})$$

$$+\frac{4\pi\alpha^2}{9sM^2}\int_{\tau}^{1.0}\frac{dx_a}{x_a}\int_{\tau/x_a}^{1.0}\frac{dx_b}{x_b}\sum_{i=1}^{n_f} e_q^2$$

$$[G_{p\to q_i}(x_a,M^2)G_{p\to g}(x_b,M^2)+(x_a\leftrightarrow x_b)]\alpha_s\Delta f_g^{DY}(\hat{\tau})$$

$$+\frac{4\pi\alpha^2}{9sM^2}\int_{\tau}^{1.0}\frac{dx_a}{x_a}\int_{\tau/x_a}^{1.0}\frac{dx_b}{x_b}\sum_{i=1}^{n_f} e_q^2$$

$$[G_{p\to\bar{q}_i}(x_a,M^2)G_{p\to g}(x_b,M^2)+(x_a\leftrightarrow x_b)]\alpha_s\Delta f_g^{DY}(\hat{\tau}). \tag{6.107}$$

The first term is Politzer's[35] leading order result that the Drell-Yan rate is given by the usual parton model formula but with the "renormalization" group improved distributions $G(x,M^2)$ with Q^2 replaced by M^2 [36]. The second, third and fourth terms are the order α_s corrections to this result. Let us now examine some of the phenomenology of the production of large mass muon pairs in pp collisions.

VII. LARGE-MASS MUON-PAIR PRODUCTION

A. QCD Factorization and the Total Muon Pair Rate

We begin our investigation of the production of large mass muon pairs in hadron-hadron collisions by examining first the leading order QCD predictions. Then we will look at some of the order α_s corrections. As we saw in the preceding section, to leading order, the quark and gluon distributions, $G_i(x,Q^2)$, determined in Section IV.B from an analysis of electroproduction are processes independent. Parton distributions determined from one process (like $ep \to e + X$) can, to leading order, be used to make predictions elsewhere (like for $pp \to \mu^+\mu^- + X$ or $pp \to \pi^o + X$). Recently several groups[37-40] have shown that all divergent perturbative contributions to these processes can be summed and absorbed into universal quark and gluon distributions. These divergences arise from, for example, the parallel emission of a massless gluon by a massless incoming or outgoing quark. (The "soft" divergences that arise as the gluon energy becomes small are canceled by other diagrams containing virtual gluon corrections.) These "parallel" divergences are a property of the incoming (or outgoing) quark line and can be "factored" out from the basic hard subprocess (this hard subprocess is, of course, different for different reactions). They are precisely the terms in Fig. 4.1 that are responsible for the Q^2 dependence of the parton distributions.

The total muon pair cross section (integrated over all muon pair $p_\perp$) is given to leading order (i.e., order $\alpha_s(Q^2)\log Q^2$) by[35,36]

$$\sigma_{tot}(s,M^2,y) \equiv \frac{d\sigma}{dM^2dy}(s,M^2,y)$$

$$= \frac{4\pi\alpha^2}{3sM^2}\left(\frac{1}{3}\right)\sum_i e_i^2\{G_{p\to q_i}(x_a,Q^2)G_{p\to\bar{q}_i}(x_b,Q^2)$$

$$+ (q_i \leftrightarrow \bar{q}_i)\}, \qquad (7.1)$$

where y is the rapidity of the muon pair of mass M and

$$x_a = \sqrt{\tau}\exp(y), \qquad x_b = \sqrt{\tau}\exp(-y), \qquad (7.2)$$

and where the parton distributions $G(x,Q^2 = M^2)$ are the "renormalization group improved" functions given in Section IV. One does not see the "Compton" term, $g + q \to q + \gamma^*$ (see Fig. 7.1a,b), explicitly in (7.1) since to leading order it is included in the probability of finding an antiquark in the proton. As illustrated in Fig. 7.2, the divergent pieces (behaving like $\alpha_s(Q^2) \log Q^2$) of gluon Bremsstrahlung from the incoming antiquark and incoming gluon quark-antiquark pair production are absorbed into and generate the Q^2 dependence of $G_{p\to\bar{q}}(x,Q^2)$[35,36,41]. Figure 7.3 shows the data on $d\sigma/dMdy$ for $pN \to \mu^+\mu^- + X$ at $W = 27.4$ GeV and $y = 0$ together with the predictions from eq. (7.1) using the parton distributions $G_i(x,Q^2)$ determined in Section IV. The agreement of the leading order prediction with the magnitude of the muon-pair cross section is quite satisfactory, although it does depend sensitively on the antiquark distributions which are not determined too well.

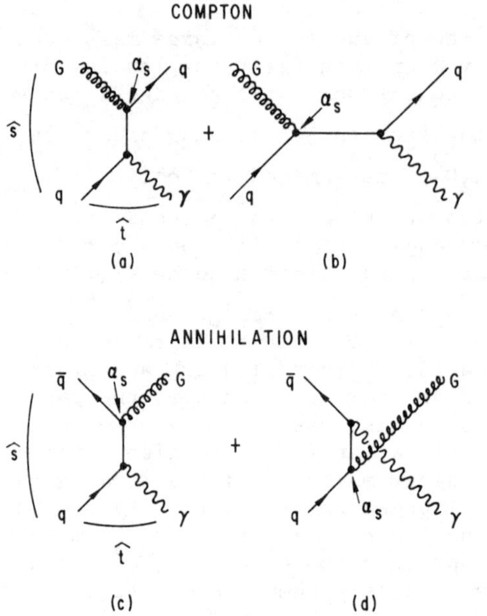

Fig. 7.1. QCD processes to first order in the strong coupling constant α_s. Diagrams (a) and (b) represent quark-gluon "Compton" scattering to yield a quark and a virtual photon. Diagrams (c) and (d) represent quark-antiquark annihilation into a gluon and a virtual photon.

B. Muon Pairs Produced at Large Mass and Large $p_\perp$

Effects due to the transverse momentum, $k_\perp$, of quarks and gluons within hadrons can sometimes be very important. In QCD, transverse momentum of partons can arise in two ways. Firstly, in, for example, a proton beam, quarks are confined in the transverse direction to within the proton radius. Therefore, from the uncertainty principle, they must have some transverse momentum. This momentum, called primordial, is intrinsic to the basic parton wave function inside the proton. It involves small Q^2 values and thus cannot be calculated using perturbation theory. At present, it must be viewed as unknown but bounded (falling off like an exponential or Gaussian in $k_\perp$). Secondly, in QCD, one expects to receive an "effective" $k_\perp$ of quarks in protons due to the hard Bremsstrahlung of gluons which can be calculated perturbatively if the momentum transfers are large.

A particularly nice place to study the interplay between these two components of transverse momentum is in the production of large mass muon pairs in pp collisions, $pp \to \mu^+\mu^- + X$. Many people have analyzed this process in terms of QCD[36,41-46]. The analysis presented here follows closely the work of Altarelli, Parisi and Petronzio[41].

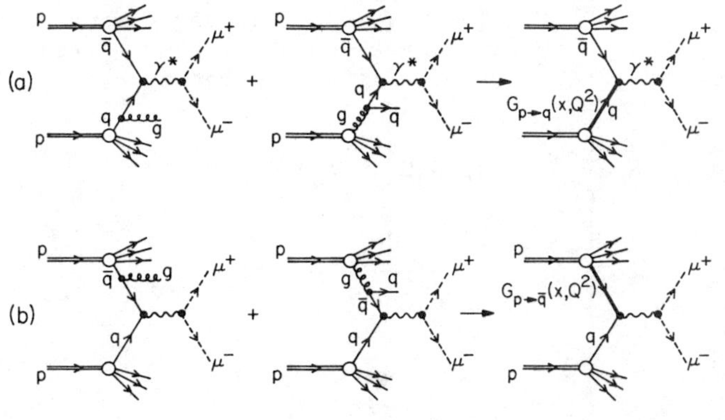

Fig. 7.2 - (a) Illustration of two diagrams for $pp \to \mu^+\mu^- + X$ which are included (divergent parts) in the "renormalization group" improved quark distributions, $G_{p\to q}(x,Q^2)$, when one calculates the contribution from the subprocess $q\bar{q} \to \gamma^* \to \mu^+\mu^-$.

(b) Illustration of two diagrams which are included (divergent parts) in the "renormalization group" improved antiquark distributions, $G_{p\to\bar{q}}(x,Q^2)$.

The perturbative component of the transverse momentum of muon pairs is generated by the two-to-two constituent subprocess $q\bar{q} \to \gamma^* +$ gluon and gluon $+ q \to \gamma^* + q$ shown in Fig. 7.1, where the virtual photon, γ^*, then decays into a $\mu^+\mu^-$ pair. Other graphs are higher order in α_s and have been neglected[47]. The cross sections for these processes are given by

$$\frac{d\sigma_A}{dM^2 d\hat{t}} = \left(\frac{8}{27}\right)\left(\frac{\alpha^2\alpha_s e_q^2}{M^2\hat{s}^2}\right)\left(\frac{2M^2\hat{s} + \hat{u}^2 + \hat{t}^2}{\hat{t}\hat{u}}\right) \tag{7.3a}$$

$$\frac{d\sigma_C}{dM^2 d\hat{t}} = \left(\frac{1}{9}\right)\left(\frac{\alpha^2\alpha_s e_q^2}{M^2\hat{s}^2}\right)\left(\frac{2M^2\hat{u} + \hat{s}^2 + \hat{t}^2}{-\hat{t}\hat{s}}\right) , \tag{7.3b}$$

where $\hat{s}$, $\hat{t}$ and $\hat{u}$ are the usual Mandelstam invariants and M^2 is the invariant mass squared of the muon pair and where A and C refer to annihilation and Compton, respectively. The cross section for

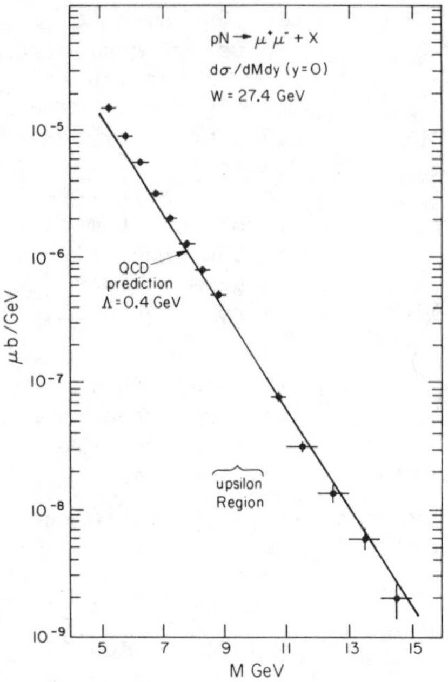

Fig. 7.3. Leading order QCD prediction from eq. (7.1) of the dimuon mass spectrum at $y = 0$ and $W = \sqrt{s} = 27.4$ GeV using the "renormalization group" improved quark and antiquark distributions $G_i(x,Q^2)$ from eq. (4.40a) and (4.40b). The data are from Ref. 52.

producing muon pairs of mass M^2, rapidity y, and transverse momentum $p_\perp$ at a center of mass energy squared s is then given by

$$\sigma_p(s,M^2,y,p_\perp) \equiv \frac{d\sigma_p}{dM^2 dy d^2 p_\perp}(s,M^2,y,p_\perp)$$

$$= \sum_{q=u,d,s} \int_{x_a^{min}}^{1.0} dx_a \{G_{p\to q}(x_a,M^2) G_{p\to \bar{q}}(x_b,M^2) + (\bar{q}\leftrightarrow q)\} \frac{x_a x_b}{(x_a - x_1)} \left(\frac{1}{\pi}\frac{d\sigma_A}{dM^2 d\hat{t}}\right)$$

$$+ \sum_{\substack{q=u,d,s,\\ \bar{u},\bar{d},\bar{s}}} \int_{x_a^{min}}^{1.0} dx_a \{G_{p\to q}(x_a,M^2) G_{p\to g}(x_b,M^2) + (g\leftrightarrow q)\} \frac{x_a x_b}{(x_a - x_1)} \left(\frac{1}{\pi}\frac{d\sigma_C}{dM^2 d\hat{t}}\right), \quad (7.4)$$

where $x_a^{min} = (x_1 - \tau)/(1-x_2)$ and $x_b = (x_a x_2 - \tau)/(x_a - x_1)$. The quantities x_1 and x_2 are given by

$$x_1 = \frac{1}{2}(x_\perp^2 + 4\tau)^{1/2} e^{y} \quad (7.5a)$$

$$x_2 = \frac{1}{2}(x_\perp^2 + 4\tau)^{1/2} e^{-y}, \quad (7.5b)$$

where $x_\perp = 2p_\perp/\sqrt{s}$ and $\tau = M^2/s$[48]. The subprocess invariants are given by

$$\hat{s} = x_a x_b s \tag{7.6a}$$

$$\hat{t} = -x_a s x_2 + M^2 \tag{7.6b}$$

$$\hat{u} = -x_b s x_1 + M^2. \tag{7.6c}$$

In (7.4), the label P refers to "perturbative contribution" and where the "renormalization improved" parton distributions $G(x,Q^2 = M^2)$ from (4.29) and the running coupling constant $\alpha_s(M^2)$ from (2.12) are used. It is clear from the work of Politzer[35] and Sachrajda[36] that here one should use the "renormalization improved" $G(x,Q^2)$ and $\alpha_s(Q^2)$ functions. However, here there are two large invariants M^2 and $p_\perp^2$ and it is not clear whether the Q^2 dependence evolves according to $G(x,Q^2{=}M^2)$ or $G(x,Q^2{=}p_\perp^2)$ or some other combination. It, of course, does not matter in leading order and at very large values of M^2 and $p_\perp^2$ but it certainly matters for the phenomenology at existing M^2 and $p_\perp^2$ values. Using $Q^2 = M^2$ in (7.4) is an assumption that is probably not precisely correct[49,50].

The perturbative contributions in (7.4) are shown in Fig. 7.4. They are absolutely normalized and agree roughly with the data at large $p_\perp$. They, however, have the wrong shape at small $p_\perp$ and diverge at $p_\perp = 0$. This infrared difficulty besets other QCD perturbative calculations. (The gluon Bremsstrahlung in Fig. 4.1b also produces divergences as Δ in (4.8) goes to zero.) One expects that non-perturbative phenomena at small $p_\perp$ will regularize this singularity leaving a smooth $p_\perp$ distribution.

The soft, non-perturbative, component of $k_\perp$ (the primordial $k_\perp$) can be used to regularize $\sigma_P(s,M^2,y,p_\perp)$ and produce a finite distribution in $p_\perp$ at all $p_\perp$. This can be done crudely by convoluting $\sigma_P(p_\perp)$ with a quark primordial motion given, for instance, by

$$f(k_\perp^2) = \exp[-k_\perp^2/(4\sigma_q^2)]/(4\pi\sigma_q^2), \tag{7.7}$$

where for a single constituent in a proton, one has

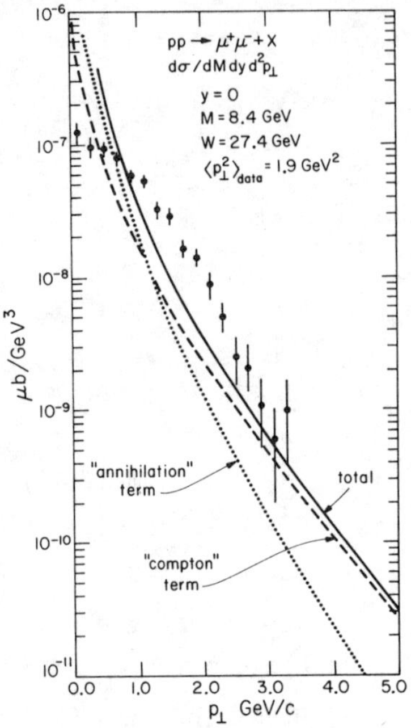

Fig. 7.4. The distribution in transverse momentum, $p_\perp$, of the $\mu^+\mu^-$ pair produced in pp collisions at W = 27.4 GeV together with the QCD perturbative predictions. The "Compton" (Fig. 7.1a,b) and "annihilation" (Fig. 7.1c,d) are given by the dashed and dotted curves, respectively.

$$\langle k_\perp^2\rangle_{\text{primordial}} = 2\sigma_q^2. \tag{7.8}$$

The result is

$$\sigma(s,M^2,y,p_\perp^2) = \int d^2k_\perp \sigma_P(M^2,s,y,(\vec{p}_\perp-\vec{k}_\perp)^2) f(k_\perp^2)$$
$$= \int d^2k_\perp \sigma_P(M^2,s,y,(\vec{p}_\perp-\vec{k}_\perp)^2) [f(k_\perp^2) - f(p_\perp^2)] + f(p_\perp^2) \int d^2k_\perp \sigma_P(M^2,s,y,(\vec{p}_\perp-\vec{k}_\perp)^2), \tag{7.9}$$

where I have added and subtracted the second term in (7.9). Actually, in doing this convolution, I should have added to σ_P a contribution of order α_s arising from the vertex correction to the subprocess $q\bar{q} \to \gamma^*$. This contribution also diverges but only contributes at $p_\perp = 0$ (i.e., has a $\delta(p_\perp)$). It does not contribute to the first term in (7.9) since $[f(k_\perp^2) - f(p_\perp^2)]\,\delta(k_\perp^2)$ is zero, but when added to the second term yields

$$\sigma(s,M^2,y,p_\perp^2) = \int d^2q_\perp \sigma_P(M^2,s,y,q_\perp^2)[f((\vec{p}_\perp-\vec{q}_\perp)^2) - f(p_\perp^2)] + f(p_\perp^2)\sigma_{tot}(s,M^2,y), \tag{7.10}$$

where I have defined $\vec{q}_\perp = \vec{p}_\perp - \vec{k}_\perp$ (one must keep track of the vector direction) and where $\sigma_{tot}(s,M^2,y)$ is given by definition by (7.1).

Both terms in (7.10) are now finite at all $p_\perp$ and one is left with one additional parameter, the primodrial $\langle k_\perp\rangle_{p\to q}$ in (7.8), to be adjusted to fit the data. The fit to the data is shown in Fig.

7.5 and yields σ_q = 0.48 GeV or

$$\langle k_\perp \rangle_{\text{primordial}} = \sqrt{\pi/2}\,\sigma_q \approx 600 \text{ MeV}. \quad (7.11)$$

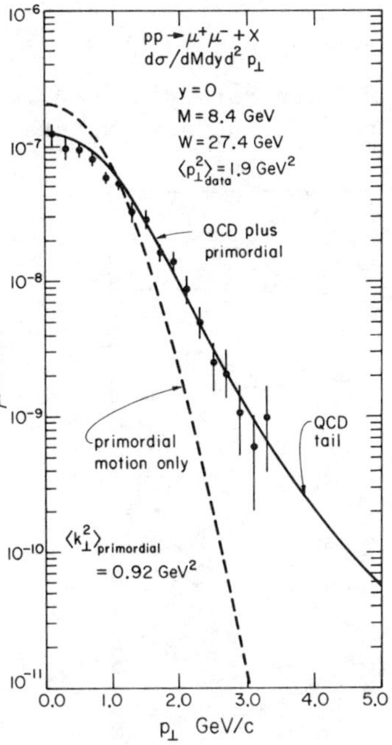

Fig. 7.5. The same data as Fig. 7.4 together with the perturbative QCD contributions of Fig. 7.1 folded with a Gaussian momentum spectrum with $\langle k_\perp \rangle_{h\to q} = \langle k_\perp \rangle_{h\to \bar{q}}$ = 600 MeV (solid curve). The dashed curve results from the primordial motion only with no perturbative QCD terms.

This means that the mean $p_\perp^2$ of the $\mu^+\mu^-$ data, which at this energy and y = 0 is about 1.9 GeV^2, results from about 0.9 GeV^2 due to primordial motion and about 1.0 GeV^2 due to the hard QCD subprocesses. Figure 7.5 shows also the second term in (7.10) which would be the prediction if only primordial motion (with σ_q = 0.48 GeV) were present.

Clearly, this smearing procedure which is used to regulate the divergences is a bit ad hoc[51] and the fit to the shape of the $p_\perp$ spectrum shown in Fig. 7.5 can not be viewed as a success of QCD. One could have fit the same data with a Gaussian with σ_q = 0.677 GeV. The real test of the presence of the QCD component to the effective $k_\perp$ of partons comes from examining the energy dependence of the muon pair $p_\perp$ spectrum. Predictions for this are shown in Fig. 7.6. One expects to see a flatter $p_\perp$ spectra as the energy increases (and M^2 is fixed). Recent data on $pN \to \mu^+\mu^- + X$ at 200, 300 and 400 GeV[52] do show a mean $p_\perp$ that increases with increasing energy in agreement with the QCD expectations.

C. "Scale" Breaking in $pp \to \mu^+\mu^- + X$

It is interesting to look at the "scale-breaking" expected in QCD for observables in $pp \to \mu^+\mu^- + X$. In the old parton model, one expected

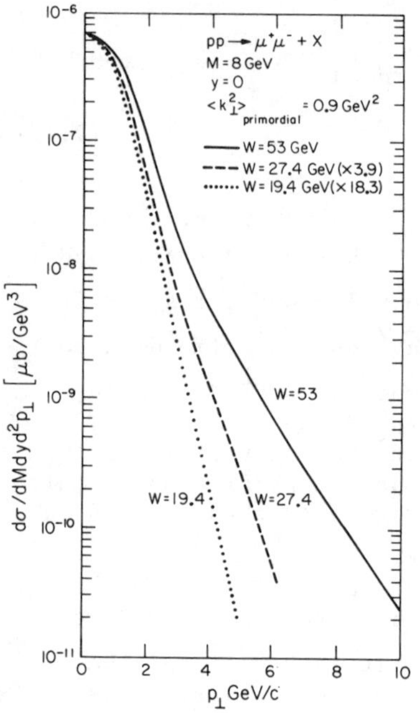

Fig. 7.6. Energy dependence of the large $p_\perp$ tail expected for $pp \to \mu^+\mu^- + X$ from the QCD perturbative contributions shown in Fig. 7.1 folded with a primordial transverse momentum with $\langle k_\perp \rangle_{h\to q} = 600$ MeV for each parton.

$$M^4 \frac{d\sigma}{dM^2 dy}(x,M^2,y) = f(\tau,y), \qquad (7.12)$$

to be only a function of $\tau = M^2/s$ and y and not to depend separately on $W = \sqrt{s}$. As shown in Fig. 7.7, one now expects

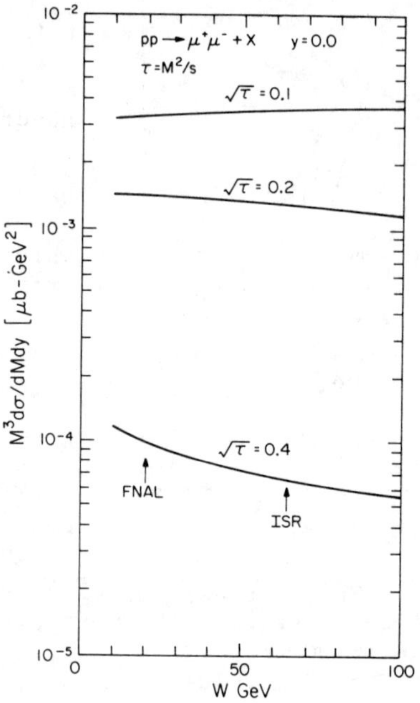

Fig. 7.7. Expected "scale breaking" of the quantity $M^3 d\sigma/dMdy$ for $pp \to \mu^+\mu^- + X$ at $y = 0$ from the QCD non-scaling structure functions with $\Lambda = 0.4$ GeV/c. Scaling would predict this quantity to be independent of W at fixed τ.

small scale breaking effects. At small τ one sees a slight rise with increasing W and at large τ a decrease with increasing W. The effects are small. They are comparable to the breaking of $\nu W_2(x,Q^2)$ in Fig. 4.3 and Fig. 4.6 and will probably not be seen experimentally for quite some time.

Other muon pair observables show larger scale breaking effects. For example, for the annihilation and Compton subprocesses in Fig. 7.1 dimensional counting yields

$$M^4 p_\perp^2 \frac{d\sigma}{dM^2 dy d^2 p_\perp} = F(\tau, y, x_\perp), \tag{7.13a}$$

or where $x_\perp = 2p_\perp/W$ or

$$W^5 \frac{d\sigma}{dM dy d^2 p_\perp} = \tilde{F}(\tau, y, x_\perp). \tag{7.13b}$$

For asymptotically free theories, (7.13) does not scale. Figure 7.8 shows that for fixed $x_\perp$ and τ, $W^5 d\sigma/dMdyd^2p_\perp$ decreases as W increases and approaches a constant ("scaling" result) asymptotically[41,44]. As can be seen in this figure, the primordial motion produces an additional "scale breaking" term. (If there were only primordial motion, then $W^5 d\sigma/dMdyd^2p_\perp$ would go to zero at increasing W (fixed τ and $x_\perp$) like $1/W^2$.) If one could observe experimentally this decrease and approach to a constant for $\tilde{F}$ in (7.13b), it would certainly be support for QCD. However, such measurements are a long way off. They require, for example, at $x_\perp = 0.2$ and $\sqrt{\tau} = 0.2$, comparing a point at FNAL W = 19.4, M = 3.9 GeV and $p_\perp = 1.94$ GeV/c with a point at ISR W = 53 GeV, M = 10.6 GeV, and $p_\perp = 5.3$ GeV/c!

D. Away-Side Jet in $A + B \to \mu^+\mu^- + X$

Several authors[42-44,46] have pointed out the usefulness of observing the "away-side" jet of hadrons that balances the momentum of a large $p_\perp$ muon pair in $A + B \to \mu^+\mu^- + X$. For pp collisions, large $p_\perp$ muon pairs occur predominantly by the "Compton" subprocess, $q + g \to q + \gamma^*$ as seen in Fig. 7.4. This means that for pp collisions, the "away-side" jet is predominantly a quark jet. On the other hand, for $\pi p \to \mu^+\mu^- + X$ (or $\bar{p}p \to \mu^+\mu^- + X$) the "annihilation" term, $\bar{q}q \to \gamma^* + g$, is the dominant subprocess for producing large $p_\perp$ muon pairs. This means that for these processes, the away jet is more likely to be initiated by a gluon than by a quark. By comparing the away-side distributions of hadrons in large $p_\perp$ muon production in $pp \to \mu^+\mu^- + X$ and $\pi p \to \mu^+\mu^- + X$, one can in principle elucidate the differences between gluon and quark fragmentation.

E. Order α_s Corrections to the Total Muon Pair Rate[28,53-54]

We have seen that the leading order QCD predictions agree qualitatively with the data on large-mass muon pair production. However, perturbative predictions to a given order are only meaningful

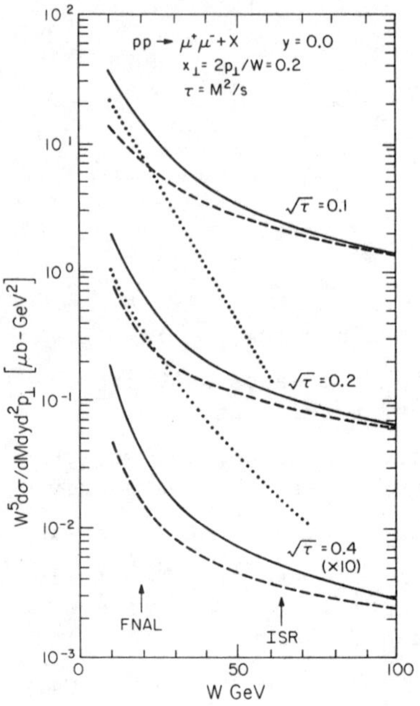

Fig. 7.8. Expected "scale breaking" of the quantity $W^5 d\sigma/dMdyd^2p_\perp$ for $pp \to \mu^+\mu^- + X$ at $y = 0$ and $x_\perp = 0.2$ from the QCD diagrams in Fig. 7.1 with $\Lambda = 0.4$ GeV/c. The solid (dashed) curves are the results after (before) smearing with a primordial parton transverse momentum with $\langle k_\perp \rangle_{h\to q} = 600$ MeV for each parton. The dotted curves would be expected if primordial motion alone were responsible for the muon pair $p_\perp$. Scaling would predict this quantity to be independent of W at fixed τ and $x_\perp$.

if the next order (and all higher order) corrections are small. Since we are dealing with an expansion in $\alpha_s(Q^2)$ (or $1/\log Q^2$) and since $\log Q^2/\Lambda^2$ is not very large, we might worry that higher order corrections may be quite important. We shall now examine the order α_s corrections to the total muon pair cross section, $d\sigma_{DY}/dM^2$, using the formalism developed in Section VI. The results are somewhat disturbing. The order α_s corrections are, in certain kinematic regions, equal to or greater than the leading order term. This probably means that one is not justified in dropping order α_s^2 terms (etc.) and that, at present, the normalization of the muon pair cross section cannot be predicted from the parton distributions determined from the electroproduction data. I would like to say, however, that there are some subtleties involved in calculating the δ-function term in (6.106b) that I don't yet completely understand[55). Therefore, I would like the reader to view this section as "work in progress" and not the finished product.

To leading order the total muon pair rate is

$$\frac{d\sigma_{DY}}{dM^2}(s,M^2) = \frac{4\pi}{9}\frac{\alpha^2}{sM^2}\int_\tau^1 \frac{dx_a}{x_a}\sum_{i=1}^{n_f} e_{q_i}^2 \left[G_{p\to q_i}(x_a,M^2)G_{p\to\bar{q}_i}(x_b,M^2) + (x_a\leftrightarrow x_b)\right], \qquad (7.14)$$

where $x_b = \sqrt{\tau}/x_a$ and where $G(x,M^2)$ are the "renormalized improved" parton distributions $G(x,Q^2{=}M^2)$ from (4.29). To order α_s, $d\sigma_{DY}/dM^2$

receives corrections from the annihilation and virtual diagrams in Fig. 6.3 and Fig. 6.4 (i.e., $\Delta f_q^{DY}(z)$) and from the Compton diagram in Fig. 6.1 (i.e., $\Delta f_g^{DY}(z)$) and is given by (6.107). Fig. 7.9 compares the leading order predictions with the full order α_s results for $d\sigma_{DY}/dM$. For $0.1 \leqslant \sqrt{\tau} \leqslant 0.6$, the order α_s results are roughly a factor of 2 larger than the leading order predictions. This can be seen more clearly in Fig. 7.10 where I have plotted the ratio of the full order α_s to the leading order predictions. As can be seen, the factor of 2 comes from the Δf_q^{DY} term in (6.107). The gluon corrections Δf_g^{DY} actually reduce the leading order predictions slightly. For $0.1 \leqslant \sqrt{\tau} \leqslant 0.6$, the large correction from the Δf_q^{DY} term comes from the δ-function piece in (6.106b). Namely,

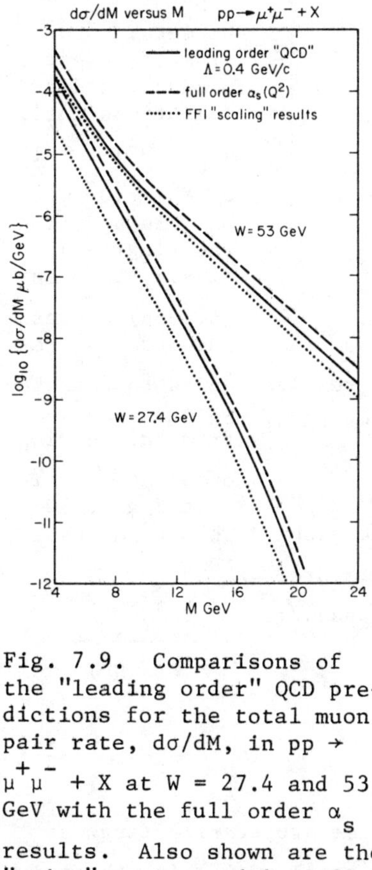

Fig. 7.9. Comparisons of the "leading order" QCD predictions for the total muon pair rate, $d\sigma/dM$, in $pp \to \mu^+\mu^- + X$ at W = 27.4 and 53 GeV with the full order α_s results. Also shown are the "naive" parton model predictions using the scaling parton distributions of FF1[26].

$$[\alpha_s \Delta f_q^{DY}(z)]_{\delta\text{-function}} = \frac{\alpha_s}{2\pi}\left(\frac{2}{3}\right)$$

$$\left(\frac{4\pi^2}{3} + \frac{11}{2}\right)\delta(z-1) \approx 2\,\alpha_s . \quad (7.15)$$

As shown in Fig. 7.11, this term alone results in

$$\frac{G^{DY}(x,Q^2)}{G(x,Q^2)} \approx 1 + 2\,\alpha_s(Q^2), \quad (7.16a)$$

or

$$\frac{d\sigma_{DY}/dM(\text{order } \alpha_s)}{d\sigma_{DY}/dM(\text{leading})} \approx 1 + 4\,\alpha_s(M^2) \quad (7.16b)$$

and since $\alpha_s(Q^2{=}M^2) \approx 1/3$, this is quite a large effect (see Fig. 7.10). It probably means that unless one can sum these contributions

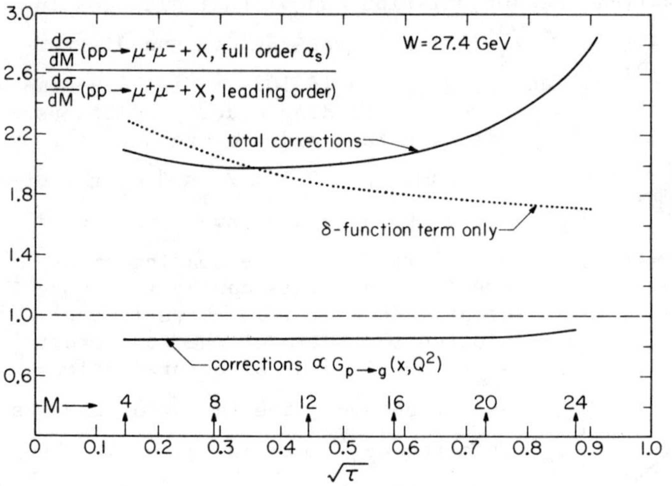

Fig. 7.10. Ratio of the complete order α_s results to the leading order QCD predictions for the total muon pair rate, $d\sigma/dM$, in $pp \to \mu^+\mu^- + X$ at W = 27.4 GeV. The contributions from the δ-function term in Δf_q^{DY} and the contributions from the gluon term Δf_g^{DY} are shown separately. The results are plotted versus both the muon pair mass, M, and $\sqrt{\tau}$, where $\tau = M^2/s$.

to all orders in α_s then the total muon pair cross section, $d\sigma_{DY}/dM^2$, cannot be determined from the electroproduction distributions until $\alpha_s(Q^2=M^2) \approx 1/10$ or $M^2 \approx 10^5$ GeV^2!

In addition, the $(\log(1-z)/(1-z))_+$ term in (6.106b) causes the order α_s results to deviate more and more from the leading order as $\sqrt{\tau} \to 1$ (see Fig. 7.10). This term results in

$$\frac{G^{DY}(x,Q^2)}{G(x,Q^2)} \underset{x\to 1}{\approx} 1 + \frac{\alpha_s(Q^2)}{2\pi}\left(\frac{4}{3}\right)\log^2(1-x) \tag{7.17}$$

and hence the order $\alpha_s(Q^2)$ corrections become arbitrarily large as $x \to 1$ as seen in Fig. 7.11. No matter how small $\alpha_s(Q^2)$ is, the order α_s term can be made to dominate over the leading term by taking x close to one. This is not quite as disturbing as at first it appears. It simply reflects the fact that $G^{DY}(x,Q^2)$ has a different shape as $x \to 1$ than does $G(x,Q^2)$. The former does not approach zero as fast as the latter. If one views the order α_s corrections as a slight change in the shape as $x \to 1$, then maybe the perturbation result is meaningful in spite of the large ratio in (7.17).

In fact, one can view the order α_s corrections in a different,

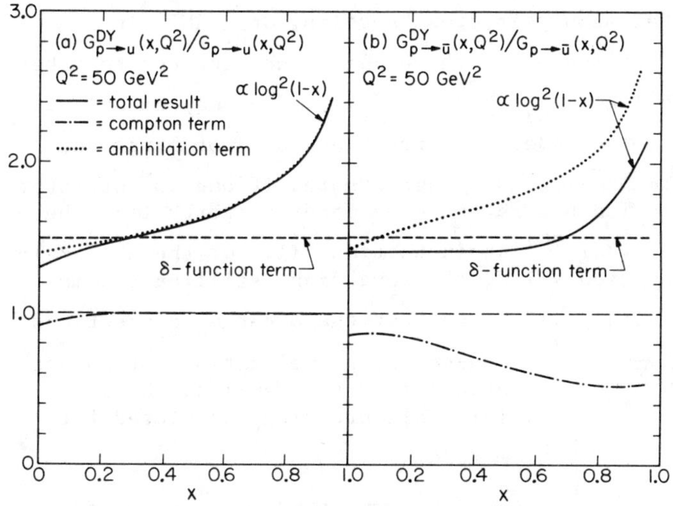

Fig. 7.11. Ratio of $G^{DY}_{p\to u}(x,Q^2)/G_{p\to u}(x,Q^2)$ and $G^{DY}_{p\to \bar{u}}(x,Q^2)/G_{p\to \bar{u}}(x,Q^2)$ versus x at $Q^2 = 50$ GeV^2. The functions $G^{DY}(x,Q^2)$ are the full order α_s distributions defined by (6.106a) for the total muon pair rate given by (6.104) and $G(x,Q^2)$ are the leading order QCD parton distributions defined for electroproduction by (6.62). The contributions from the annihilation term, Δf^{DY}_q, in (6.106b) and the Compton term, Δf^{DY}_g, in (6.106c) are shown separately. Also shown is the contribution from the δ-function term alone in Δf^{DY}_q. The functions $G^{DY}_{p\to u}$ and $G^{DY}_{p\to \bar{u}}$ are not separate observables and only have real significance when convoluted together to produce the total rate, $d\sigma_{DY}/dM^2$, in (6.104).

and somewhat less disturbing, manner. Instead of plotting the ratio of the corrected to the uncorrected predictions as I have done in Fig. 7.10 and Fig. 7.11, one might ask how much one must change M in Fig. 7.9 to get from the leading order to the order α_s results. Since $d\sigma_{DY}/dM$ is a steeply falling function, the change in M is not great, only about 7% at M = 12 GeV. Thus we can predict the rate $d\sigma_{DY}/dM$ but with about a 7 - 10% uncertainty as to the precise value of M at which the rate is actually attained. Viewed in this way, the results of this section don't seem as disturbing. However, until we really understand the physics behind these large order α_s corrections and perhaps learn to sum them to all orders or find that the order α_s^2 corrections are small, I feel we cannot predict $d\sigma_{DY}/dM$ from the parton distributions determined in electroproduction!

I should say, however, that the $G^{DY}(x,Q^2)$ functions in (6.106) refer only to the total muon pair cross section, $d\sigma_{DY}/dM^2$, in (6.104). I have not yet looked at the order α_s corrections to other muon pair observables like $d\sigma_{DY}/dM^2dy$ in (7.1). One may be able to find observables where the order α_s corrections are not large and hence make an accurate prediction. Furthermore, if one is not interested in the normalization but rather some other property like the $p_\perp$ spectrum of the muon pair[56] as in Section VII.B or the M^2 "scaling" behavior as in Section VII.C, then one might still have some predictive power[57]. Fig. 7.12 shows that the order α_s corrections change the normalization but do not affect much the "scaling" behavior (or scale breaking) predicted for $M^3 d\sigma/dM$.

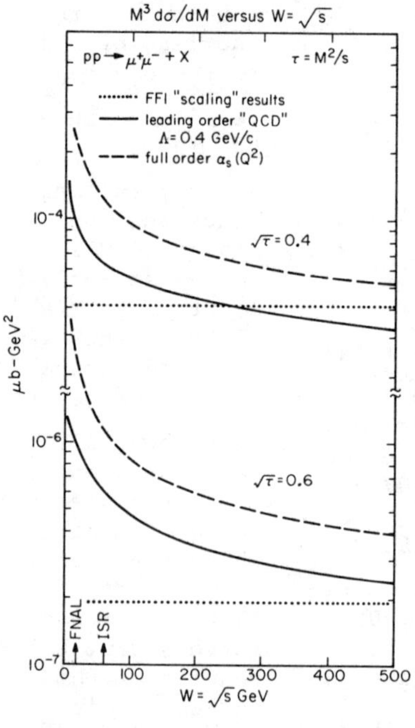

Fig. 7.12. Scaling behavior of $M^3 d\sigma/dM$ for large mass muon pair production in $pp \to \mu^+\mu^- + X$ at $\sqrt{\tau} = 0.4$ and 0.6 versus $W = \sqrt{s}$. The scaling results of the "naive" parton model[26], the leading order QCD predictions, and the complete order α_s calculations are compared.

F. <u>The Production of $W^\pm$ Bosons in Hadron-Hadron Collisions</u>[58,59]

Before leaving this section, I cannot resist predicting the total cross section for producing $W^\pm$ bosons in hadron-hadron collisions. To leading order, the cross section for $A + B \to W^\pm + X$ is given by

$$\sigma_{W^\pm}(s) = \sqrt{2}\pi G\tau \int_\tau^1 \frac{dx}{x} F^\pm_{AB}(x,\tau/x), \tag{7.18a}$$

where G is the weak coupling constant ($G \approx 10^{-5}/M_p^2$) and $\tau = M_W^2/s$ and where

$$F^{+}_{AB}(x_a,x_b) = \frac{1}{3}\{[G_{A\to u}(x_a,Q^2)G_{B\to\bar{d}}(x_b,Q^2)$$

$$+ G_{A\to\bar{d}}(x_a,Q^2)G_{B\to u}(x_b,Q^2)]\cos^2\theta_c$$

$$+ [G_{A\to u}(x_a,Q^2)G_{B\to\bar{s}}(x_b,Q^2)$$

$$+ G_{A\to\bar{s}}(x_a,Q^2)G_{B\to u}(x_b,Q^2)]\sin^2\theta_c \qquad (7.18b)$$

$$+ [G_{A\to c}(x_a,Q^2)G_{B\to\bar{s}}(x_b,Q^2)$$

$$+ G_{A\to\bar{s}}(x_a,Q^2)G_{B\to c}(x_b,Q^2)]\cos^2\theta_c$$

$$+ [G_{A\to c}(x_a,Q^2)G_{B\to\bar{d}}(x_b,Q^2)$$

$$+ G_{A\to\bar{d}}(x_a,Q^2)G_{B\to c}(x_b,Q^2)]\sin^2\theta_c\},$$

where θ_c is the Cabibbo angle and $Q^2 = M_W^2$ with

$$M_W^2 = \pi\alpha/(\sqrt{2}G\sin^2\theta_W) \approx (37.3\text{ GeV})^2/\sin^2\theta_W \qquad (7.18c)$$

and where I have taken the Weinberg angle to be

$$\sin^2\theta_W = 0.23. \qquad (7.18d)$$

The expression for $F^{-}_{AB}(x_a,x_b)$ is similar and I won't bother to write it down.

The leading order QCD predictions are given in Fig. 7.13 and Fig. 7.14 for $pp \to W^{\pm} + X$ and $\bar{p}p \to W^{\pm} + X$ together with the "naive" parton model results (using the scaling distributions of FF1[26]). The QCD distributions have changed significantly in going from low Q^2 to $Q^2 = M_W^2 \approx (77.8\text{ GeV})^2$. The number of quarks at high x has decreased and the number at low x has increased. This results in a leading order QCD prediction that is lower than the naive parton model predictions at low s/M_W^2 and larger at high s/M_W^2.

As for the muon pair cross section, these predictions cannot be

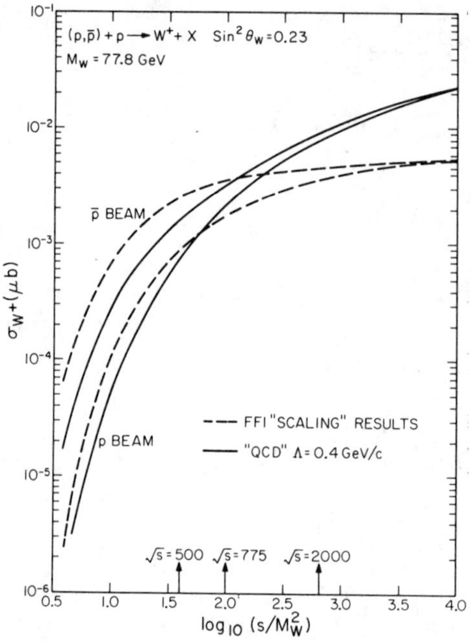

Fig. 7.13. Predictions for the total cross section for $pp \to W^+ + X$ and $\bar{p}p \to W^+ + X$ versus $\log_{10}(s/M_W^2)$ and where $M_W = 77.8$ GeV. The dashed and solid curves are the "naive" parton model predictions[26] and the leading order QCD results, respectively.

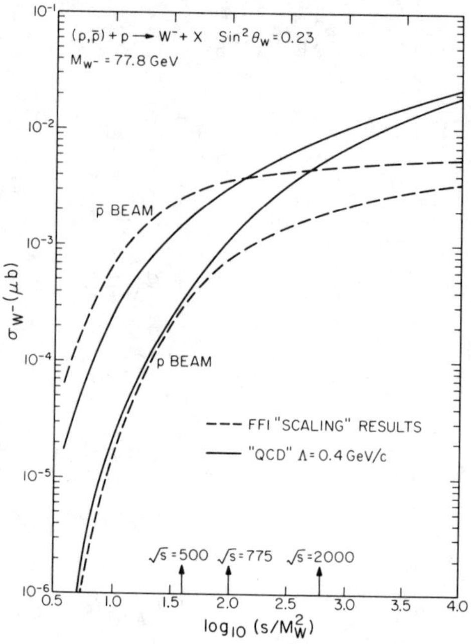

Fig. 7.14. Same as Fig. 7.13 but for $pp \to W^- + X$ and $\bar{p}p \to W^- + X$.

trusted unless the higher order corrections are small. The order α_s results for $W^\pm$ production are obtained by the use of (7.18) but with $G(x,Q^2)$ replaced by $G^{DY}(x,Q^2)$ defined by (6.106) and dropping α_s^2 terms. The ratio of the order α_s results to the leading order predictions for $pp \to W^+ + X$ is shown in Fig. 7.15. Here $\alpha_s(Q^2=M_W^2)$ is a bit smaller than in Fig. 7.10, but the order α_s corrections are still quite large. At $\sqrt{s} = 500$ GeV, they increase the leading order result by about 50%. One is tempted to say that the result is meaningful and that probably α_s^2 corrections are still smaller. However,

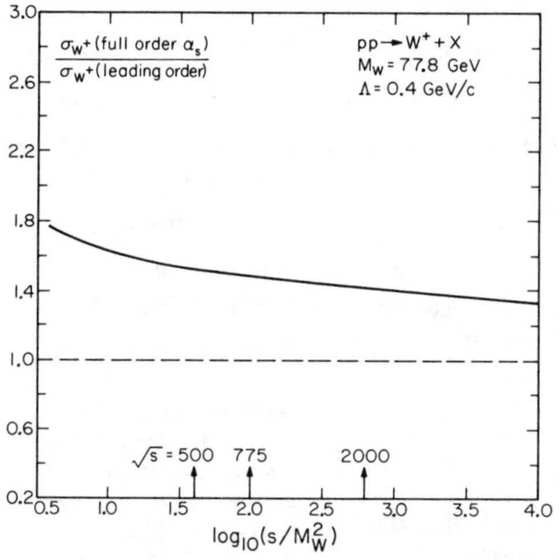

Fig. 7.15. Ratio of the full order α_s predictions to the "leading order" QCD results for the total pp → W^+ + X cross section.

a perturbation series of the form 1 + 0.5 is still quite dangerous and so the predictions in Fig. 7.13 and Fig. 7.14 must be viewed with caution[60].

VIII. QUARK AND GLUON "JETS": FRAGMENTATION FUNCTIONS

Gluon radiation from an outgoing quark results in scale breaking of the quark fragmentation function $D_i^h(z)$. As in (4.8), the probability of radiating a hard gluon goes as $\alpha_s(Q^2)\log^2(Q^2/\Delta^2)$ where Δ is, say, the invariant mass of the resulting quark-gluon pair. If Δ is large enough then this quark and gluon will appear as two distinct jets (each of which may radiate hard gluons and split again). If Δ is small then it will merely "appear" as if the original quark jet is becoming fatter as Q^2 increases. If one chooses to describe this in terms of a single fragmentation function of the original quark, $D(z,Q^2)$, (which may not be the best way to handle this), then clearly as Q^2 increases, the probability of finding a large z hadron decreases. As Q^2 increases, the momentum of the original quark is shared among the radiated gluon jets[61].

One can analyze the Q^2 dependence of the fragmentation functions, $D_i^h(z,Q^2)$, by the methods developed by Altarelli and Parisi[13] for the quark distributions[61-63]. One defines "non-singlet" functions for partons to fragment into a specific hadron h as follows

$$D_{NS}^h(z,Q^2) = D_{q_i}^h(z,Q^2) - D_{\bar{q}_i}^h(z,Q^2). \tag{8.1}$$

The change of these functions w.r.t. τ is then given in an analogous manner to (4.14). Namely,

$$\frac{dD^h_{NS}(z,\tau)}{d\tau} = \frac{\alpha(\tau)}{2\pi}\int_z^1 \frac{dy}{y} P_{q\leftarrow q}(y) D^h_{NS}(\frac{z}{y},\tau). \tag{8.2}$$

As illustrated in Fig. 8.1, in general, there is mixing of the gluon fragmentation functions $D^h_g(z,\tau)$ and the quark fragmentation functions $D^h_{q_i}(z,\tau)$, since, for example, an observed hadron h could come from an outgoing quark or from a hard gluon that was radiated by the initial quark. Thus, in general,

$$\frac{dD^h_{q_i}(z,\tau)}{d\tau} = \frac{\alpha(\tau)}{2\pi}\int_z^1 \frac{dy}{y} [P_{q\leftarrow q}(y) D^h_{q_i}(\frac{z}{y},\tau) + P_{g\leftarrow q}(y) D^h_g(\frac{z}{y},\tau)], \tag{8.3a}$$

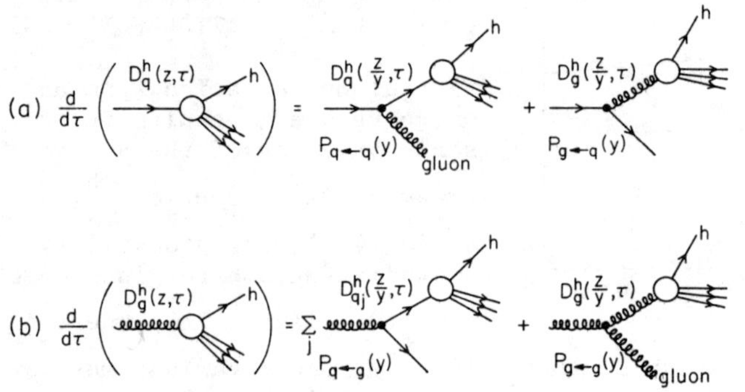

Fig. 8.1 - (a) Illustrates that the change of the quark fragmentation function, $D^h_q(z,\tau)$, w.r.t. $\tau = \log(Q^2/Q_o^2)$ is due to the Bremsstrahlung radiation of a gluon, $P_{q\leftarrow q}(y)$, and to the production of a gluon, $P_{g\leftarrow q}(y)$, which fragments into the observed hadrons.

(b) Illustrates that the change of the gluon fragmentation function, $D^h_g(z,\tau)$, w.r.t. τ is due to the production of a quark-antiquark pair, $P_{q\leftarrow g}(y)$, and to the Bremsstrahlung radiation of a gluon, $P_{g\leftarrow g}(y)$. This figure can be compared to Fig. 4.2 which illustrates the causes for the Q^2 change of the quark and gluon distributions. To leading order, the functions $P_{a\leftarrow b}(z)$ are the same.

$$\frac{dD_g^h(z,\tau)}{d\tau} = \frac{\alpha(\tau)}{2\pi}\int_z^1 \frac{dy}{y}\,[\sum_{j=1}^{2n_f} P_{q_j\leftarrow g}(y)D_{q_j}^h(\frac{z}{y},\tau)$$

$$+ P_{g\leftarrow g}(y)D_g^h(\frac{z}{y},\tau)], \tag{8.3b}$$

where the probabilities $P_{a\leftarrow b}(z)$ are the same as in Fig. 4.2 and given in (4.22) and (4.23). The moments of the fragmentation function for a constituent of type i to a given hadron, h, given by

$$\bar{M}_i^h(n,Q^2) = \int_0^1 z^{n-1}D_i^h(z,Q^2)dz \tag{8.4a}$$

are given in terms of the moments at some reference momentum Q_0 by an equation similar to (4.40b). Namely,

$$\bar{M}_j^h(n,Q^2) = \sum_{i=1}^{9} \bar{M}_i^h(n,Q_0^2)\,\bar{R}_{ij}(n,Q^2,Q^2,\Lambda^2), \tag{8.4b}$$

where the matrix $\bar{R}_{ij}$ is simply related to R_{ij}.

Care must be taken in choosing $D_i^h(z,Q_0^2)$ since one must use in (8.3) the distribution of "primary" mesons before decay. In addition, one must guess at the distribution of hadrons in a gluon jet, $D_g^h(z,Q_0^2)$. Figure 8.2 shows the resulting Q^2 dependence of $D_u^{\pi^o}(z,Q^2)$ and $D_g^{\pi^o}(z,Q^2)$ for the particular reference momentum choices discussed in Refs. 25 and 64. The distribution of charged hadrons at Q_0^2 is adjusted to fit the data shown in Fig. 8.3. The gluon fragmentation function at Q_0^2 has been chosen to be steeper than the quark fragmentation function. That is, it is assumed that gluons fragment into fewer high z hadrons and are assumed to have a higher rapidity plateau than do quarks[61,65]. Notice that although $D_u^{\pi^o}(z,Q^2)$ decreases at large z as Q^2 increases, the amount of "scale breaking" is not predicted to be as great as for $\nu W_2(x,Q^2)$. This is because the shape of the z distribution of $D_u^{primary}(z,Q_0^2)$ at Q_0^2 is not as steep at large z as the x distribution of $\nu W_2(x,Q_0^2)$ is at large x.

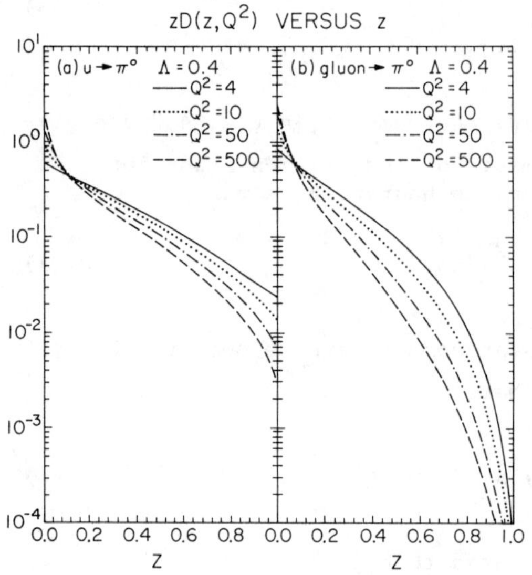

Fig. 8.2 - (a) The Q^2 dependence of the fragmentation function for a u-quark to a π^o, $D_u^{\pi^o}(z,Q^2)$, expected from QCD. The distributions at high Q^2 are calculated from the distribution at the reference momentum Q_o^2 = 4 GeV^2 using Λ = 0.4 GeV/c, where $D_q^h(z,Q_o^2)$ is taken from the analysis in Ref. 64.
(b) Same as (a) but for the gluon fragmentation function $D_g^{\pi^o}(z,Q^2)$.

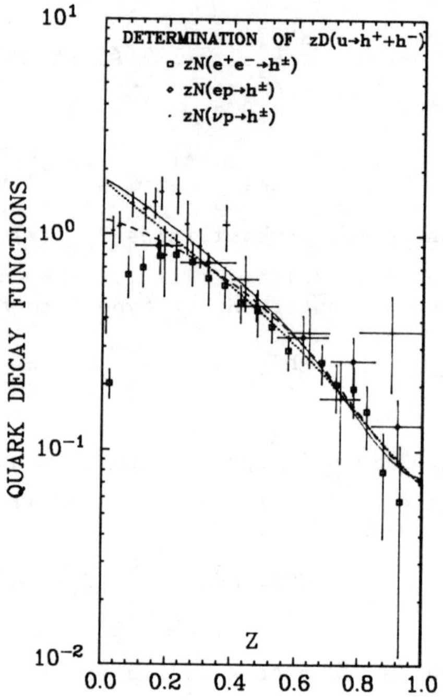

Fig. 8.3. Comparison of the charged particle distributions $zD_u^{h^+}(z,Q_o^2) + zD_u^{h^-}(z,Q_o^2)$ at $Q_o^2 = 4\ GeV^2$ with data from $e^+e^- \to h^\pm + X$, $ep \to h^\pm + X$ and $\nu p \to h^\pm + X$. (The data used are described in Fig. 3 of Ref. 64).

IX. LARGE $p_\perp$ MESON AND "JET" PRODUCTION IN HADRON-HADRON COLLISIONS

A. The QCD Approach

In the naive parton model, the large $p_\perp$ production of hadrons in the process $A + B \to h_1 + h_2 + X$ is described by the diagram in

Fig. 9.1b. The large-transverse-momentum reaction is assumed to occur as the result of a single large-angle scattering $a + b \to c + d$

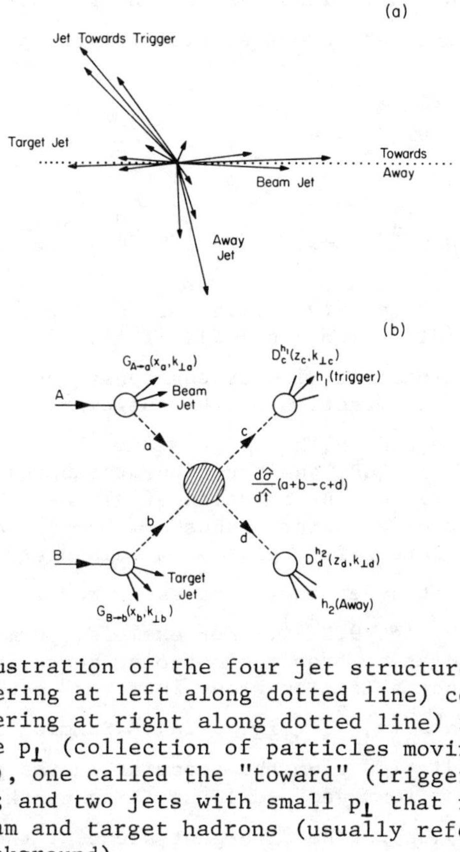

Fig. 9.1 - (a) Illustration of the four jet structure resulting from a beam hadron (entering at left along dotted line) colliding with a target hadron (entering at right along dotted line) in the CM frame: two jets with large $p_\perp$ (collection of particles moving roughly in the same direction), one called the "toward" (trigger) side and one on the "away" side; and two jets with small $p_\perp$ that result from the break up of the beam and target hadrons (usually referred to as the "soft hadronic" background).

(b) Illustration of the underlying structure of the large $p_\perp$ process $A + B \to h_1 + h_2 + X$. The large $p_\perp$ trigger hadron h_1 occurs as the result of a large angle scattering of constituents ($q_a + q_b \to q_c + q_d$), followed by the decay of fragmentation of constituent c into a towards side jet of hadrons (one being the trigger h_1) and constituent d into an away side jet of hadrons (one being h_2). The quantities x_a, x_b, $k_{\perp a}$, $k_{\perp b}$ are the longitudinal fraction of the incoming hadrons A, B momentum and perpendicular momentum of constituents a, b and z_c, z_d, $k_{\perp c}$, $k_{\perp d}$ are the fraction of the outgoing constituents longitudinal momentum and perpendicular momentum carried by the detected hadrons h_1 and h_2.

of constituents a and b, followed by the decay or fragmentation of constituent c into the trigger hadron h_1 and constituent d into the "away-side" hadrons, h_2. This results in the four jet structure in Fig. 9.1a. The invariant cross section for the process $A + B \to h + X$ is given by

$$E \frac{d\sigma}{d^3p} (s, p_\perp, \theta_{cm}) = \int dx_a \int dx_b G_{A\to a}(x_a) G_{B\to b}(x_b) D_c^h(z_c) \frac{1}{z_c} \frac{1}{\pi} \frac{d\hat{\sigma}}{d\hat{t}} (\hat{s}, \hat{t}), \tag{9.1}$$

where $\hat{s}$, $\hat{t}$ are the usual invariants but for the constituent two-to-two subprocess $d\hat{\sigma}/d\hat{t}$ $(a + b \to c + d)$. The quantities x_a and x_b are the fractional momentum carried by the constituents a and b, respectively, and z_c is the fraction of the outgoing constituent momentum that appears in the hadron, h.

In the theory of QCD, the constituent subprocess, $a + b \to c + d$, must be corrected for the emission of gluons. That is one must include the higher order subprocesses $a + b \to c + d + g$, $a + b \to c + d + g + g$, etc., where g is a gluon. As discussed by C. Sachrajda[66], to leading order these processes modify (9.1) in the manner illustrated in Fig. 9.2[67]. For example, summing all the (divergent) gluon radiation from the outgoing quark in $q + q \to q + q$ generates the "renormalization group improved" scale breaking fragmentation function $D_q^h(z,Q^2)$ (Fig. 9.2a). Summing over all the (divergent) gluon radiation from the incoming quarks generates the "renormalization group improved" scale breaking quark distributions $G_{p\to q}(x,Q^2)$ (Fig. 9.2b). The divergent piece of the diagram

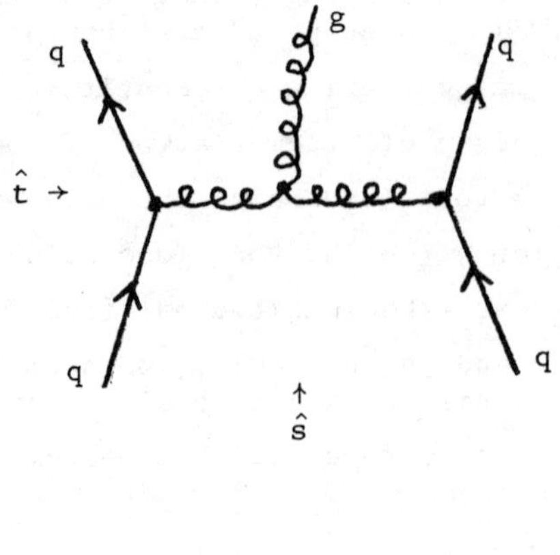

(9.2)

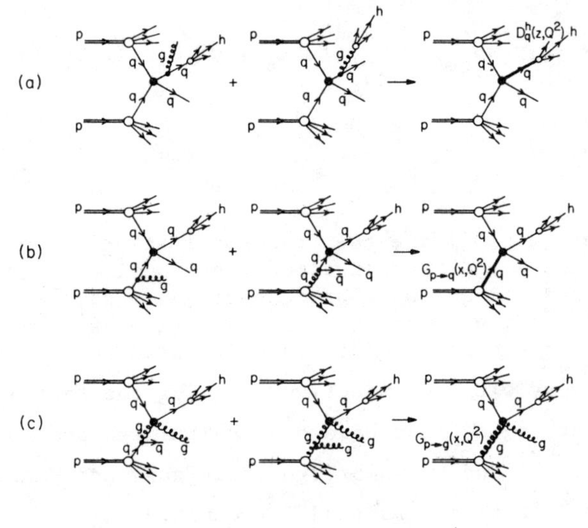

Fig. 9.2 - (a) Illustration of two diagrams for $pp \to h + X$ which are included (divergent parts) in the "renormalization group" improved quark fragmentation function, $D_q^h(z,Q^2)$, when one calculates the contribution from the subprocess $qq \to qq$.

(b) Illustration of two diagrams for $pp \to h + X$ which are included (divergent parts) in the "renormalization group" improved quark distribution, $G_{p\to q}(x,Q^2)$, when one calculates the contribution from the subprocess $qq \to qq$.

(c) Illustration of two diagrams for $pp \to h + X$ which are included (divergent parts) in the "renormalization group" improved gluon distribution, $G_{p\to g}(x,Q^2)$, when one calculates the contribution from the subprocess $qg \to qg$.

is contained in the "renormalization improved" gluon distribution $G_{p\to g}(x,Q^2)$ (Fig. 9.2c), where the constituent subprocess is now $g + q \to g + q$. Other higher order corrections to the gluon propagators and to the vertices generate the "renormalized" coupling $\alpha_s(Q^2)$ as discussed in Section II. Thus to leading order in QCD, eq. (9.1) becomes

$$E \frac{d\sigma}{d^3p}(s,p_\perp,\theta_{cm}) = \sum_{a,b} \int dx_a \int dx_b G_{A\to a}(x_a,Q^2)$$

$$G_{B\to b}(x_b,Q^2) D_c^h(z_c,Q^2) \frac{1}{z_c} \frac{1}{\pi} \frac{d\hat{\sigma}}{d\hat{t}}(\hat{s},\hat{t}), \tag{9.3}$$

where $G(x,Q^2)$ and $D(z,Q^2)$ are the "renormalization group improved" parton distributions and fragmentation functions, respectively, and where one includes all seven subprocesses: $qq \to qq$, $\bar{q}q \to \bar{q}q$, $\bar{q}\bar{q} \to \bar{q}\bar{q}$, $gq \to gq$, $g\bar{q} \to g\bar{q}$, $gg \to \bar{q}q$, $\bar{q}q \to gg$, and $gg \to gg$. Each $2 \to 2$ differential cross section, $d\hat{\sigma}/d\hat{t}$, is calculated to lowest order in perturbation theory with an effective coupling constant $\alpha_s(Q^2)$ as in (2.12). These cross sections have been calculated previously by Cutler and Sivers[68] and by Combridge, Kripfganz and Ranft[69] and all behave as $\hat{s}^{-2}$ at fixed $\hat{t}/\hat{s}$ (and for constant α_s).

For ep collisions, Q is the 4-momentum transfer from the electron to the quark and for $pp \to \mu^+\mu^- + X$, it is the mass of the muon pair. On the other hand, the correct kinematic quantity to use for Q^2 in the constituent subprocesses contributing to large $p_\perp$ meson production in pp collisions is not at present known. (As I pointed out previously, changing the definition of Q^2 is a higher than leading order effect.)

For our analyses of high $p_\perp$[23-25], we have taken for definiteness

$$Q^2 = 2\hat{s}\hat{t}\hat{u}/(\hat{s}^2 + \hat{t}^2 + \hat{u}^2), \tag{9.4}$$

where $\hat{s}$, $\hat{t}$ and $\hat{u}$ are the usual Mandelstam invariants but for the constituent subprocesses. This arbitrariness in the form for Q^2 makes predictions at low Q^2 (i.e., low $p_\perp$) in hadron-hadron collisions uncertain.

B. Smearing

There is considerable experimental evidence that the constituents inside the proton have a large effective internal transverse momentum[20,70,71]. Effects due to the transverse momentum of quarks within hadrons, $(k_\perp)_{h\to q}$, and of hadrons within the outgoing jets, $(k_\perp)_{q\to h}$, called "smearing" effects are particularly important for large $p_\perp$ calculations. This is due to the "trigger bias" which selects the configuration in which the initial quarks (or gluons) are already moving toward the trigger. As for the muon pairs, in QCD, this transverse momentum of the partons can arise from two sources (illustrated in Fig. 9.3).

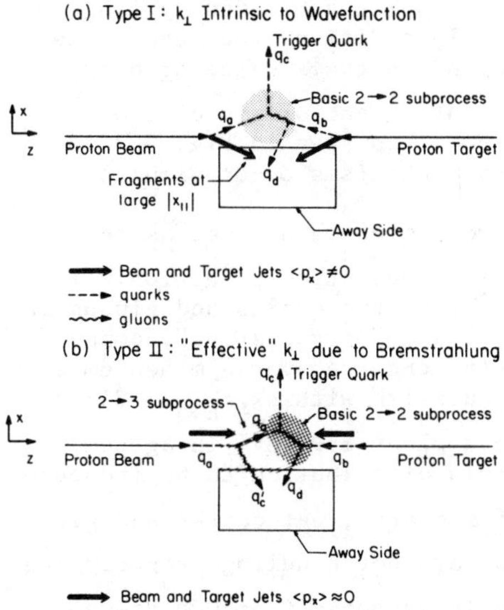

Fig. 9.3 - (a) Illustration of the non-perturbative ("primordial") component of the transverse momentum of quarks within proton that is intrinsic to the wave function of the proton. One expects this transverse momentum to be balanced by the remaining constituents in the proton which can, in turn, fragment into particles at high $x_\parallel$. The away-side consists of the recoiling quark, q_d, and two slightly shifted jets, one from the beam and one from the target.

(b) Illustration of a perturbative component to the transverse momentum of a quark with a hadron which is due to the Bremsstrahlung of a gluon before the basic $2 \to 2$ scattering occurs. In this case, the trigger quark is balanced by two away-side jets, one from the quark q_d and one from the radiated gluon q_c'.

Firstly, in a proton beam, quarks are confined in the transverse direction to within the proton radius. Therefore, from the uncertainty principle, they must have some transverse momentum. This "primordial" momentum is intrinsic to the basic parton "wave function" inside the proton. As illustrated in Fig. 9.3a, one might expect the wave function to have a term where the trigger parton $k_\perp$ is balanced by another constituent (or constituents) which has the opposite $k_\perp$ and most of the remaining longitudinal momentum. Consider now the plane formed by the beam, target and a 90° trigger hadron (called the x-z plane in Fig. 9.3). Typically, the trigger arises from the fragmentation of a constituent with $k_{\perp x} > 0$ which is balanced by the remaining constituents having $k_{\perp x} < 0$. One expects to see this negative $k_{\perp x}$ as a shift in the beam and target jets at large $|x_\parallel|$. This shift (i.e., nonzero $\langle k_{\perp x}\rangle$) of the beam jet as one increases the $p_\perp$ of a 90° trigger has recently been observed by the BFS group at ISR[72] (see Fig. 7.5 of Ref. 24).

Secondly, in QCD, one expects to receive an "effective" $k_\perp$ of quarks in protons due to the Bremsstrahlung of gluons. This perturbative term, which arises from diagrams like those in Fig. 9.2, is illustrated in Fig. 9.3b.

It corresponds to including two particle to three or more particle processes ($2 \to 3$) rather than just the two particle to two particle $2 \to 2$ scatterings. For such subprocesses, the $k_\perp$ of the quark q_a is balanced by a gluon jet on the away-side which subsequently fragments into many low momentum hadrons. In addition, the mean value of the effective $k_\perp$ is expected to depend on the x value of quark q_a and on the Q^2 for the processes. Separating the origin of the transverse momenta into Type I and Type II as seen in Fig. 9.3 is, of course, a bit artificial since both mechanisms occur simultaneously.

The analysis of large $p_\perp$ meson production is not yet as complete as the discussion of $pp \to \mu^+\mu^- + X$ presented in Section VII. We have not separated the "effective" $k_\perp$ of the quarks and gluons in the initial hadrons in the "primordial" and "perturbative" components. For the present, we parameterize the transverse momentum of the partons (quarks and gluons) by a Gaussian with $\langle k_\perp \rangle_{h\to q} = 848$ MeV which produces for $pp \to \mu^+\mu^- + X$ a mean $p_\perp^2$ of 1.9 GeV^2 in agreement with the data in Fig. 7.5. We take this distribution to be independent of x and Q^2 and to be the same for quarks, antiquarks and gluons in the proton[73]. In so doing, we are <u>not</u> handling properly the x and Q^2 dependence of the high $k_\perp$ tails expected from QCD Bremsstrahlung. The next step would, of course, be to calculate and include explicitly the $2 \to 3$ subprocess expected by QCD (like $qq \to qqg$, etc.) and smear the results with the "primordial" $k_\perp$ only (which is presumably smaller than the effective 848 MeV we now use). Care would have to be taken since one would include only the non-divergent parts of the $2 \to 3$ subprocesses. The divergent parts have already been included by the use of the "renormalization improved" distributions $G(x,Q^2)$ and fragmentation functions $D(z,Q^2)$. For the present, however, we merely use the data in Fig. 7.5 to give an "effective" $k_\perp$ distribution and include explicitly only $2 \to 2$ subprocesses.

It is clear from examining data that smearing effects are present in nature. However, no one yet really knows how to handle the primordial $\langle k_\perp \rangle$ properly theoretically[75]. Different models for smearing give quite differing results[76-79]. Surely our crude way of handling it is not precisely correct. Present day calculations are uncertain in the low $p_\perp$ ($p_\perp \leqslant 4$ GeV/c) region where these effects are important.

The emission of gluons after the hard scattering ($2 \to 2$) subprocesses induces an "effective" $k_\perp$ of the hadrons that fragment from the outgoing quarks because one is sometimes really seeing two jets rather than one. As for the quark distributions in the proton, we do not include these effects (we also neglect the interferences that arise between the amplitude for emitting gluon before and after the hard 2 → 2 process) and for the present take the transverse momentum distribution of hadrons from outgoing quarks (and gluons) to be a Gaussian with $\langle k_\perp \rangle_{q\to h}$ = 439 MeV independent of z or Q^2. Again, this is not precisely correct and should be improved upon in later work.

C. The Single Particle Cross Section

In the naive parton model, the quark distribution and fragmentation function in (9.1) scale. In this case, the invariant cross section for producing large $p_\perp$ mesons directly reflects the energy dependence of the constituent cross section, $d\hat{\sigma}/d\hat{t}$. If this latter cross section behaves as

$$d\hat{\sigma}/d\hat{t} = h(\hat{t}/\hat{s})/\hat{s}^n, \tag{9.5a}$$

then the former behaves as

$$E d\sigma/d^3p = f(x_\perp, \theta_{cm})/p_\perp^N, \tag{9.5b}$$

where $N = 2n$, $x_\perp = 2p_\perp/W$ and $W = \sqrt{s}$.

The experimental determination of the effective $p_\perp$ power index N in (9.5b) involves the comparison of data at different energies W and different $p_\perp$ but at the same ratio $x_\perp = 2p_\perp/W$ and the same angle θ_{cm}. For example, Fig. 9.4 illustrates how one determines the $p_\perp$ dependence at $x_\perp = 0.2$. Data at the lowest Fermilab energy W = 19.4 GeV and $p_\perp$ = 1.94 GeV/c are compared with the ISR energy W = 53 GeV at $p_\perp$ = 5.3 GeV/c. As can be seen in the lower part of this figure, where I have multiplied the cross section by $p_\perp^8$, the W = 53 GeV, $p_\perp$ = 5.3 GeV/c point is about a factor of $(1.94/5.3)^8$ times the W = 19.4 GeV, $p_\perp$ = 1.94 GeV/c point. By combining these two points with points from other energies, one finds that the data at $x_\perp = 0.2$ behave roughly like $p_\perp^{-8}$ over the range $2 \leqslant p_\perp \leqslant 6$ GeV/c and $\theta_{cm} = 90°$. Thus the naive expectation from field theory that $n \approx 2$ is not seen in the data.

However, one cannot use dimensional counting in QCD. There is an intrinsic mass scale Λ in (2.20) that is generated by the interaction of quarks and gluons. Figure 9.5 shows the behavior of the single particle invariant cross section, $pp \to \pi^0 + X$, arising from the subprocess $qq \to qq$. The cross section, $E d\sigma/d^3p$, is plotted at $\theta_{cm} = 90°$ and $x_\perp$ = 0.2 and 0.4 versus $p_\perp$ and is multiplied by $p_\perp^4$ which would be independent of $p_\perp$ if "scaling" were valid (dotted curves). The dot-dashed and dot-dot-dashed curves illustrate the

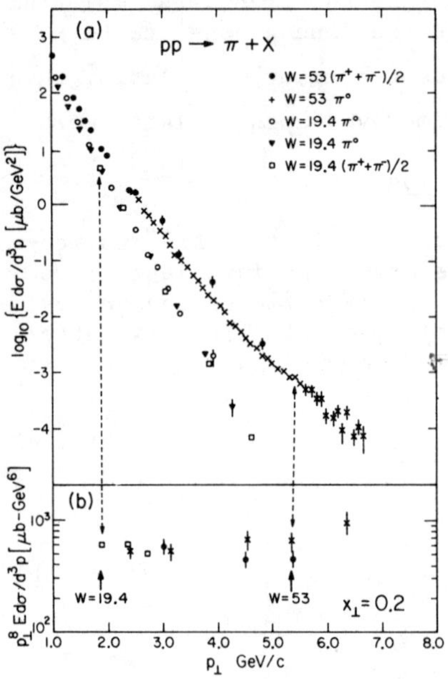

Fig. 9.4 - (a) The invariant cross section for pp → π + X at $W = \sqrt{s} = 19.4$ and 53 GeV/c with $\theta_{cm} = 90°$ (open squares = Ref. 92, solid dots = Ref. 93, crosses = Ref. 94, solid triangles = Ref. 95, open circles = Ref. 96).

(b) Plot of the invariant cross section for pp → π + X at $x_{\perp} = 0.2$ and $\theta_{cm} = 90°$ times $p_{\perp}^8$ versus $p_{\perp}$.

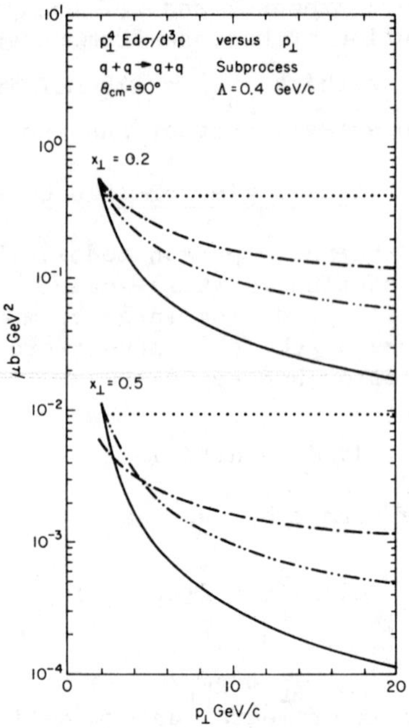

Fig. 9.5. Behavior of the quantity $p_{\perp}^4\, Ed\sigma/d^3p$ for $pp \to \pi^0 + X$ at $\theta_{cm} = 90°$ and $x_{\perp} = 0.2$ and 0.5 arising from the QCD subprocess $q + q \to q + q$. The dot-dashed and dot-dot-dashed curves result when one includes a running coupling constant $\alpha_s(Q^2)$ and when one includes both $\alpha_s(Q^2)$ plus the non-scaling structure functions $G(x, Q^2)$, respectively. The solid curves arise from including a running coupling constant $\alpha_s(Q^2)$ plus $G(x,Q^2)$ and in addition include the scale breaking fragmentation functions $D(z,Q^2)$. The dotted curves indicate perfect "scaling."

effects of including the running coupling constant $\alpha_s(Q^2)$ and $\alpha_s(Q^2)$ plus the scale breaking quark distributions $G(x,Q^2)$, respectively.

The solid curves are the results after one includes a running coupling constant $\alpha_s(Q^2)$ plus scale breaking quark distributions $G(x,Q^2)$ plus scale breaking quark fragmentation functions $D(z,Q^2)$. The curves decrease with increasing $p_\perp$ eventually approaching a $p_\perp^{-4}$ behavior asymptotically at very large $p_\perp$.

Figure 9.6 shows the final results for $p_\perp^4\, Ed\sigma/d^3p$ for $pp \rightarrow \pi^0 + X$ at $\theta_{cm} = 90°$ and $x_\perp = 0.2$ where now all seven subprocesses discussed in Section IX.A are included. The dash-dot and solid curves are the results before and after the addition scale breaking due to the transverse momentum of the quarks and gluons within the initial protons (smearing) have been included with $\langle k_\perp \rangle_{h\rightarrow q} = 848$ MeV and $\Lambda = 0.4$ GeV. The dashed curve is the final result (after smearing) with $\Lambda = 0.6$ GeV and the dotted curve shows a $p_\perp^{-8}$ behavior. The final result exhibits a rough $p_\perp^{-8}$ behavior over the range $2 \leq p_\perp \leq 6$ with an approach to $p_\perp^{-4}$ at very high $p_\perp$ ($p_\perp \gtrsim 14$ GeV/c).

The QCD results for $p_\perp^4\, Ed\sigma/d^3p$ for $pp \rightarrow \pi^0 + X$ in Fig. 9.6 can be compared to the prediction for $W^5\, d\sigma/dMdyd^2p_\perp$ for $pp \rightarrow \mu^+\mu^- + X$ in Fig. 7.8. The nature of the expected scale breaking is similar except the "breaking" is somewhat larger in the $pp \rightarrow \pi^0 + X$ case. This is due to the additional "scale breaking" from the fragmentation functions and from a trigger bias effect which makes $pp \rightarrow \pi^0 + X$ more sensitive to transverse momentum effects. When one performs a high $p_\perp$ experiment, one sits at large $p_\perp$ and waits for an event. This biases one in favor of the configuration where the initial partons are already moving toward the trigger and so smearing makes a large effect in these experiments.

The data on $Ed\sigma/d^3p$ at <u>fixed</u> W = 19.4 and 53 GeV versus $p_\perp$ are compared with the theory in Fig. 9.7. The agreement is quite good. The results before smearing are shown by the dot-dashed curves. Smearing has little effect for $p_\perp \geq 4.0$ GeV/c at W = 53 GeV but has a sizable effect (even at $p_\perp = 6.0$ GeV/c) at W = 19.4 GeV due to the steepness of the cross section at this low energy. At $p_\perp = 2.0$ and W = 19.4 GeV, smearing increases the cross section by about a factor of 10. The contributions to the total invariant cross section from quark-quark elastic scattering (plus $q\bar{q} \rightarrow q\bar{q}$ and $\bar{q}\bar{q} \rightarrow \bar{q}\bar{q}$) are shown by the dotted curves. As noted by several authors, gluons make important contributions to the cross section at small $x_\perp$ ($x_\perp \leq 0.4$)[68,69].

The disagreement in the normalization of the theory seen in Figs. 9.6 and 9.7 at low $x_\perp$ (the $\Lambda = 0.4$ GeV/c solution is about a factor of 2 low at $p_\perp = 2$ GeV/c and W = 53 GeV) is not significant. The $\Lambda = 0.6$ GeV solution agrees better at low $p_\perp$ but the theory at present cannot be calculated precisely at these low $p_\perp$ values. At

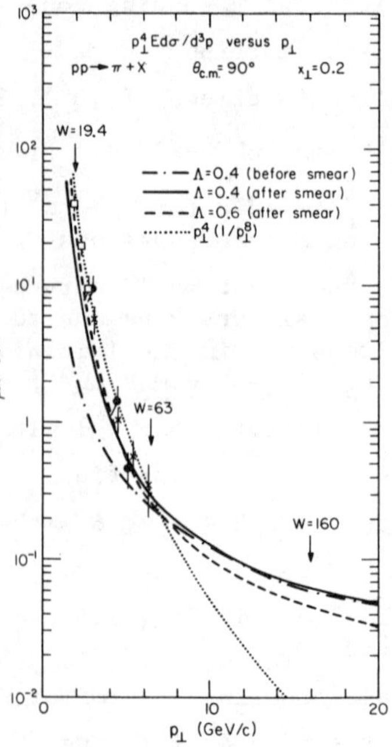

Fig. 9.6. The data on $p_\perp^4$ times $Ed\sigma/d^3p$ for large $p_\perp$ pion production at $\theta_{cm} = 90°$ and <u>fixed</u> $x_\perp = 0.2$ versus $p_\perp$ (open squares = Ref. 92, solid dots = Ref. 93, crosses = Ref. 94) compared with the predictions (with absolute normalization) of a model that incorporates all the features expected from QCD. The dot-dashed and solid curves are the results before and after smearing with $\langle k_\perp \rangle_{h\to q} = 848$ MeV, respectively, using $\Lambda = 0.4$ GeV/c and the dashed curves are the results using $\Lambda = 0.6$ GeV/c (after smearing). The dotted curve is $p_\perp^4(1/p_\perp^8)$.

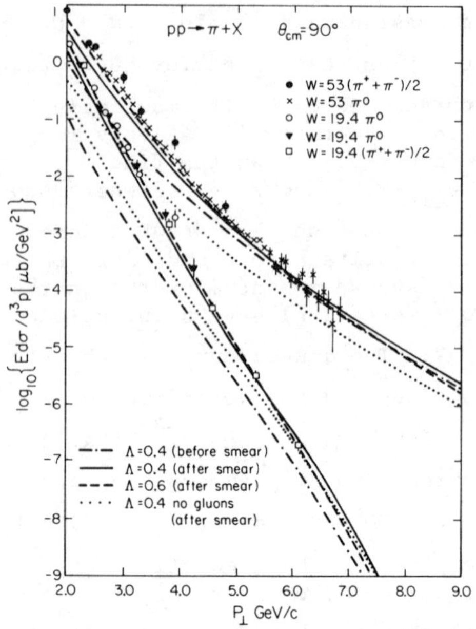

Fig. 9.7. Comparison of a QCD model (normalized absolutely) with data on large $p_\perp$ pion production in proton proton collisions at $W = \sqrt{s} = 19.4$ and 53 GeV/c with $\theta_{cm} = 90°$ (open squares = Ref. 92, solid dots = Ref. 93, crosses = Ref. 94, solid triangles = Ref. 95, open circles = Ref. 96). The dot-dashed and solid curves are the results before and after smearing, respectively, using $\Lambda = 0.4$ GeV/c and $\langle k_\perp \rangle_{h\to q} = 838$ MeV and the dashed curves for $\Lambda = 0.6$ GeV/c (after smearing). The contribution arising from quark-quark, quark-antiquark and antiquark-antiquark scattering (i.e., no gluons) is shown by the dotted curves (after smearing).

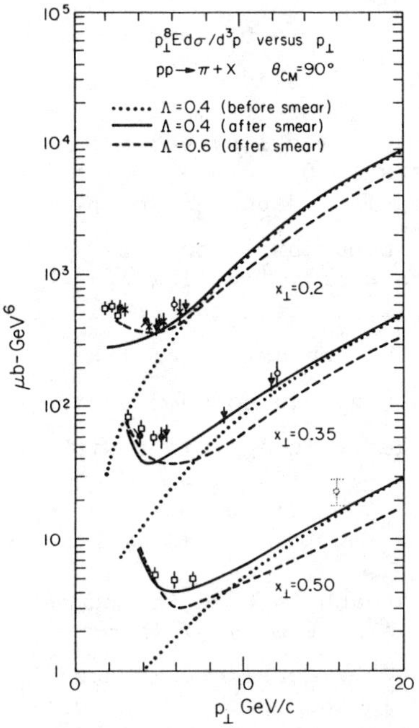

Fig. 9.8. The data on $p_\perp^8$ times $Ed\sigma/d^3p$ for large $p_\perp$ pion production at $\theta_{cm} = 90°$ and fixed $x_\perp = 0.2$, 0.35 and 0.5 versus $p_\perp$ (open squares = Ref. 92, solid dots = Ref. 93, crosses = Ref. 94) compared with the predictions (with absolute normalization) of a model that incorporates all the features expected from QCD. The dot-dashed and solid curves are the results before and after smearing, respectively, using $\Lambda = 0.4$ GeV/c and the dashed curves are the results using $\Lambda = 0.6$ GeV/c (after smearing). Recent data (triangles = Ref. 83, open circles = Ref. 82) from the ISR do show a deviation from a straight line ($1/p_\perp^8$) behavior as expected from QCD.

low $p_\perp$, the results depend too sensitivly on things like the unknown gluon distributions, the precise shape of the transverse momentum distributions, the low $\hat{s}$ and $\hat{t}$ cut-off employed[76-79], the choice for Q^2, and possibly higher order QCD corrections. (For $\alpha_s(Q^2) \gtrsim 0.3$ non-perturbative effects may begin to play a role.) It may well be that the scattering of quarks and gluons as described by the leading QCD subprocesses is responsible for all the cross section down to $p_\perp$'s as low as 1.0 or 2.0 GeV/c; one simply cannot say at present.

On the other hand, other non-leading subprocesses may play a role at low $p_\perp$ for single particle triggers particularly if these subprocesses produce single particles without being suppressed by a fragmentation function. One process that behaves like $1/p_\perp^6$ (before scale breaking) and must occur at some level is $q + g \to \pi + q$. Here one only needs to assume that quarks and gluons within the incident protons are important but estimates give a contribution from this subprocess that is only about 1/100 the size of the π cross section at $W = 53$ GeV, $\theta_{cm} = 90°$ and $p_\perp = 2.0$ GeV/c. Other CIM constituent subprocesses like, $\pi + q \to \pi + q$, require knowledge of the probability of finding a π within a proton, $G_{p\to\pi}(x)$, and are thus difficult to estimate. However, recent calculations by Blankenbecler, Brodsky and Gunion[80] and by D. Jones and J. F. Gunion[81] indicate that CIM terms may be important at low $p_\perp$ with the leading QCD processes dominating for $p_\perp \geq 5$ GeV/c.

Figure 9.8 shows a comparison of the predicted and experimental behavior of $p_\perp^8$ times $Ed\sigma/d^3p$ for $pp \to \pi + X$ at $\theta_{cm} = 90°$ and $x_\perp$ = 0.2, 0.35 and 0.5 versus $p_\perp$. The dot-dashed and solid curves are the final results before and after smearing, respectively, with Λ = 0.4. The dashed curves are the results (after smearing) using Λ = 0.6. For the range $2.0 \le p_\perp \le 6.0$ GeV/c at $x_\perp = 0.2$, and $4.0 \le p_\perp \le 10.0$ GeV/c at $x_\perp = 0.5$, the results are roughly independent of $p_\perp$ (when multiplied by $p_\perp^8$). However, this $p_\perp^{-8}$ behavior is only a "local" effect. It holds only over a small range of $p_\perp$ (at low $p_\perp$); the region depending somewhat on $x_\perp$. As $p_\perp$ increases, the predictions approach the expected $p_\perp^{-4}$ behavior. The new data from ISR (triangles[82] and open circles[83]) shown in Fig. 9.8 do show an increase from the flat ($1/p_\perp^8$) behavior at large $p_\perp$ in agreement with the QCD expectations.

D. Away-Side Correlations

An important consequence of the QCD approach is that the number of away-side hadrons with large $p_\perp$ (or $-p_x$ as in Fig. 9.9) is predicted to be considerably smaller than in the quark-quark scattering approach. Figure 9.10 shows that the number of away hadrons carrying a certain fraction, z_p, of the trigger momentum is predicted to be 3 to 4 times less than the FFF results[20], and now agree quite well with experiment. This reduction in the away-side multiplicity for large z_p is due to three factors. First, $\langle k_\perp \rangle_{h\to q}$ has been increased from 500 MeV to 848 MeV. Second, the fragmentation functions $D_i^h(z,Q^2)$ decrease at large z as Q^2 increases (see Fig. 8.2). Finally, in the QCD approach, the away-side constituent is quite often a gluon which produces on the average fewer hadrons at large z than do the quarks.

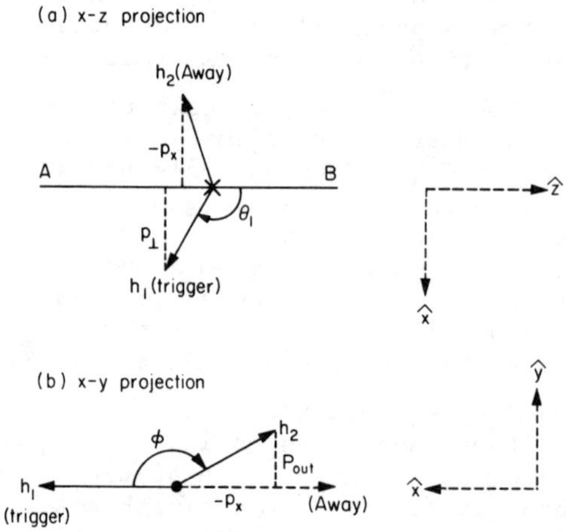

Fig. 9.9. Definition of kinematic variables used in describing the process $A + B \to h_1 + h_2 + X$: (a) x - z projection, where the beam, target and trigger hadron h_1 form this plane: (b) x - y projection.

Demanding a high $p_\perp$ trigger means that the toward-side constituent is predominantly a quark (72% of the time at W = 53 GeV, $p_\perp$

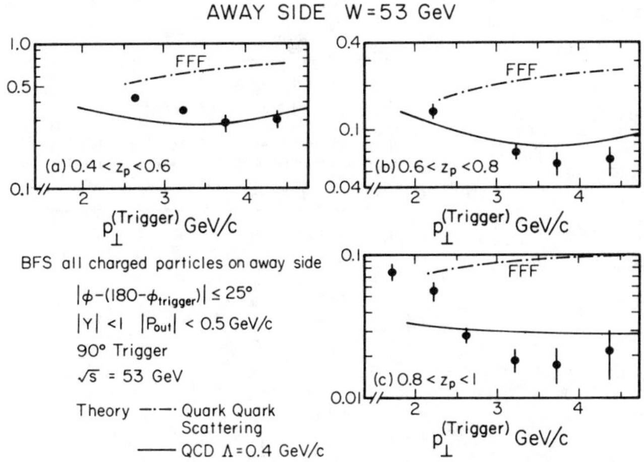

Fig. 9.10. The dependence on the trigger $p_\perp$ of the away-side hadron multiplicity $n(z_p) = (1/\sigma)d\sigma/dz_p$, where $z_p = -p_x(\text{away})/p_\perp(\text{trig})$ from the British-French-Scandinavian collaboration[84] on $pp \to h_1^\pm + h_2^\pm + X$ at W = 53 GeV, $\theta_1 = 90°$ and with an away-side acceptance of 25° in ϕ and $|Y_2| < 1$, $|P_{out}| < 0.5$ GeV/c. The predictions from the QCD approach with Λ = 0.4 GeV/c (solid curves) and the results of the quark-quark "black-box" model of FFF[20] (dash-dot curves) are shown. Background contributions from the fragmentation of the beam and target (see Fig. 9.1) which might be important for low $p_\perp$ triggers have <u>not</u> been included in either the QCD or FFF predictions.

= 4.0 GeV/c, θ_{cm} = 90°) whereas the recoiling away-side constituent will quite often be a gluon (62% at the above energy). This is illustrated in Fig. 9.11 for $pp \to \pi^o + X$ at W = 53 GeV, $p_\perp$ = 4.0 GeV/c and θ_{cm} = 90°. These away-side gluons produce equal numbers of positives and negatives so that the away-side plus to minus ratio is, at the ISR (low $x_\perp$), predicted to be considerably smaller than in FFF where all away-side partons were quarks. Figure 9.12 shows that the QCD approach yields almost equal numbers of positives and negatives for $p_\perp(\text{away})$ = 1.5 GeV/c at W = 53 GeV and $3.0 \le p_\perp(\text{trig}) \le 4.0$ GeV/c in agreement with recent ISR data[84]. In FFF[20], this ratio was predicted to be about 1.5 in gross disagreement with the experiment. At present, neither the QCD approach nor the FFF model can explain the apparently large increase in the away-side positive to negative ratio when triggering on K^- as observed by R-413[84] (Fig. 9.12). This increase is not seen in the E494 FNAL data[85] which are in reasonable agreement with the QCD contributions (Fig. 9.13).

E. <u>P-out Distributions</u>

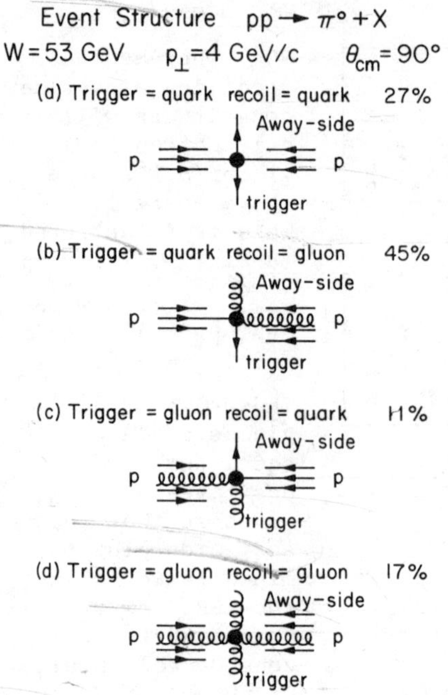

Fig. 9.11. Illustration of the underlying event structure expected from QCD at W = 53 GeV, $p_\perp$ = 4 GeV/c and θ_{cm} = 90° for the process $pp \to \pi^0 + X$.

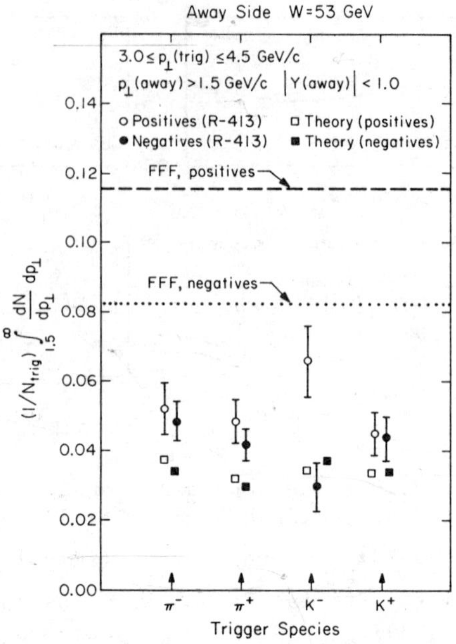

Fig. 9.12. The number of away-side positive and negative hadrons with $p_\perp$(away) > 1.5 GeV/c per trigger with 3.0 ≤ $p_\perp$(trig) ≤ 4.5 GeV/c from the BFS collaboration (R-413)[84] on $pp \to h_1 + h^\pm + X$ where W = 53 GeV/c and θ_1 = 90° and $|Y_2| < 1.0$. The results for π^-, π^+, K^- and K^+ triggers are shown and compared to the predictions of the quark-quark "black-box" model of FFF[20] and the QCD approach with Λ = 0.4 GeV/c (open and solid squares). Background contributions from the beam and target jets (see Fig. 9.1) have not been included in either the QCD or FFF predictions.

The use of an effective transverse momentum $\langle k_\perp \rangle_{h\to q}$ = 848 MeV as determined from the muon pair data results in mean P-out values that are considerably larger than the FFF results where $\langle k_\perp \rangle_{h\to q}$ was 500 MeV. (P-out is defined in Fig. 9.9.) As can be seen in Fig. 9.14, the new results are in much better agreement with the hadron data, although the predicted <P-out> is still a bit too small. Some of the discrepancy may be due to contributions from the beam and target jets which have not been included. Also, it may

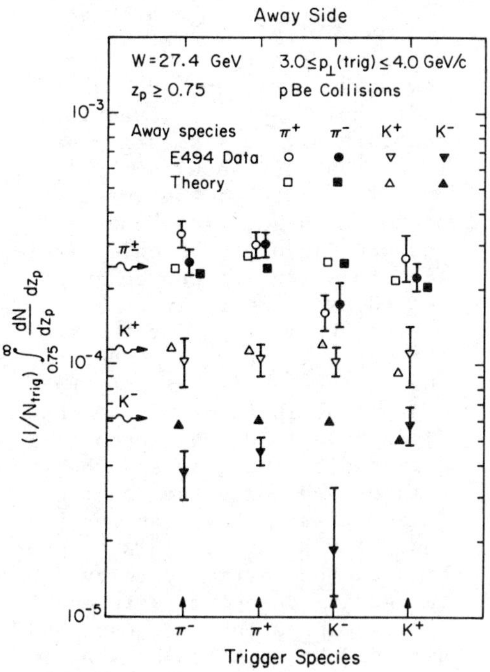

Fig. 9.13. The number of away-side mesons (of type π^+, π^-, K^+, K^-) with $z_p \geq 0.75$ per trigger (of type π^+, π^-, K^+, K^-) from the Fermilab experiment E494[85)]. The data are taken at W = 27.4 GeV with $3.0 \leq p_\perp(\text{trig}) \leq 4.0$ for proton beryllium collisions and are compared with the prediction of the QCD approach (for proton-proton collisions) with Λ = 0.4 GeV/c. No A-dependence corrections have been made to the theory or the data.

be that one should be using a slightly larger $\langle k_\perp \rangle_{h\to q}$ in the ISR range W = 53 GeV since the effective $\langle k_\perp \rangle$ distribution should increase slightly with increasing energy as seen in Fig. 7.6.

We have not really done a proper analysis of the P-out distribution. What one should do (and many are undoubtedly working on this) is to include 2 → 3 process explicitly in the large $p_\perp$ analysis similar to what was done for $pp \to \mu^+\mu^- + X$. The primordial motion could then be set at the value determined from $pp \to \mu^+\mu^- + X$ (i.e., $\langle k_\perp \rangle_{\text{primordial}}$ = 600 MeV or perhaps less). One would then predict a large momentum tail to the P-out distribution. It would not be expected to be bounded (like a Gaussian) and the tail would increase at increasing energies. There is, at present, no experimental evidence for a large momentum tail to the P-out distribution, although <P-out> is large. As Fig. 9.15 shows, it looks Gaussian, but so does the $\mu^+\mu^-$ spectrum in Fig. 7.5.

F. Scaling of $n(z_p) = (1/\sigma)d\sigma/dz_p$ and the Back-to-Back Cross Section

As seen in Fig. 9.6 and Fig. 9.8, the basic QCD subprocesses (before smearing) behave roughly like $1/p_\perp^6$ at fixed $x_\perp$ for $2 \leq p_\perp \leq 10$ GeV/c. To get a $1/p_\perp^8$ behavior, one must include large smearing effects that raise the small $p_\perp$ prediction considerably while leaving the large $p_\perp$ region essentially unchanged. There is considerable controversy over the question of whether smearing really

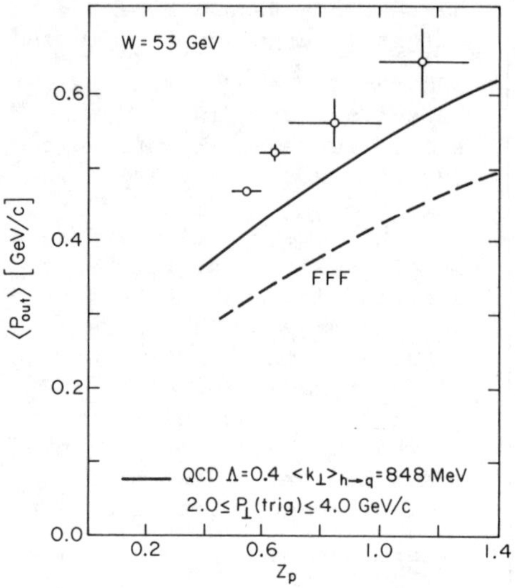

Fig. 9.14. The dependence on z_p of the mean value of the $|P_{out}|$ of away-side charged hadrons at W = 53 GeV and $2.0 \le p_\perp(\text{trig}) \le 4.0$ GeV/c with θ_1 averaged over 45° and 20° from the CCHK collaboration[70] on $pp \to h_1^\pm + h_2^\pm + X$, where $z_p = -p_x(\text{away})/p_\perp(\text{trig})$ (see Fig. 9.9). The predictions from the QCD approach at $\theta_1 = 45°$ with $\Lambda = 0.4$ GeV/c, and $\langle k_\perp \rangle_{h\to q} = 848$ MeV, and $\langle k_\perp \rangle_{q\to h} = 439$ MeV (solid curve) and the results of FFF[20] (dashed curve) curve are shown.

produces enough scale breaking to change $1/p_\perp^6$ to $1/p_\perp^8$ [76-79]. Other methods for handling the low $\hat{s}$ and $\hat{t}$ region, such as the "off-shell" approach, result in considerably less smearing effects than our results[77-79].

Fortunately, the question of how much smearing "breaks" the scaling of the single particle cross section can be answered experimentally. The increase at small $p_\perp$ for the single particle trigger is due to the "trigger bias" effect that biases one in favor of the configuration in which the initial quarks are moving toward the trigger and hence the basic constituent subprocess can occur at a smaller $\hat{s}$ and $\hat{t}$. This trigger bias can be partially removed by triggering on events with equally large $p_\perp$'s on the toward and away side. This can be seen in Fig. 9.16 where the predictions for $n(z_p) = (1/N(\text{trig}))dN/dz_p$ for $pp \to h_1 + h_2 + X$ where $\theta_1 = 90°$ and $x_{\perp 1} = 2p_{\perp 1}/\sqrt{s} = 0.35$ and where $z_p = -p_{x_2}/p_{\perp 1}$ as shown in Fig. 9.9. This quantity does not "scale." As $p_{\perp 1}$ increases from 3.4 GeV/c (W = 19.4 GeV) to 9.3 GeV/c (W = 53 GeV), $n(z_p \simeq 1)$ rises by about a factor of 4.6. This is a reflection of the fact that the numerator, $d\sigma/dz_p(z_p \simeq 1)$, decreases with increasing $p_{\perp 1}$ (and fixed $x_{\perp 1}$) less rapidly than does the denominator (i.e., the single particle cross section σ). (Remember at fixed W = 53 GeV $n(z_p)$ was roughly independent of $p_\perp$ as shown in Fig. 9.10.)

We expect the $p_\perp$ dependence of the two-particle back-to-back cross section to differ (in the region where smearing is an important effect) from that of the single particle cross section. This is also seen in Fig. 9.17 where I plot the two particle back-to-back

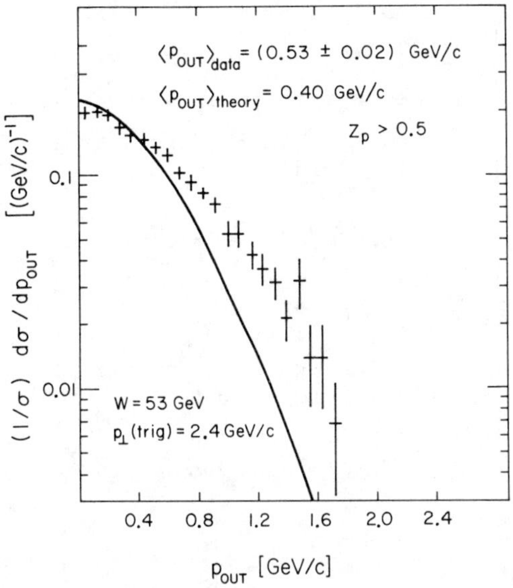

Fig. 9.15. The P-out spectrum (see Fig. 9.9) for away-side charged hadrons with $z_p \geq 0.5$ at W = 53, $2.0 \leq p_\perp(\text{trig}) \leq 4.0$ GeV/c with θ_1 averaged over 45° and 20° from the CCHK collaboration[70] on $pp \to h_1^\pm + h_2^\pm + X$. The prediction from the QCD approach at $\theta_1 = 45°$ with $\Lambda = 0.4$ GeV/c, $\langle k_\perp \rangle_{h\to q} = 848$ MeV and $\langle k_\perp \rangle_{q\to h} = 439$ MeV is shown by the solid curve.

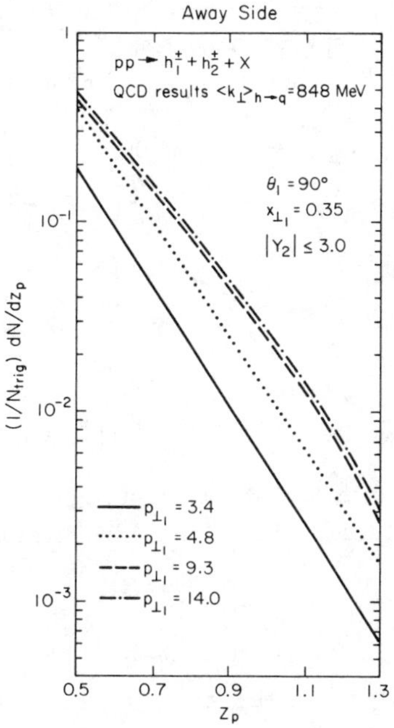

Fig. 9.16. Predicted behavior of the distribution of away-side hadrons $n(z_p) \equiv (1/N_{trig})dN/dz_p$ for $pp \to h_1^\pm + h_2^\pm + X$ at $\theta_1 = 90°$ and at fixed $x_{\perp_1} = 0.35$ as a function of $p_{\perp_1}$ (or energy), where $z_p = -p_x(h_2)/p_\perp(h_1)$ (see Fig. 9.9).

cross section $E_1 d\sigma/d^3p_1 dz_p dy_2 d\phi_2$ at $z_p = 1$ (times $p_\perp^8$) versus $p_\perp$ at $x_\perp = 0.35$. It behaves roughly like $1/p_\perp^6$ over the range $4 \leq p_\perp \leq 6.0$ GeV/c whereas the single particle cross section results, when multiplied by $p_\perp^8$, are roughly independent of $p_\perp$ over the range. The two particle back-to-back cross section reflects more closely the dependence on $p_\perp$ of the basic subprocess without the additional scale breaking due to smearing.

New data from a Columbia, Fermilab, Stony Brook

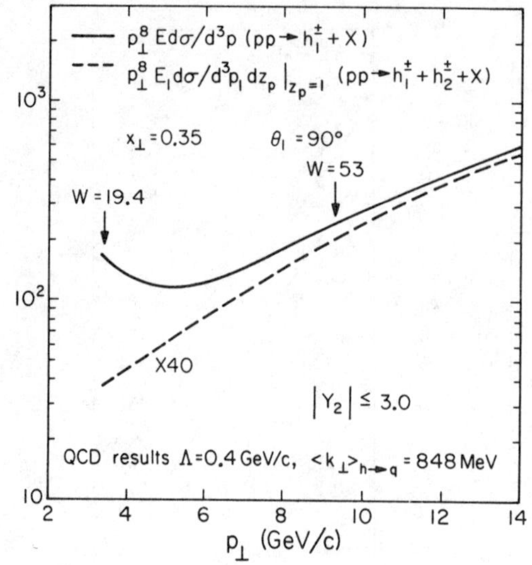

Fig. 9.17. Comparison of the behavior of $p_\perp^8$ times the single charged particle cross section, $Ed\sigma/d^3p$ $(pp \to h_1^\pm + X)$, and $p_\perp^8$ times the two particle back-to-back cross section, $E_1 d\sigma/d^3p_1 dz_p\big|_{z_p=1}$ $(pp \to h_1^\pm + h_2^\pm + X)$ at fixed $x_\perp$ = 0.35 (times 40). The predictions are calculated at θ_1 = 90° with Λ = 0.4 GeV/c and $\langle k_\perp\rangle_{h\to q}$ = 848 MeV and with an away-side acceptance of $135° \leq \phi_2 \leq 225°$ and $|Y_2| \leq 3.0$.

collaboration[85] at FNAL are shown in Fig. 9.18. The data clearly show a rise in $N(z_p) = \int n(z_p)dz_p$ at fixed $x_{\perp 1}$ as the energy (or $p_{\perp 1}$) increases. Their fits to the data yield

$$Ed\sigma/d^3p(pp \to \pi + X) \propto p_\perp^{-8.2 \pm 0.2} \qquad (9.6a)$$

at fixed $x_\perp$ for a single particle trigger and

$$E_1 d\sigma/d^3p_1 dz_p dy_2 d\phi_2(pp \to h_1 + h_2 + X)\Big|_{\xi \approx 1,\ y_2 \approx 0,\ \phi_2 \approx 180°} \propto p_\perp^{-6.4 \pm 0.2} \qquad (9.6b)$$

at fixed $x_\perp$ ($x_\perp > 0.25$) for the back-to-back trigger. This is an important result. It means that even at FNAL energies and $p_\perp$ values, one is seeing a basic subprocess behaving more like $1/p_\perp^6$ (after the scale breaking due to smearing has been removed) which can be explained by a basic $1/p_\perp^4$ subprocess plus the scale breaking due to $\alpha_s(Q^2)$, $G(x,Q^2)$ and $D(z,Q^2)$ in the amount expected by QCD[86].

G. The "Jet" Cross Section

A dramatic prediction of the QCD parton approach is the size of the cross section for producing a jet (parton = quark, antiquark or gluon) of momentum $p_\perp$ compared to that for producing a single

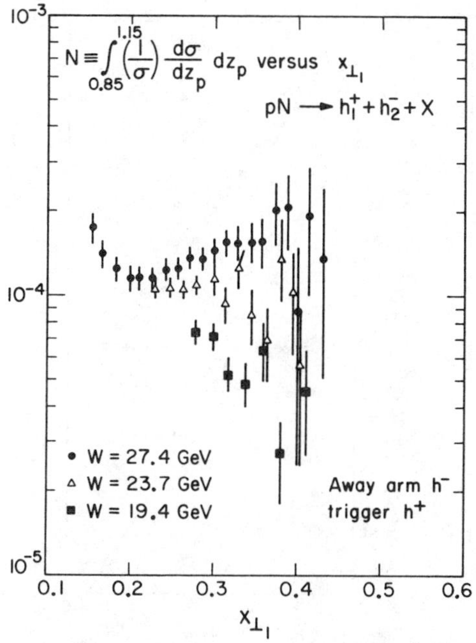

Fig. 9.18. Data on the number of away-hadrons (charge negative) per trigger hadron (charge positive) with $0.85 \leq z_p \leq 1.15$, $N \equiv \int_{0.85}^{1.15} (1/\sigma)(d\sigma/dz_p)dz_p$, for $pN \rightarrow h_1^+ + h_2^- + X$ at various energies W versus $x_{\perp 1}$. The data are from a recent Columbia-Fermilab-Stony Brook collaboration at FNAL (Ref. 85).

particle at the same $p_\perp$. In this approach, the single particle trigger always comes from a parton carrying more $p_\perp$ (typically about 15% more for quarks and greater for gluons) than the trigger particle. Furthermore, the chance of a parton fragmenting into hadrons in such a way that one particle carries almost all the momentum is small (only a few percent) as can be seen in Fig. 8.2. These two effects combine to give the large $\sigma(pp \rightarrow \text{jet} + X)/\sigma(pp \rightarrow \pi^0 + X)$ ratio shown in Fig. 9.19[87]. In the QCD approach, this ratio does not scale (i.e., it is a function of $x_\perp$, θ_{cm} and W).

The cross section for producing a "jet" of particles whose transverse momentum sum to give $p_\perp$ has been measured now by two groups[71,88] and is shown in Fig. 9.20. The measured jet rate is several orders of magnitude greater than the π^0 rate and is in qualitative agreement with the QCD predictions.

It is extremely difficult to make precise quantitative comparisons with the jet data in Fig. 9.20. Theoretically what is shown in Fig. 9.19 and Fig. 9.20 is the cross section for producing a quark (or gluon) with a given momentum (in Fig. 9.19 it is divided by the π^0 cross section at the same momentum). However, as discussed in Ref. 64, quarks of a given momentum (equal to their energy) cannot produce jets with the momentum of all particles equal to the energy of all particles. Our jet model[64] gives $E_{tot} - p_{z_{tot}} \approx 1.2$ GeV for quark jets. Since the cross section for producing jets falls so steeply, the cross section for producing a jet with a given $p_{z_{tot}}$ is considerably smaller than that for producing one with a given E_{tot}. As

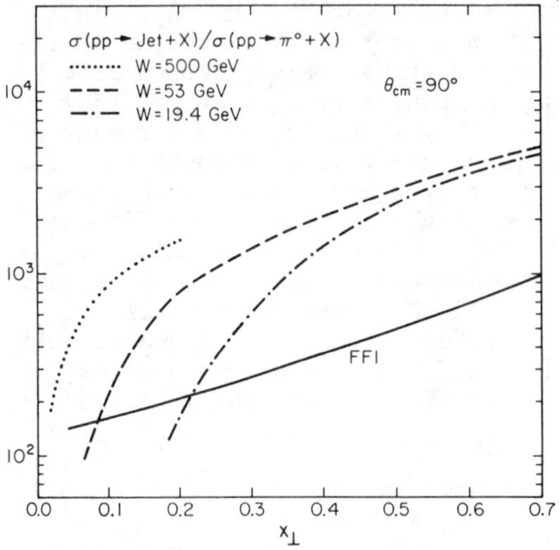

Fig. 9.19. Prediction of the jet to single π^o ratio at θ_{cm} = 90° versus $x_\perp$ for W = 500, 53 and 19.4 GeV from the QCD approach using Λ = 0.4 GeV/c. The jet cross section is defined as the cross section for producing a parton (quark, antiquark and glue) with the given $x_\perp = 2\ p_\perp/W$. Also shown is the prediction from the quark scattering model of FFF[20] which is independent of W at fixed $x_\perp$ and θ_{cm}.

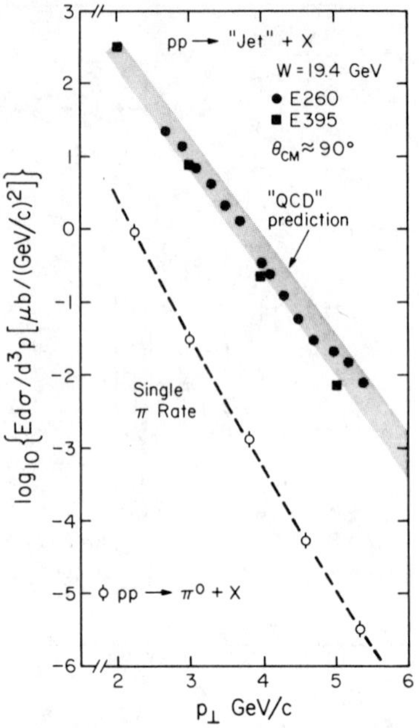

Fig. 9.20. Comparison of the jet and single π^o cross sections measured at 200 GeV/c (W = 19.4 GeV) and $\theta_{cm} \approx 90°$. The jet data are from two FNAL experiments, E260[88] and E395[71], where a jet is defined as the sum of all particles into their respective detectors. Also shown is the QCD prediction for the cross section of producing a parton (quark, antiquark or gluon) at θ_{cm} = 90° and W = 19.4 GeV with a given energy equal $p_\perp$.

explained in Ref. 89, it is the former that is more closely connected to what is measured experimentally. At W = 19.4 GeV/c, the cross section to produce a jet where $p_{z_{tot}}$ = 5 GeV/c at 90° is about 10 times smaller than the cross section to produce a jet whose E_{tot} = 5 GeV/c (see Fig. 9.21). The difference between E_{tot} and $p_{z_{tot}}$ of a jet arises, of course, from low momentum particles that have energy due to their mass (or $k_\perp$) but have little momentum p_z. This is tangled with the experimental uncertainty in all hadron jet experiments

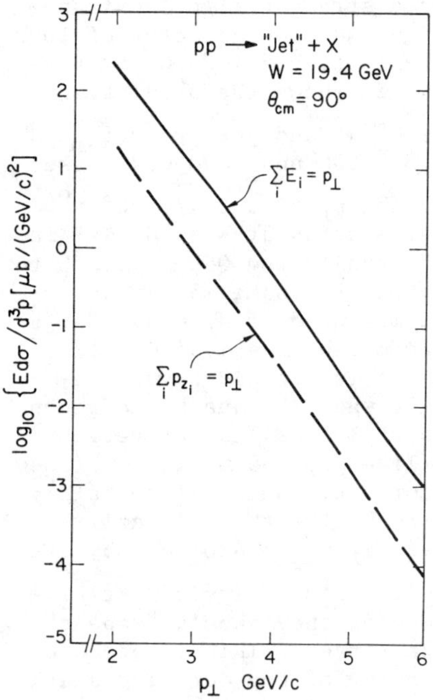

Fig. 9.21. Comparison of two different definitions of the cross section for producing a "jet" at $\theta = 90°$ and W = 19.4 GeV. The first is the cross section for producing a jet of particles whose total energy is $p_\perp$; the second is the cross section for producing a jet of particles whose $\hat{z}$-component (component along the quark direction of momentum sums to give $p_\perp$. The quark jet model in Ref. 64 was used to calculate the difference between the total energy and total p_z of all hadrons in a jet.

concerning low $p_\perp$ particles. One cannot be sure that one is not losing the low $p_\perp$ jet particles that are not well collimated or gaining low $p_\perp$ background from the beam and target jets in Fig. 9.1. Only by doing a very careful analysis, including the precise acceptances of a given experiment, can one make any quantitative statements.

The "bias" in favor of toward-side quarks, discussed in Section IX.D and illustrated in Fig. 9.11, does not occur when one triggers on jets rather than on single particles and thus gluons make up a sizable fraction of the jet cross section. With our guesses for the gluon distributions, gluons are responsible for 73% of the jet triggers at $p_\perp$ = 4 GeV/c, W = 53 GeV and $\theta_{cm} = 90°$. Even at higher $x_\perp$ values like $p_\perp$ = 6.0 GeV/c, W = 19.4 GeV, $\theta_{cm} = 90°$, gluons still make up 45% of the jets. One might hope someday to distinguish experimentally between gluon and quark jets. The gluon jets are assumed to have a higher multiplicity of particles each with lower momentum on the average. In addition, unlike the quark jets discussed in Ref. 64, gluon jets will carry on the average no net charge (or strangeness, etc.).

H. Very High Energy Expectations

Figure 9.8 shows that the QCD predictions begin to deviate from a $1/p_\perp^8$ behavior (at fixed $x_\perp$) as $p_\perp$ increases yielding a much larger cross section than expected from a $p_\perp^{-8}$ model. This is also seen in Fig. 9.22 where the QCD predictions for $p_\perp^8$ times $E d\sigma/d^3p$ versus $p_\perp$ at $x_\perp = 0.05$ and $\theta_{cm} = 90°$ are plotted. At W = 500 GeV, the QCD results are a factor of 100 greater

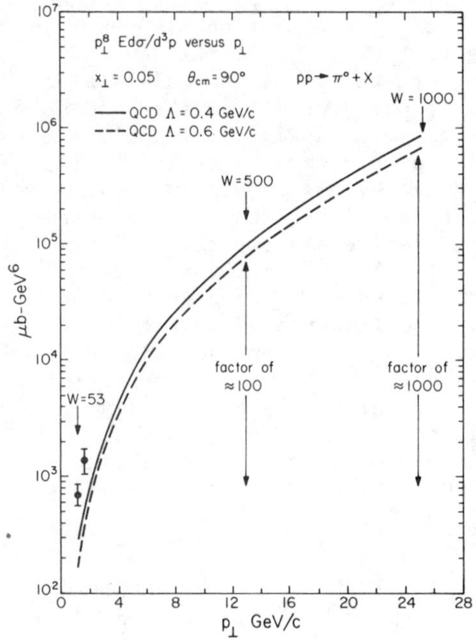

Fig. 9.22. The behavior of $p_\perp^8$ times the 90° single π^o cross section, $Ed\sigma/d^3p$, at $x_\perp = 0.05$ versus $p_\perp$ calculated from the QCD approach with $\Lambda = 0.4$ GeV/c (solid curve) and $\Lambda = 0.6$ GeV/c (dashed curve). The two low $p_\perp$ data points are at W = 53 and 63 GeV[93]. The predictions are a factor of 100 (1000) times larger than the flat ($p_\perp^{-8}$) extrapolation to W = 500 GeV (1000 GeV).

than a straight ($1/p_\perp^8$) extrapolation and show a factor of 1000 increase at W = 1000 GeV. Figure 9.23 shows the predictions for 90° π^o and jet production at W = 53, 500 and 1000 GeV versus $p_\perp$. The $p_\perp$ = 30 GeV/c 90° π^o cross section at W = 500 GeV is predicted in the QCD approach to be about the same magnitude as that measured at $p_\perp$ = 6.0 GeV/c at Fermilab (W = 19.4 GeV)!

It is not clear yet precisely what the quark and gluon jets will look like at very high $p_\perp$ (like $p_\perp$ = 30 GeV/c). If QCD is correct, they will certainly not look like the well collimated $\langle k_\perp \rangle_{q\to h}$ = 430 MeV objects used in this analysis. At $p_\perp$ = 30 GeV/c, they should "appear" to be fatter. This is because as the $p_\perp$ of the outgoing quark increases, it becomes increasingly likely that it radiate a hard gluon and become two jets (one quark and one gluon). Then, this quark or gluon might radiate producing still more subjets. This is the same mechanism that is responsible for the scale breaking of the fragmentation functions $D(z,Q^2)$ discussed in Section VIII. The net result is that most of the time it will look as if there is one somewhat fatter jet (with the fatness increasing with increasing momentum); however, occasionally when the radiation is hard enough, one will see two or three subjets. I will discuss this more in a later section.

Figure 9.24 shows the leading order QCD predictions for $pp \to \pi^o + X$ and $pp \to$ Jet + X at W = 500, 1000, 2000 and 40,000 GeV. One should not take these predictions too seriously since I have extrapolated the leading order formulas quite a long way. Higher order corrections could be important for such long extrapolations (i.e.,

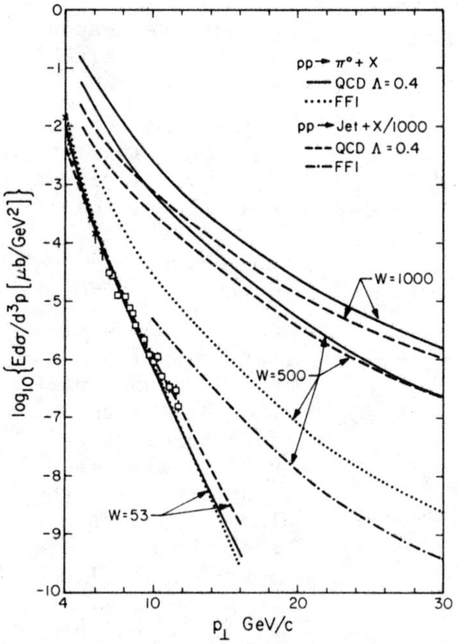

Fig. 9.23. Comparison of the results on the 90° π^o cross section, $Ed\sigma/d^3p$, from the QCD approach with Λ = 0.4 GeV/c (solid curve) and the quark-quark "black-box" model of FFF (dotted curves). Both models agree with the data at W = 53 GeV (crosses = Ref. 94) where the open squares are the "preliminary" data from the CCOR collaboration[83)] normalized to agree with the lower $p_\perp$ experiments. The QCD approach results in much larger cross sections than the FFF model at W = 500 and 1000 GeV. The FFF results at 1000 GeV (not shown) are only slightly larger than the results at 500 GeV. Also shown are the cross sections for producing a jet at 90° (divided by 1000) as predicted by the QCD approach (dashed curves) and the FFF model (dot-dashed curve).

$\alpha_s(Q^2) - \alpha_s(Q_o^2)$ is large - see (6.22)). Nevertheless, it is clear that QCD predicts quite large cross sections at these energies. This is also seen in Fig. 9.25 and Fig. 9.26 where I have plotted the integrated spectrum for pp → Jet + X.

I. Charm Production at Large $p_\perp$ [90,91)]

An interesting consequence of the presence of gluons within the proton is that one can now produce heavy quarks (like charm) at large $p_\perp$ by the process $gg \to c\bar{c}$ as illustrated in Fig. 9.27a. This subprocess is somewhat analogous to $e^+e^- \to q\bar{q}$ that results in heavy quark production in e^+e^- collisions. Figure 9.28 shows the estimated total cross section for producing a charm quark, c, with $p_\perp(c) \geq 2.0$ GeV/c in proton-proton collisions from this subprocess. Since each charm quark must fragment into at least one charmed hadron (these hadrons won't necessarily have $p_\perp \geq 2.0$ GeV/c), this can be viewed as the total cross section for charm that arises from charm quarks having $p_\perp \geq 2.0$ GeV/c. The large $p_\perp$ charm production is not negligible. From this subprocess alone, it reaches about 10 μb at the highest ISR energies[97)].

J. Direct Photons at Large $p_\perp$ [98,102)]

As shown in Fig. 9.7,

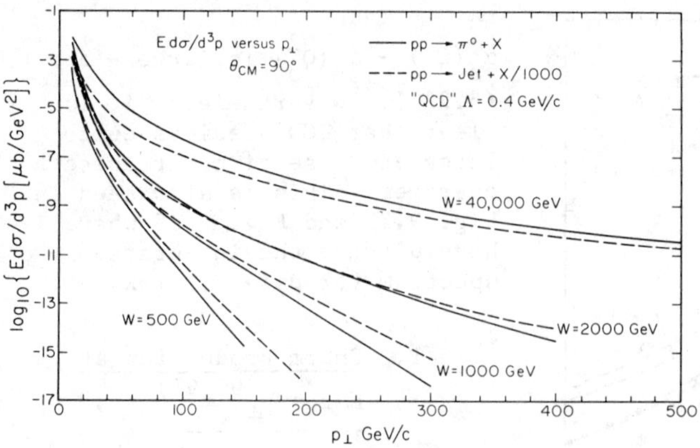

Fig. 9.24. Leading order QCD predictions for $pp \to \pi^0 + X$ (solid curves) and $pp \to$ Jet + X (divided by 1000, dashed curves) at 90° using $\Lambda = 0.4$ GeV/c. The predictions are plotted versus $p_\perp$ and given at $W = \sqrt{s} = 500$, 1000, 2000 and 40,000 GeV.

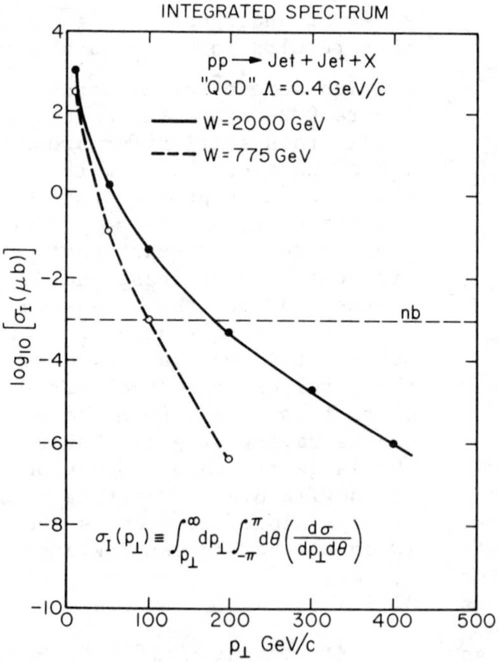

Fig. 9.25. Leading order QCD predictions for the integrated spectrum in $pp \to$ Jet + X at $W = \sqrt{s} = 775$ and 2000 GeV.

gluons are responsible for a sizable portion of the large $p_\perp$ π^0 cross section. In fact, Fig. 9.11 indicates that the dominant subprocess at W = 53 GeV, $p_\perp = 4$ GeV/c, $\theta_{cm} = 90°$ is quark-gluon scattering, $g + q \to g + q$. If gluons participate in this subprocess, then necessarily they must produce direct large $p_\perp$ photons by the process, $g + q \to \gamma + q$, as illustrated in Fig. 9.27b. Even though the process $g + q \to \gamma + q$ is down by α_{QED}/α_s relative to $g + q \to g + q$, when comparing the rate for large $p_\perp$ photons to that for producing say, π^0's it is enhanced since this latter must proceed via a quark or gluon fragmentation function.

Figure 9.29 shows the ratio of photons produced by the subprocess $g + q \to \gamma + q$ to the total QCD π^0 rate[103] and Fig. 9.30 shows the predicted invariant cross section,

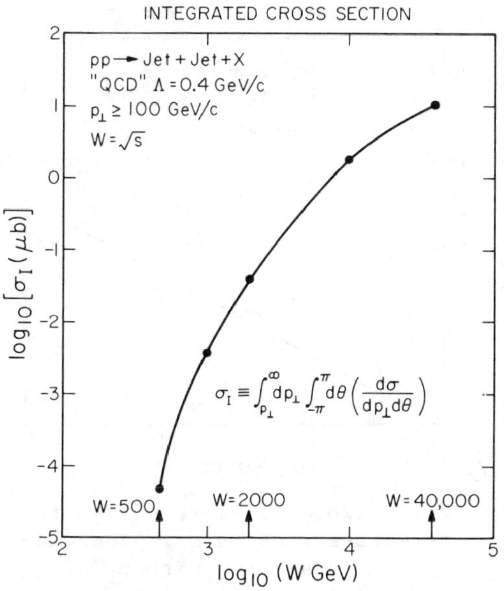

$$\sigma_I \equiv \int_{p_\perp}^{\infty} dp_\perp \int_{-\pi}^{\pi} d\theta \left(\frac{d\sigma}{dp_\perp d\theta}\right)$$

Fig. 9.26. Leading order QCD predictions for the total rate for producing jets with $p_\perp$ greater than 100 GeV/c versus $W = \sqrt{s}$.

$Ed\sigma/d^3p$, for $pp \to \gamma + X$ at $W = 53$ and 19.4 GeV and $\theta_{cm} = 90°$ (before and after smearing with $\langle k_\perp \rangle_{h\to q} = 848$ MeV) compared to the observed π^o rate. At $p_\perp = 14$ GeV/c, $\theta_{cm} = 90°$ and $W = 53$ GeV, one expects about as many direct photons as π^o's! These photon events are quite distinctive. They occur with a photon at large $p_\perp$ on the trigger side with no accompanying toward-side hadrons. The away-side hadrons come from the fragmentation of a quark. On the other hand, the production of photons due to Bremsstrahlung from, say, $q + q \to q + q$ results in events where the photon is produced in association with other trigger-side hadrons. In addition, the γ/π^o rate for photons produced via Bremsstrahlung is only about 2 - 5% as illustrated in Fig. 9.29[93,101].

It is interesting that in ep collisions one probes the quark distributions with an incoming virtual photon, while studying large $p_\perp$ real photons in $pp \to \gamma + X$, one can probe the distribution of gluons within the proton through the subprocess $g + q \to \gamma + q$. If one does not find a reasonable rate for producing γ's at large $p_\perp$, the QCD approach as I have outlined it will be in trouble.

X. THE SEARCH FOR THREE-JET EVENTS

A. Analysis of Event Shapes in e^+e^- Annihilation

Experiments have shown that the final states in $e^+e^- \to$ hadrons consist predominantly of two jets of hadrons presumably resulting from the process $e^+e^- \to q\bar{q}$. The theory of QCD expects this basic two-jet structure, but predicts that occasionally one of the outgoing quarks should emit a hard gluon, g, resulting in a three-jet final state[104,105]. As discussed in Section IV.B and Section VIII,

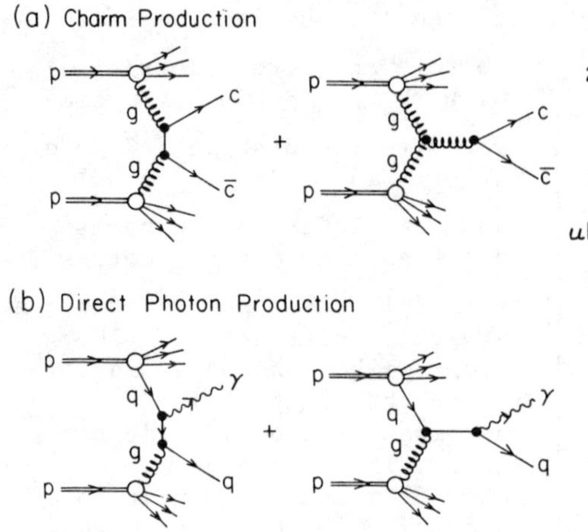

Fig. 9.27 - (a) Two diagrams for the production of charm at large $p_\perp$ in pp collisions from the subprocess $gg \to c\bar{c}$, where g is a gluon.

(b) Two diagrams for the production of direct photons, γ, at large $p_\perp$ in pp collisions from the "Compton" subprocess $gq \to \gamma q$.

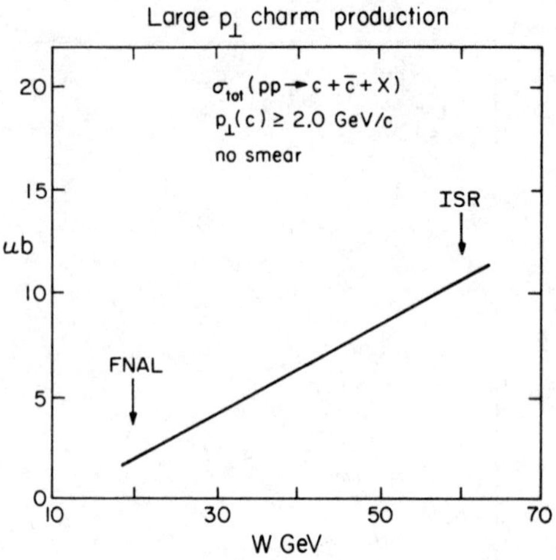

Fig. 9.28. Estimate of the total cross section for producing a charm quark, c, with $p_\perp(c) \geq$ 2 GeV/c in pp collisions from the $gg \to c\bar{c}$ subprocess illustrated in Fig. 9.27b.

the probability of emitting a hard gluon is proportional to $\alpha_s(Q^2)$ $\log^2(Q^2/\Delta^2)$ and thus increases logarithmically with increasing Q^2. The observation of such "three-jet" events would provide strong support for QCD.

Unfortunately, two-jet events can from time to time fragment into a configuration of final state hadrons that looks like a three-jet configuration; so the question is one of rate. One looks for observables to describe event structure that can be calculated to order $\alpha_s(Q^2)$ in QCD and are "infrared finite." Presumably observables that do not discriminate between final states differing by the inclusion of a very low energy particle or by the replacement of one particle by two collinear particles with the same total momentum are infrared finite in QCD perturbation theory.

Two such observables recently examined by De Rújula, Ellis, Floratos and Gaillard[106] are the "thrust" defined by

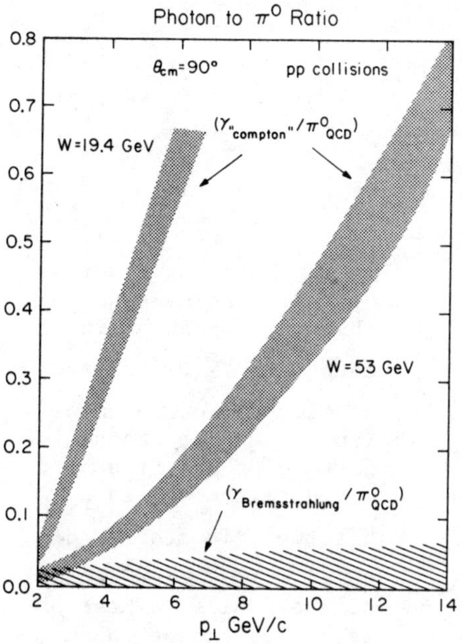

Fig. 9.29. Estimate of the rate of large $p_\perp$ direct photon production at $\theta_{cm} = 90°$ to the rate for producing π^0 at W = 19.4 and 53 GeV, where the photons are produced by the "compton" subprocess $gq \to \gamma q$ illustrated in Fig. 9.27b and the π^0 rate is given by the QCD predictions in Fig. 9.7 (Λ = 0.4 GeV/c). Also shown is the estimate for γ's produced by Bremsstrahlung from quark-quark and quark-gluon scattering.

$$T = 2 \max \frac{\tilde{\sum_i} p_\parallel^i}{\sum_i |\vec{p}_i|} \qquad (10.1)$$

and the "spherocity"

$$S = \left(\frac{4}{\pi}\right)^2 \min \left(\frac{\sum_i |\vec{p}_\perp^{\,i}|}{\sum_i |\vec{p}_i|}\right)^2 , \qquad (10.2)$$

where the sum in the denominator runs over all observed particles and the sum $\tilde{\sum}$ in (10.1) runs over all particles in a hemisphere. The momenta $p_\parallel$ are parallel to a "jet" axis, normal to the plane defining the hemispheres, and chosen to maximize T. The thrust and spherocity lie in the range $1 \geq T \geq 0.5$ and $0 \leq S \leq 1$, respectively. One expects the number of events with S or 1 - T large to be considerably different if the QCD perturbative $e^+e^- \to q\bar{q}g$ state exists than if it does not. This is shown in Fig. 10.1a,b. The perturbative QCD $q\bar{q}g$ contribution results in a large S and large 1 - T tail to the $d\sigma/dS$ and $d\sigma/dT$ distributions, respectively, that are considerably larger (at $\sqrt{s}$ = 18 GeV) than the expected background from the two-jet $q\bar{q}$ contribution. However, as these distributions indicate, the great majority of events still have small S and small 1 - T and look like two jets[107].

Since there is no natural axis defined for the final states in e^+e^- annihilation, it would be convenient to have a set of observables that characterizes the shape of each event that are rotationally invariant. Recently G. Fox and S. Wolfram[108] have come up with just such a set of observables. They define

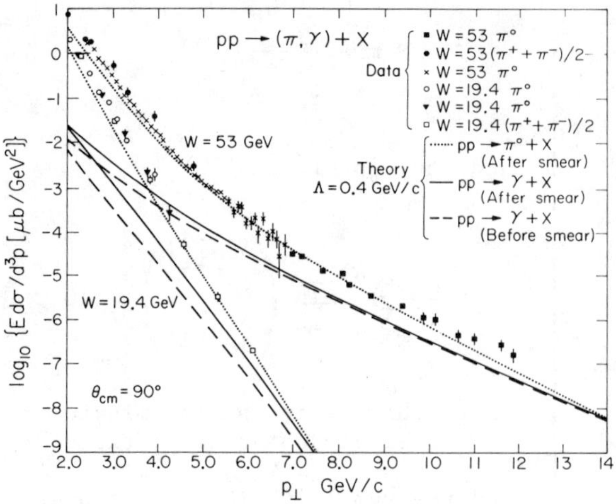

Fig. 9.30. Predicted cross section, $E d\sigma/d^3p$, for producing direct photons, $pp \to \gamma + X$, at $\theta_{cm} = 90°$ and $W = 19.4$ and 53 GeV, where the photons are produced by the "Compton" subprocess $gq \to \gamma q$ illustrated in Fig. 9.27b. The results are shown before and after smearing with $\langle k_\perp\rangle_{h\to q} = \langle k_\perp\rangle_{h\to g} = 848$ MeV and are compared with the data on $pp \to \pi^0 + X$ from Figs. 9.7 and 9.23.

$$H_\ell = \left(\frac{4\pi}{2\ell+1}\right) \sum_{m=-\ell}^{+\ell} \left| \sum_i Y_\ell^m(\Omega_i) \frac{|\vec{p}_i|}{\sqrt{s}} \right|^2, \qquad (10.3)$$

where the inner sum is over all the hadrons which are produced in the event and $Y_\ell^m(\Omega)$ are the usual spherical harmonics. One must choose a particular set of axes to evaluate the angles Ω_i, but the values of H_ℓ deduced will be independent of the choice. These observables can be calculated to order $\alpha_s(Q^2)$ in QCD and have the advantage that one need not find any "jet-axis" by minimization.

Energy and momentum conservation requires $H_1 = 0$ and $H_0 \approx 1$. In principle, all the other H_ℓ carry independent information. A convenient measure of the event shape is provided by the mean value of H_ℓ. For example, the processes $e^+e^- \to q\bar{q}$ and $e^+e^- \to q\bar{q}g$ calculated to lowest order in α_s give

$$\langle H_2\rangle = 1 + \frac{2\alpha_s}{3\pi}(33 - 4\pi^2). \qquad (10.4)$$

Figures 10.1c and 10.1d show the distribution $d\sigma/dH_2$ and $d\sigma/dH_3$, respectively, for the process $e^+e^- \to q\bar{q}$ and for the final sum of this term plus the first order QCD $e^+e^- \to q\bar{q}g$ contribution at $\sqrt{s} = 20$ GeV. The fragmentation of quarks into hadrons results in a large modification of the idealized two-jet productions (i.e., $H_2 = 1$). Nevertheless, one could clearly establish (at $\sqrt{s} \gtrsim 10$ GeV) whether

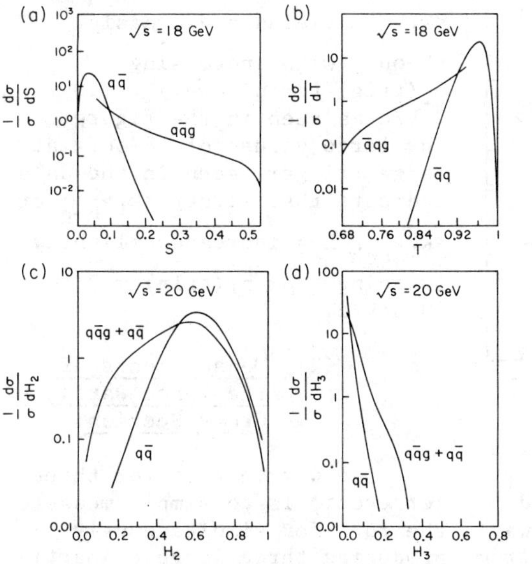

Fig. 10.1 - (a) Distribution of sphericity S, defined by eq. (10.2) for the process $e^+e^- \to$ hadrons at $\sqrt{s}$ = 18 GeV from Ref. 106. The curves are the "two jet" $q\bar{q}$ and the QCD perturbative "three jet" $q\bar{q}g$ contributions.

(b) Same as Fig. (a) but for the thrust T, defined by eq. (10.1).

(c) Distribution of the quantity H_2, defined by eq. (10.3), for the process $e^+e^- \to$ hadrons at $\sqrt{s}$ = 20 GeV from Ref. 108. The curves are the "two jet" $q\bar{q}$ contribution and the total ($q\bar{q}$ plus the QCD perturbative "three jet" $q\bar{q}g$) result.

(d) Same as (c) but for the quantity H_3.

$e^+e^- \to q\bar{q}g$ events exist by observing the rate at which small H_2 (or large H_3) events occur. (Fox and Wolfram use the jet model of Ref. 64 to "smear" their idealized perturbative results and make predictions concerning the outgoing hadrons. The $q\bar{q}g$ contributions in Fig. 10.1a,b have not been "smeared" over the final state quark and gluon fragmentations.)

B. Three Large $p_\perp$ Jets in pp Collisions

(1) Measurements of P-out

In the QCD approach, one expects a broad P-out distribution for mesons produced out of the production plane in pp collisions like that observed in Fig. 9.15. This lack of coplanarity is due to the presence of two-to-three subprocess like $qq \to qqg$ as illustrated in Fig. 9.2 and Fig. 9.3b. One expects a large momentum tail of the P-out distribution due, for example, to the Bremsstrahlung radiation of a hard gluon from an outgoing quark in $qq \to qq$. The tail should increase with increasing trigger $p_\perp$.

There is some evidence for an increasing <P-out> with increasing trigger $p_\perp$ from the CCOR ISR experimental results shown in Fig. 10.2. In this figure, <P-out> is plotted for all away particles with $p_\perp$(away) > 1.0 GeV/c so that $z_p = -p_x(\text{away})/p_\perp(\text{trig})$ is decreasing as $p_\perp$(trig) increases.

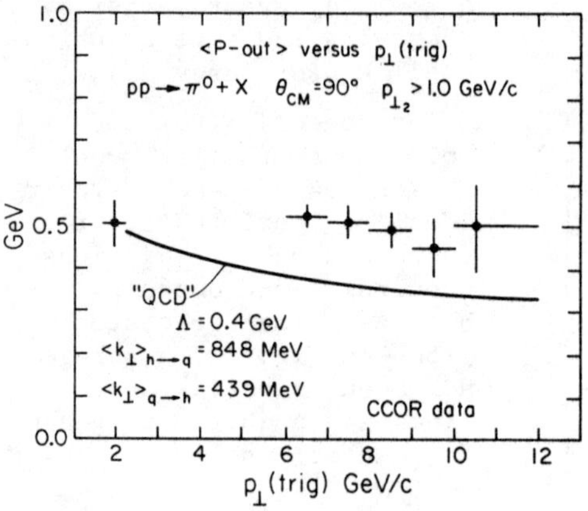

Fig. 10.2. Data on the mean value of P-out (see Fig. 9.9 for the definition) of the away-side hadrons in pp collisions with $p_\perp$(away) > 1.0 GeV/c versus $p_\perp$(trigger) from the CCOR collaboration[83]. The curve is the prediction of the QCD approach with $\langle k_\perp \rangle_{h\to q}$ = 848 MeV and $\langle k_\perp \rangle_{q\to h}$ = 439 MeV, where $\langle k_\perp \rangle_{h\to q}$ and $\langle k_\perp \rangle_{q\to h}$ are taken to be independent of Q^2 (or $p_\perp$(trigger)). (In QCD, one would expect $\langle k_\perp \rangle_{h\to q}$ and $\langle k_\perp \rangle_{q\to h}$ to increase with increasing Q^2.)

A model with fixed $\langle k_\perp \rangle_{h\to q}$ and $\langle k_\perp \rangle_{q\to h}$ yields a decreasing <P-out> with increasing $p_\perp$(trig) for $p_\perp$(away) > 1.0 GeV/c as seen in the figure. The large values of <P-out> at large triggers seen in the data indicate that either $\langle k_\perp \rangle_{h\to q}$ or $\langle k_\perp \rangle_{q\to h}$ has increased slightly in going from $p_\perp$(trig) = 2 to 10 GeV/c.

(2) Measurements of the Three Particle Cross Section

One way to test for three jet events is to simply measure the rate for simultaneously producing three large $p_\perp$ particles all at large angles to each other[109]. For example, one could measure the rate for producing three particles all with $p_\perp$ > 3 GeV/c and all at 90° to the beam and all 120° degrees apart. The rate will, of course, be small, but it will be orders of magnitude larger if three jet events exist than if they don't. Efforts to estimate these rates are in progress.

(3) Measuring Event Shapes in Large $p_\perp$ Jet Production

Fox and Wolfram[108] have suggested that one can analyze large $p_\perp$ events in a manner analogous to $e^+e^- \to$ hadrons by defining (now in two dimensions)

$$C_\ell = \left| \sum_i |\vec{p}_{\perp i}| \exp(i\ell\phi_i) \right|^2, \tag{10.5}$$

where $\vec{p}_{\perp i}$ are the perpendicular momenta of the resulting hadrons and ϕ_i is measured relative to an <u>arbitrary</u> $\hat{\xi}$-axis chosen in the plane perpendicular to the beam direction (the $\hat{x}$-$\hat{y}$ plane in Fig. 9.9). The choice of the $\hat{\xi}$-axis does not matter since the C_ℓ are rotationally invariant. Again these observables are infrared finite and can be calculated in QCD perturbation theory. They are distinctly different for two and three-jet processes. (For idealized two jet events $C_\ell/C_o = 1$ for ℓ = even and $C_\ell/C_o = 0$ for ℓ = odd. For three jet events $C_\ell/C_o < 1$, ℓ = even.) Future measurements of these observables will establish if, as expected by QCD, three jet events really exist.

XI. SUMMARY AND CONCLUSIONS

Many of the calculations discussed here should be considered as first, "crude", phenomenological attempts to examine experimental consequences of QCD. The theory of QCD is, however, more than a phenomenological model. It is a precise and complete theory purporting to be an ultimate explanation of all hadronic experiments at all energies, high and low. There are many reasons to hope and expect it to be right. The question is, is it indeed right? Mathematical complexity has, so far, prevented quantitatively testing its correctness. The theory itself is remarkably simple and beautiful; however, what it predicts is not yet clearly known. Nevertheless, its property of asymptotic freedom leads one to expect that phenomena of high momentum transfer should be analyzable (by perturbation) and some applications of the theory have been examined here[110]. Unfortunately, most processes involve both low and high energy aspects and ways of separating the low energy (or Q^2) pieces that cannot be calculated by perturbation from the high Q^2 perturbative corrections are just now becoming understood.

One reason to view many of the present day applications of QCD as preliminary is because calculations to leading order are only meaningful if one shows that the next order corrections are indeed small. For example, although to leading order the antiquark distributions as measured in neutrino and antineutrino interactions are the same as the antiquark distribution one should use in the Drell-Yan calculation (7.1), there are higher order corrections that vitiate this direct connection. In fact, my preliminary findings are that for the total muon pair rate, $d\sigma_{DY}/dM^2$, the order α_s corrections are quite large. So large that one must go to $M^2 \approx 10{,}000$ GeV^2 before one could believe the leading order result. This may be true for other observables and other processes, and, if so, it represents an important new problem for perturbative QCD. Since we are making an expansion in $1/\log Q^2$, one must go to large $\log Q^2$ (not large Q^2) before one can neglect higher terms and trust the leading order results (in leading order, we have already summed all terms of the form $(\alpha(Q^2)\log Q^2)^N$). I believe that for those cases where the coefficient of the order α_s term is large, one must develop means of summing the perturbation series if one is going to make precise predictions at present or foreseeable Q^2 values.

One place where the next order corrections are not too important (for $0.2 \le x \le 0.8$) is in the QCD equations for the Q^2 dependence for the parton distributions in lepton scattering experiments[111]. Recent data from ep, μp and νp experiments do show clearly the "scale" breaking effects expected from QCD. However, since most of the data are at relatively low Q^2, the results are sensitive to $1/Q^2$ contributions that are difficult, if not impossible, to calculate (they involve knowledge of, for example, the primordial transverse momentum).

The transverse momentum of muon pairs is certainly larger than one would have expected from the naive parton model; however, the experiments have not really seen the high $p_\perp$ tail predicted by QCD. There is some evidence to suggest that $\langle p_\perp \rangle_{\mu^+\mu^-}$ does increase with increasing energy (fixed M^2) as expected. But, there are no data to check the approach to a constant of $W^5\, d\sigma/dMdyd^2k_\perp$ as W increases, shown in Fig. 7.8. Finally, there is the question of why the primordial transverse momentum still comes out as large as 600 MeV even after one includes the first order QCD perturbative corrections[112].

At one time it was thought that the experimentally observed $p_\perp^{-8}$ behavior of large $p_\perp$ meson production in hadron-hadron collisions might pose a problem for QCD. However, we now see that there is no problem[113]. The energy ($p_\perp$) of existing experiments is too low and there are too many non-asymptotic effects acting. All the scale breaking effects act in the <u>same</u> direction to produce an effective apparent $p_\perp$ power that is roughly eight at low $p_\perp$. In addition, the predicted size of the invariant cross section is just about right. Results closer to a $p_\perp^{-4}$ fall off should appear only at much higher $p_\perp$. Furthermore, one has indirect evidence from large $p_\perp$ correlations that gluons as well as quarks must be included in a description of the data.

We conclude that there is no evidence from eN, μN, νN, $\bar{\nu}$N interactions <u>or</u> $pp \to \mu^+\mu^- + X$, <u>or</u> large $p_\perp$ production in hadron-hadron experiments against QCD. On the contrary, the overall picture favors a QCD approach. However, most of the conclusive and exciting predictions of the theory have not yet been seen. The next generation proton-proton machines should see hundreds or even thousands of times more mesons at large $p_\perp$ than expected from extrapolations of existing data. One should occasionally see three distinct jets in

e^+e^- collisions and at large $p_\perp$ in pp collisions. One should see gluon jets as well as quark jets.

QCD is not just "another theory." If it is not the correct description of nature then it will be quite some time before another candidate theory emerges.

Acknowledgments

It is a pleasure to thank my collaborators R. P. Feynman and G. C. Fox without whom this work could not have been performed. In addition, I would like to acknowledge useful discussions with S. Brodsky, H. Georgi, J. F. Owens, H. D. Politzer, R. Raitio, D. Ross and S. Wolfram. Finally, let me congratulate Adolf Hochstim and Bill Frazer on a most enjoyable and stimulating summer workshop. If the students learned half as much as the lecturers did, then it was a most profitable three weeks for all.

Footnotes and References

1. H. D. Politzer, Physics Reports 14C (1974).
2. D. Gross and F. Wilczek, Phys. Rev. D8, 3633 (1973); Phys. Rev. D9, 980 (1974).
3. M. Gell-Mann, Phys. Rev. 125, 1067 (1962).
4. T. Kinoshita, J. Math. Phys. 3, 650 (1962).
5. T. D. Lee and M. Nauenberg, Phys. Rev. 133B, 1547 (1964).
6. R. G. Moorhouse, M. R. Pennington and G. G. Ross, Nucl. Phys. B124, 285 (1977).
7. Gail Hanson, Talk given at the 13th Rencontre de Moriond on High Energy Leptonic Interactions, Les Arcs, Savoit, France, 1978.
8. R. P. Feynman, "Photon-Hadron Interactions," (Benjamin, Reading, Mass., 1972).
9. This is explained in a clear way by H. Georgi in "The Use and Misuse of the Parton Model," Harvard preprint HUTP-78/A003.
10. H. D. Politzer, Phys. Lett. 70B, 430 (1977); H. Georgi and H. D. Politzer, Phys. Rev. Lett. 40, 3 (1978).
11. H. Georgi and H. D. Politzer, Phys. Rev. D14, 1829 (1976); A. De Rújula, H. Georgi and H. D. Politzer, Ann. Phys. (N. Y.) 103, 351 (1977).
12. A. J. Buras, E. G. Floratos, D. A. Ross and G. T. Sachrajda, Nucl. Phys. B131, 308 (1977); A. J. Buras and K. J. F. Gaemers, Nucl. Phys. B132, 249 (1978).
13. G. Altarelli and G. Parisi, Nucl. Phys. B126, 298 (1977).
14. G. C. Fox, Nucl. Phys. B131, 107 (1977).
15. T. Quirk, private communication (1978); and T. Quirk invited talk at the symposium on Jets in High Energy Collisions, Copenhagen, July 1978.
16. H. Anderson, H. S. Matis and L. C. Myrianthopoulos, Phys. Rev. Letters 40, 1061 (1978).

17. E. D. Bloom and F. J. Gilman, Phys. Rev. Letters 25, 1140 (1970).
18. O. Nachtmann, Nucl. Phys. B63, 237 (1973).
19. G. Altarelli and G. Martinelli, Phys. Letters 76B, 89 (1978).
20. R. P. Feynman, R. D. Field and G. C. Fox, Nucl. Phys. B128, 1 (1977).
21. G. C. Fox, "The Physics of Inclusive Charged Current Neutrino Reactions," CALT-68-658, invited talk at the International Conference on Neutrino Physics, Purdue University 1978.
22. G. C. Fox, Nucl. Phys. B134, 269 (1978).
23. R. D. Field, Phys. Rev. Letters 40, 997 (1978).
24. G. C. Fox, "Application of Quantum Chromodynamics to High Transverse Momentum Hadron Production," invited talk at the Orbis Scientiae 1978 (Coral Gables) CALT-68-643.
25. R. P. Feynman, R. D. Field and G. C. Fox, Phys. Rev. D18, 3320 (1978).
26. R. D. Field and R. P. Feynman, Phys. Rev. D15, 2590 (1977).
27. Aachen-Bonn-CERN-London-Oxford-Saclay Collaboration, "Analysis of Nucleon Structure Functions in CERN Bubble Chamber Neutrino Experiments," Oxford preprint 16/78 (1978).
28. G. Altarelli, R. K. Ellis and G. Martinelli, "Leptoproduction and Drell-Yan Processes Beyond the Leading Approximation in Chromodynamics," MIT preprint CTP #723 (1978).
29. W. Bardeen, A. J. Buras, D. W. Duke and T. Muta, "Deep Inelastic Scattering Beyond Leading Order in Asymptotically Free Gauge Theories," Fermilab-PUB-78/42 THY.
30. H. Wahl, invited talk presented at the International Conference on Neutrino Physics, Purdue University (1978).
31. There is an additional term in $d\sigma^{\Sigma}_{DIS}/d\hat{t}$ of the form $m_q^2 Q^2/\hat{s}^2$ that produces a finite contribution when integrated over $\hat{s}$. This piece is included when calculating the $\delta(z-1)$ contribution to $f_{\Sigma,q}(z)$.
32. I am finding it difficult to calculate the δ-function piece of $f_{\Sigma,q}(z)$ in eq. (6.52) with the off-shell technique. To calculate the δ-function term, one must integrate the cross section $\hat{\sigma}^{\Sigma}_{DIS}$ over all $\hat{s}$ in the range $-m_q^2 \leq \hat{s} \leq \infty$. For $P_q^2 = -m_q^2$, $\hat{s}$ can be negative! Thus, although the technique removes the problem with $1/\hat{t}$, it does not remove the $1/\hat{s}$ pole in (6.47a). Because of this difficulty, I am not at present able to deduce the δ-function in (6.52) with this technique. What I have done instead is to fix it by requiring that $\int_0^1 f_{2,q}(z)dz = 0$, where $f_{2,q}(z)$ is given by (6.58a). This condition is necessary to insure that eq. (4.33) is satisfied even to order α_s.

33. Again, as discussed in footnote 32, I am having trouble deducing the δ-function piece of $f_q^{DY}(\hat{\tau})$ in (6.101b) using the off-shell method. Instead what I have done is to calculate the δ-function piece of the difference $\alpha_s(f_q^{DY}-f_{2,q})$ using another regularization scheme and adjusted the δ-function piece of f_q^{DY} in (6.101b) to agree with the result. The regularization scheme I used was to give the gluon a mass μ and leave the quarks massless. At this order in α_s, this method is explicitly gauge invariant and does not suffer from the problems of the off-shell method (now $\hat{s}_{min} = \mu^2$ in DIS). The method will not work if one is calculating diagrams with initial gluons.
34. The δ-function contribution to $\Delta f_q^{DY}(z)$ in (6.106b) was deduced using the massive gluon regularization scheme discussed in footnote 33.

35. H. D. Politzer, Nucl. Phys. B129, 301 (1977) and CALT-68-628 (1977).
36. C. T. Sachrajda, Phys. Letters 73B, 185 (1978).
37. R. K. Ellis, H. Georgi, M. Machacek, H. D. Politzer and G. Ross, "Factorization and the Parton Model in QCD," MIT preprint #718 (1978); Caltech preprint CALT-68-684.
38. D. Amati, R. Petronzio and G. Veneziano, "Relating Hard QCD Processes Through Universality of Mass Singularities," CERN preprint TH-2470 and TH-2527 (1978).
39. S. Libby and G. Sterman, ITP Stony Brook preprint ITP-SB-78-42.
40. A. V. Efremor and A. V. Radyoshicin, "High Momentum Transfer Processes in QCD," JINR preprint, Dubna (1978).
41. G. Altarelli, G. Parisi and R. Petronzio, Phys. Lett. 76B, 351 (1978); Phys. Lett. 76B, 356 (1978); R. Petronzio, CERN preprint TH-2495 (1978).
42. H. Fritzsch and P. Minkowski, Phys. Lett. 73B, 80 (1978).
43. C. Michael and T. Weiler, contribution to the XIIIth Rencontre de Moriand, Les Arcs, France (1978).
44. K. Kajantie and R. Raitio, Helsinki report HU-TFT-77-21; K. Kajantie, J. Lindfors and R. Raitio, Helsinki report HU-TFT-78-18; K. Kajantie and J. Lindfors, Helsinki report HU-TFT-78-33; also see the talk by K. Kajantie at this symposium.
45. F. Halzen and D. Scott, University of Wisconsin report COO-881-21 (1978).
46. E. L. Berger, "Massive Lepton Pair Production in Hadronic Collisions," invited talk at the International Conference at Vanderbilt University (1978, ANL-HEP-PR-78-12; E. L. Berger, "Tests of QCD in the Hadroproduction of Massive Lepton Pairs," ANL-HEP-PR-78-18.

47. It is a bit dangerous to neglect the subprocess $q + q \to q + q + \gamma^*$, for although this subprocess is down by α_s from $g + q \to \gamma^* + q$ and $q + \bar{q} \to \gamma^* + g$, there are many more quarks at high x in a proton than there are gluons or antiquarks. However, recently it has been shown that the subprocess $q + q \to q + q + \gamma^*$ makes a negligible contribution over most of the kinematic range. See J. Kripfganz and A. P. Contogouris, McGill University preprint (1979).
48. The region of integration in eq. (7.4) is discussed in detail in Ref. 44.
49. Y. L. Dokshitser, D. I. D'Yakonov and S. I. Troyan, "Inelastic Processes in Quantum Chromodynamics," from the XIIIth Winter School of Leningrad B. P. Konstantinov Institute of Nuclear Physics (1978), SLAC translation #183.
50. G. Pariri and R. Petronzio, "Small Transverse Momentum Distributions in Hard Processes," CERN-TH-2627 (1979).
51. This smearing of the final $\sigma_p(s,M^2,y,p_\perp^2)$ is not the same method of smearing as we used in Refs. 20, 24-25. In these papers, the initial quarks were assigned a perpendicular and parallel component of momentum and the $\hat{s}$ and $\hat{t}$ invariants were then constructed. Thus both $\hat{s}$ and $\hat{t}$ were affected by the smearing.
52. D. C. Hom et al., Phys. Rev. Lett. <u>36</u>, 1239 (1976) and <u>37</u>, 1374 (1976); S. W. Herb et al., Phys. Rev. Lett. <u>39</u>, 252 (1977); W. R. Innes et al., Phys. Rev. Lett. <u>39</u>, 1240 (1977).
53. J. Kubar-Andre and F. E. Paige, "Gluon Corrections to the Drell-Yan Model," BNL preprint BNL-24794.
54. J. Abad and B. Humpert, Phys. Letters <u>78B</u>, 627 (1978).
55. The problem is that several groups using different regularization schemes have gotten slightly different values for the δ-function in (6.106b). The authors in Ref. 28 get $(\pi^2/3 - 1)$ using the off-shell technique rather than $(4\pi^2/3 + 11/2)$ I get using the massive gluon method (see footnote 33). On the other hand, the authors of Ref. 53 find $(4\pi^2/3 + 1)$ using a massive gluon and massive quark method. Recently, the authors of Ref. 28 have also deduced the value $(4\pi^2/3 + 1)$ using dimensional regularization (K. Ellis, private communication). Perhaps the correct answer is $(4\pi^2/3 + 1)$ and I have made an error in getting $(4\pi^2/3 + 11/2)$ or perhaps there is something more subtle going on. I just cannot say at present and will continue to use the value presented in (6.106b). However, it makes little difference if one instead uses $(4\pi^2/3 + 1)$. The point is that both are large numbers.
56. This is assuming that the order α_s^2 corrections to the $p_\perp$ spectrum are not so large as to vitiate the conclusions based on

calculations of order α_s.

57. One must be careful for if the Δf_q^{DY} (or equivalently the ΔB_n^{DY}) are too large, then according to (6.22b) there will be a sizable correction to the Q^2 evolution formula.
58. C. Quigg, Rev. Mod. Phys. 49, 297 (1977).
59. R. F. Peierls, T. L. Trueman and L. L. Wang, Phys. Rev. D16, 1397 (1977).
60. Although Fig. 7.15 shows the ratio of the order α_s predictions to the "leading order" results, it does not show the difference in the W^+ rate due to the order α_o^2 corrections to the Q^2 evolution formula in (6.22b). These effects could also be important and should be examined.
61. See K. Konishi, A. Ukawa and G. Veneziano, "A Simple Algorithm for QCD Jets," CERN preprint TH2509 (1978).
62. J. F. Owens, Phys. Letters 76B, 85 (1978).
63. Tsuneo Uematsu, Kyoto University preprint RIFP-335 (1978).
64. R. D. Field and R. P. Feynman, Nucl. Phys. B136, 1 (1978).
65. S. Brodsky and J. Gunion, Phys. Rev. Letters 37, 402 (1976); S. Brodsky, invited talk at the XII Rencontre de Moriond (1977), (SLAC-PUB-1937).
66. C. T. Sachrajda, Phys. Letters 76B, 100 (1978).
67. See also, W. Furmanski, "Large $p_\perp$ Jet Cross-Section from QCD," Phys. Letters 77B, 312 (1978) and Jagellonian University preprints TPJU-10/78, TPJU-11/78 and TPJU-12/78.
68. R. Cutler and D. Sivers, Phys. Rev. D16, 679 (1977); Phys. Rev. D17, 196 (1978).
69. B. L. Combridge, J. Kripfganz and J. Ranft, Phys. Lett. 234 (1977).
70. M. Della Negra et al., (CCHK Collaboration), Nucl. Phys. B127, 1 (1977).
71. Fermilab-Lehigh-Pennsylvania-Wisconsin Collaboration, talks given by W. Selove and A. Erwin at the symposium on Jets in High Energy Collisions, Niels Bohr Institute-Nordita, Copenhagen, July 1978.
72. M. G. Albrow et al., Nucl. Phys. B135, 461 (1978).
73. It is quite possible that quarks, antiquarks and gluons do not all have the same "effective" $k_\perp$ spectra. For example, the larger mean $k_\perp$ of the Upsilon compared with the non-resonant background might be interpreted by saying that gluons have a larger effective $\langle k_\perp \rangle$ than do quarks. This approach has been adapted by the Florida State group in Ref. 74.
74. J. F. Owens, E. Reya and M. Gluck, FSU preprint 77-09-07 (1978); J. F. Owens and J. D. Kimel, FSU HEP 78-03-30 (1978).
75. Recently some attempts have been made to handle the primordial $k_\perp$ more properly. See, for example, Ref. 50 and Howard Georgi and Jon Sheiman, "Transverse Momentum Distributions in Lepton-Hadron Scattering from QCD," HUTP-78/A034.

76. F. Halzen, G. A. Ringland and R. G. Roberts, Phys. Rev. Lett. 40, 991 (1978).
77. J. F. Gunion, "The Interrelationship of the Constituent Interchange Model and Quantum Chromodynamics," presented at the discussion meeting on Large Transverse Momentum Phenomena, SLAC, January 1978; R. Horgan, W. Caswell and S. J. Brodsky, SLAC-PUB (1978).
78. K. Kinoshita and Y. Kinoshita, "Effects of Parton Transverse Momenta on Hadronic Large $p_\perp$ Reactions," preprint submitted to the XIX International Conference on High Energy Physics, Tokyo, 1978.
79. R. R. Horgan and P. N. Scharbach, "Transverse Momentum Fluctuations and High $p_\perp$ Processes in Quantum Chromodynamics," SLAC-PUB-2188 (1978).
80. R. Blankenbecler, S. J. Brodsky and J. F. Gunion, "The Magnitudes of Large Transverse Momentum Cross Sections," SLAC-PUB-2057 (1977).
81. D. Jones and J. F. Gunion, "The Transition from Constituent Interchange to QCD $p_\perp^{-4}$ Dominance at High Transverse Momentum," SLAC-PUB-2157 (1978).
82. A. G. Clark et al., Phys. Letters 74B, 267 (1978); A. G. Clark, talk presented at this symposium.
83. CERN-Columbia-Oxford-Rockefeller Experiment, reported by L. Di Lella in the Workshop on Future ISR Physics, Sept. 14-21, 1977, edited by M. Jacob; and the talk by L. Di Lella at the symposium on Jets in High Energy Collisions, Niels Bohr Institute-Nordita, Copenhagen, July 1978.
84. H. Boggild, (British-French-Scandinavian Collaboration), invited talk at the Eighth Symposium on Multiparticle Dynamics, Kaysersberg, France (1977); R. Moller, invited talk at the Moriond Conference (1977). Also see the talk by R. Moller and E. Kluge at the symposium on Jets in High Energy Collisions, Niels Bohr Institute-Nordita, Copenhagen, July 1978.
85. Columbia-Fermilab-Stony Brook Collaboration, R. J. Fisk, Ph.D. thesis, State University of New York at Stony Brook (1978); R. J. Fisk et al., Phys. Rev. Letters 40, 984 (1978).
86. The results of the Columbia-Fermilab-Stony Brook Experiment (Ref. 85) are not quite as easy to interpret as I first thought. Finite size detectors can produce (apparent) scale breaking effects so one must be quite careful to include acceptance corrections into ones with theoretical predictions. After a more careful analysis, my preliminary conclusion is that the experimental results indicate that about one power of $p_\perp$ (i.e., $p_\perp^{-6} \to p_\perp^{-7}$) is coming from primordial $k_\perp$ smearing in the $p_\perp$ range of the experiment. This is in contrast to the two powers that result from our smearing[23-25] (i.e., $p_\perp^{-6} \to p_\perp^{-8}$); however, a more careful analysis should be performed.
87. This was also predicted in a less model dependent way by D. S. Ellis, M. Jacob and P. V. Landshoff, Nucl. Phys. B108, 93

(1976); M. Jacob and P. V. Landshoff, Nucl. Phys. B113, 395 (1976).
88. C. Bromberg et al., Phys. Rev. Letters 38, 1447 (1977); C. Bromberg et al., CALT-68-613 (to be published in Nucl. Phys.).
89. G. C. Fox, "Recent Experimental Results on High Transverse Momentum Scattering from Fermilab," invited talk given at the Argonne APS Meeting (1977).
90. J. Babcock, D. Sivers and S. Wolfram, Phys. Rev. D18, 162 (1978).
91. B. Combridge, invited talk presented at the symposium on Jets in High Energy Collisions, Niels Bohr Institute-Nordita, Copenhagen, July 1978.
92. J. W. Cronin et al., (CP Collaboration), Phys. Rev. D11, 2105 (1975); D. Antreasyan et al., Phys. Rev. Lett. 38, 112 (1977); Phys. Rev. Lett. 38, 115 (1977).
93. B. Alper et al., (BS Collaboration), Nucl. Phys. B100, 237 (1975).
94. F. W. Busser et al., Nucl. Phys. B106, 1 (1976).
95. G. Donaldson et al., Phys. Rev. Lett. 36, 1110 (1976).
96. D. C. Carey et al., Fermilab Report No. FNAL-PUB-75120-EXP. (1975).
97. The subprocess $gg \to c\bar{c}$ is not the only contributor to high $p_\perp$ charm production. As discussed in Ref. 91, the subprocesses $cq \to cq$ and $cg \to cg$ are also important. In addition, G. C. Fox, S. Wolfram and I recently deduced that the production of large $p_\perp$ gluons that subsequently fragment into a charmed meson is a new and important source of high $p_\perp$ charmed mesons.
98. G. Farrar and S. Frautschi, Phys. Rev. Letters 36, 1017 (1976); G. R. Farrar, Phys. Letters 67B, 337 (1977).
99. F. Halzen and D. M. Scott, Phys. Rev. Letters 40, 1117 (1978); University of Wisconsin preprint COO-881-21 (1978).
100. C. O. Escobar, Nucl. Phys. B98, 173 (1975); Phys. Rev. D15, 355 (1977).
101. R. Ruckl, S. J. Brodsky and J. F. Gunion, "The Production of Real Photons at Large Transverse Momentum in pp Collisions," UCLA preprint (1978).
102. H. Fritzsch and P. Minkowski, CERN preprint TH2320 (1978).
103. There is, of course, also a contribution to direct photon production from the annihilation subprocess, $q + \bar{q} \to g + \gamma$, which I have neglected since the gluon content within the proton is considerably larger than the antiquark content.
104. G. Sterman and S. Weinberg, Phys. Rev. Lett. 39, 1436 (1977).
105. J. Ellis, M. K. Gaillard and G. G. Ross, Nucl. Phys. B111, 253 (1976).
106. A. De Rújula, J. Ellis, E. G. Floratos and M. K. Gaillard, Nucl. Phys. B138, 387 (1978).
107. See also, C. Basham, L. Brown, S. D. Ellis and S. T. Love, Phys. Rev. D17, 2298 (1977) and "Energy Correlations in e^+e^- Annihilations: Testing QCD," University of Washington preprint RLO-1388-759 (1978).

108. G. C. Fox and S. Wolfram, "Observables for the Analysis of Event Shapes in e^+e^- Annihilation and Other Processes," CALT-68-680 and CALT-68-678.

109. J. Kripfganz and A. Schiller, "QCD Three-Jet Production in Large $p_\perp$ Processes," Leipzig preprint KMU-HEP-78-10 (1978).

110. Due to the lack of time and energy, I have not been able to cover all applications of QCD. For example, there are interesting applications of QCD to the production of jets in photon-photon collisions. See C. H. Llewellyn Smith, Oxford preprint 56/78 (1978); S. J. Brodsky, T. A. De Grand, J. F. Gunion and J. H. Weis, SLAC-PUB-7102 (1978); C. T. Hill and G. G. Ross, CALT-68-659 (1978); and W. Frazer and J. F. Gunion, UCSD preprint UCSD-10P10-194 (1978).

111. D. A. Ross, "The Effects of Higher Order QCD Corrections in Deep Inelastic Scattering," CALT-68-699 (1979); D. A. Ross and C. T. Sachrajda, CERN preprint TH2565 (1978).

112. Some progress in understanding this has been made by the authors of Ref. 50.

113. See also, A. P. Contogouris, R. Gaskell and S. Papadopoulos, McGill University preprints; A. P. Contogouris, McGill University preprint (1978); J. Ranft and G. Ranft, Leipzig preprint (1978); R. Raitio and R. Sosnowski, University of Helsinki preprint HU-TFT-77-22, invited talk given at the Workshop on Large $p_\perp$ Phenomena, University of Bielefeld, Sept. 5-8, 1977.

THE PHASES OF A GAUGE THEORY

L. Susskind
Stanford University

I. INTRODUCTION 244

II. THE ISING MODEL IN A TRANSVERSE FIELD 245

III. Z_2 GAUGE THEORY 251

IV. ORDER AND DISORDER 255

V. LATTICE DUALITY 257

VI. NON ABELIAN GAUGE THEORY 264

ISSN: 0094-243X/79/50243-25$1.50

THE PHASES OF A GAUGE THEORY

L. Susskind
Stanford University

I. INTRODUCTION

In these lectures I am going to describe the possible phases of gauge theories. The word phase is being used in the same sense as in statistical mechanics when we speak of an ordered phase, disordered phase, ferromagnetic phase and phase transition. It refers to the global or long range behavior of the system and, unlike the small distance behavior, is undefined except for infinite systems.

How does one prove that a phase transition occurs in a given mathematical model when a parameter is varied? Indeed how does one recognize such a transition? One way is to concentrate on some thermodynamic quantity such as free energy or specific heat and look for singularities. However this is usually the hard way, especially in the case of second and higher order transitions. Often the easy way is to find an "order parameter" which vanishes in one phase and has a non vanishing value in the other. The simplest example is the spontaneous magnetization M of a ferromagnet which is exactly zero for all temperature $\geq T_c$ and is non vanishing for $T<T_c$ (see Fig. 1).

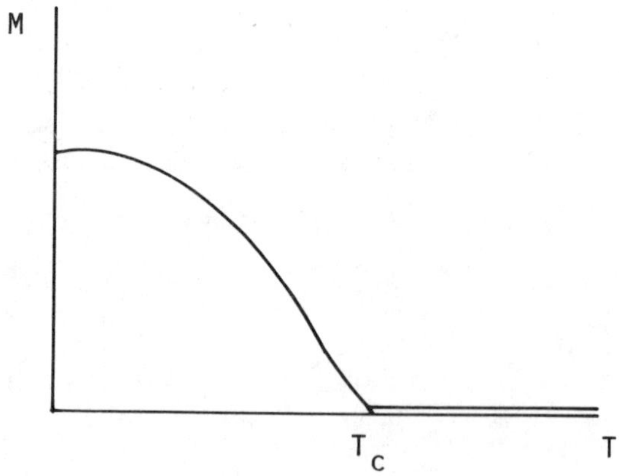

Figure 1. The spontaneous magnetization of a ferromagnet shows the phase transition at T_c.

Proving the existence of a transition involves four main steps. The first and second are showing that a T=0 and T=∞ the magnetization or order parameter is non zero and zero respectively. This is usually easy. The third and fourth steps are to prove the stability of these two behaviors for ranges of temperature around zero and infinity. This is usually done by examining the low and high temperature expansions.

The present interest in such questions of phases in quantum field theory is due to the very different behavior of quantum electrodynamics, weak interactions and quantum chromodynamics in the long wavelength limit. Indeed QED exists in a "ordinary" phase characterized by the existence of free massless photons and charges which interact at large distance via the coulomb law. To explain the weak interactions we postulate a "Higgs phenomenon" or phase transition to a phase in which gauge bosons gain a mass and the coulomb law is replaced by the Yukawa potential. Most interesting of all is the "confined phase" of QCD needed to explain the apparent absence of free quarks. In these lectures I will introduce you to the bewildering variety of phases and order parameters which are needed to completely understand the phase diagram of a gauge theory.

II. THE ISING MODEL IN A TRANSVERSE FIELD

a) The Model

We are going to proceed by studying a number of examples of increasing complexity. The first is the 1+1 dimensional Ising model in a transverse field. It is closely related to the 2 dimensional Ising model. It is obtained by making one of the dimensions (time) imaginary, and allowing it to become continuous. Thus we consider an infinite one dimensional lattice of spins $\sigma(n)$ (n=...-3,-2,-1, 0,1,2,3...). The spins σ are ordinary Pauli spins satisfying the algebra

$$\begin{aligned} [\sigma_i(n),\sigma_j(n')] &= 0 \qquad (n\neq n') \\ [\sigma_i(n),\sigma_j(n)] &= 2i\,\varepsilon_{ijk}\,\sigma_k(n) \\ \{\sigma_i(n)\sigma_j(n)\} &= 2\,\delta_{ij} \end{aligned} \tag{2.1}$$

The ordinary 1-dimensional Ising model is defined by the Hamiltonian

$$H = -\lambda \sum_n \sigma_3(n)\sigma_3(n+1) \tag{2.2}$$

where λ is a constant. Since all the $\sigma_3(n)$ commute the dynamics is entirely trivial

$$\frac{d}{dt}\sigma_3(n) = 0 \qquad \text{(all n)}$$

We now add to Eq. 2.2 an external field in the 1 direction. The additional term is defined to be $-\sum_n \sigma_1(n)$.

$$H = -\sum_n \sigma_1(n) - \lambda \sum_n \sigma_3(n)\sigma_3(n+1)$$
$$\equiv H_1 + \lambda H_3 \qquad (2.3)$$

The parameter λ replaces the temperature in the two dimensional Ising model. In other words we are interested in changes of phase as λ is varied. We will search for such a transition by using the order parameter strategy.

Let us begin with the behavior of the ground state for λ very small. For λ=0 the ground state $|0>$ is a product state satisfying

$$\sigma_1(n)|0> = |0> \qquad (2.5)$$

with no correlation between sites. Evidently the expectation values of the $\sigma_3(n)$ vanish in such a state since σ_3 anticommutes with σ_1.

$$<0|\sigma_3(n)|0> = 0 \qquad (2.6)$$

The vanishing of $<\sigma_3>$ is due to a symmetry of the Hamiltonian and its ground state. Consider the unitary operator

$$G^\dagger = G = \prod_n \sigma_1(n) \qquad (2.7)$$

This operator acts on the spin degress of freedom thusly

$$G\,\sigma_1(n)G^\dagger = \sigma_1(n)$$
$$G\,\sigma_3(n)G^\dagger = -\,\sigma_3(n) \qquad (2.8)$$

Thus G represents the operation of flipping all the Ising spins $\sigma_3(n)$. Evidently the two terms in H are spearately invariant under G.

Now suppose that the ground state $|0>$ is invariant under G. Then we easily prove 2.6:

$$<0|\sigma_3|0> = <0|G\sigma_3G^\dagger|0> = -\,<0|\sigma_3|0> \qquad (2.9)$$

Furthermore the λ=0 vacuum satisfying 2.5 is obviously invariant under G.

Let us now consider small but nonvanishing λ. (We denote the ground state for a given λ by the notation $|\lambda>$. The new ground state is given by perturbation theory to be

$$|\lambda> = |0> + \frac{\lambda}{E-H_1} H_3|0> + \ldots \qquad (2.10)$$

Since both H_3 and H_1 are G-invariant it follows that $|\lambda>$ is invariant to all orders in λ. Thus it follows that

$$<\lambda|\sigma_3|\lambda> = 0 \tag{2.11}$$

for all $\lambda<\lambda^*$ where λ^* is the radius of convergence of the perturbation series.

Now consider very large λ. We will write H as

$$H = \lambda \left\{ H_3 + \frac{1}{\lambda} H_1 \right\} \tag{2.12}$$

and think of $1/\lambda\ H_1$ as a perturbation. In this case, for $1/\lambda = 0$ the ground state is degenerate and consists of the two states in which the spins σ_3 are all parallel.

$$\sigma_3(n)|\infty>_\pm \equiv \pm|\infty>_\pm \tag{2.13}$$

and

$$<\infty|\sigma_3|\infty> = \pm 1 \tag{2.14}$$

The double degeneracy of the ground state is not special to the limit $\lambda=\infty$. By studying the perturbation expansion in powers of λ^{-1} we can prove that for small but nonzero λ^{-1} the degeneracy remains and that

$$<\lambda_\pm|\sigma_3|\lambda>_\pm = \pm M(\lambda) \tag{2.15}$$

where $|\lambda>_\pm$ are the two degenerate ground states and $M(\lambda)$ is a nonvanishing function of λ.

The expectation value of σ_3 in the ground state is the order parameter of the Ising model in a transverse field. Since it identically vanishes for $\lambda<\lambda^*$ and does not vanish for λ greater than some value, a transition must occur at some finite λ.

b) Self Duality

One of my reasons for dwelling on the ising model is that it is the simplest system which exhibits the remarkable property of duality. This property of duality underlies much of the recent work on the phase diagram of gauge theories.

Let us define the dual lattice to consist of a new lattice whose sites are identified with the links or bonds of the original as shown in Fig. 2

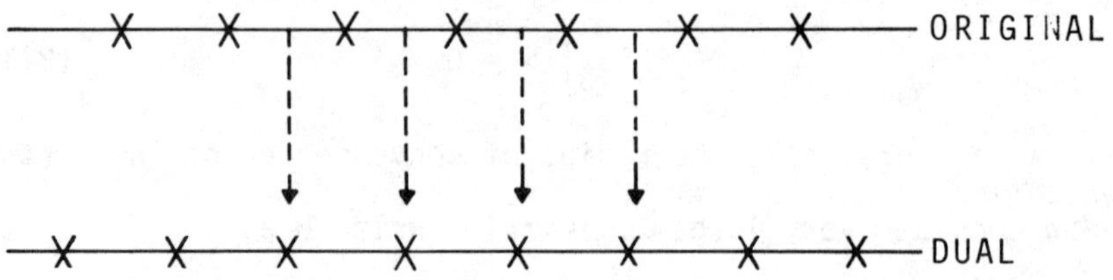

Figure 2. A one-dimensional lattice and its dual lattice

The nth site of the dual lattice is identified with the link joining sites n and n+1 of the original. Next we introduce spin variables μ_1 and μ_3 which live on the dual sites. The variable $\mu_1(n)$ tells us information about the bond (n,n+1). In particular it takes on the value +1(-1) if the spins $\sigma(n)$ and $\sigma(n+1)$ are parallel (anti-parallel). Thus

$$\mu_1(n) = \sigma_3(n)\sigma_3(n+1) \qquad (2.16)$$

The variable $\mu_3(n)$ is called the disorder variable and is defined by

$$\mu_3(n) = \prod_{m\leq n} \sigma_1(n) \qquad (2.17)$$

It has the interesting property of flipping all the $\sigma_3(m)$ to the left of the site n+1.

$$\begin{aligned} \mu_3(n)\sigma_3(m)\mu_3(n) &= \sigma_3(m) \quad (m>n) \\ \mu_3(n)\sigma_3(m)\mu_3(n) &= -\sigma_3(m) \quad (m\leq n) \end{aligned} \qquad (2.18)$$

It is easy to see that the set of algebraic relations satisfied by the variables $\mu(n)$ is isomorphic to the σ algebra. Indeed

$$\begin{aligned} \mu_1^2(n) &= \mu_3^2(n) = 1 \\ \{\mu_1(n),\mu_3(n)\}_+ &= 0 \\ [\mu_i(n),\mu_j(n')] &= 0 \qquad (n\neq n') \end{aligned} \qquad (2.19)$$

Now the amazing property of the system defined by 1.3 is that it can be reexpressed in terms of the μ in the same form. To this end note

$$\sigma_1(n) = \mu_3(n)\mu_3(n-1) \qquad (2.20)$$

Combining 2.3, 2.16 and 2.20 gives

$$H = -\lambda \sum \mu_1(n) - \sum \mu_3(n)\mu_3(n+1)$$
$$= \lambda \{-\sum \mu_1 - \frac{1}{\lambda} \sum \mu_3(n)\mu_3(n+1)\} \qquad (2.21)$$

Evidently we have derived an equivalence between two Hamiltonians. The first is expressed in terms of σ and has coupling λ. The second is of exactly the same form with μ replacing σ and λ being replaced by $1/\lambda$. This means that the small and large λ behavior of the original system are closely related by duality. In fact, solving the model for $\lambda>1$ instantly provides a solution for $\lambda<1$. For example, suppose the ground state energy $E(\lambda)$ is computed for $\lambda>1$ by solving 2.3 From 2.21 we can conclude that for $\lambda<1$ can be obtained using

$$E(\lambda) = \lambda E(1/\lambda) \ . \qquad (2.22)$$

As another illustration of duality consider the question of the magnetization of the ground state. For $\lambda>>1$ we know

$$\langle\lambda|\sigma_3|\lambda\rangle = M(\lambda) \neq 0 \qquad (2.23)$$

and for $\lambda<<1$

$$\langle\lambda|\sigma_3|\lambda\rangle = 0 \qquad (2.24)$$

(In fact it is easy to prove that the transition point between 2.23 and 2.24 must be at the symmetry point of the duality transformation where $\lambda=1$.) From 2.23 and 2.24 we can easily conclude

$$\langle\lambda|\mu_3|\lambda\rangle = M(\frac{1}{\lambda}) \qquad (2.25)$$

$$\langle\lambda|\mu_3|\lambda\rangle = 0 \qquad (2.26)$$

for $\lambda<1$ and $\lambda>1$ respectively.

The expectation value $\langle\lambda|\mu_3|\lambda\rangle$ is nonvanishing exactly when the original order parameter $\langle\sigma_3\rangle$ vanishes, i.e. the disordered phase. For this reason the quantity $\langle\mu_3\rangle$ is called the disorder variable. However it should be recognized from the outset that the question of which phase is ordered is one of convention. We can either describe the small λ phase as disordered with respect to the variables σ_3 or ordered with respect to μ_3.

The operator $\mu_3(n)$ has another significance as a creation operator for a "kink" at dual site (n). Consider the ground state for $\lambda>>1$ when the magnetization is almost complete ($\langle\sigma_3\rangle \approx 1$). The action of $\mu_3(n)$ on such a state is to reverse the magnetization over the portion of the lattice between $-\infty$ and n. We say a kink has been created. Obviously if the ground state has any nonvanishing magnetization, flipping half the lattice must lead to an orthogonal state since it differs in an observable manner over an infinite volume. Thus $\mu_3|\lambda>$ must be orthogonal to $|\lambda>$ for $\lambda>1$. The disorder parameter can only become nonzero when $\mu_3|\lambda>$ is no longer orthogonal to λ or when the addition of a kink can not be unambiguously recognized in the vacuum. This obviously requires the average magnetization to vanish.

An interesting picture of the transition from the ferromagnetic to the unmagnetized phase can be formulated as a kind of condensation of kinks. Begin with the completely magnetized vacuum for the case $\lambda=\infty$. Obviously this state is completely empty of kinks (a kink may be defined as a link with $\sigma_3(n)\sigma_3(n+1)=-1$). Such a state has a long range correlation

$$\lim_{n\to\infty} <\infty|\sigma_3(0)\sigma_3(n)|\infty> = 1$$

Now let us turn on the term $\sum\sigma_1$ as a perturbation. Since it flips a spin at some site n it creates a kink pair (see Fig. 3)

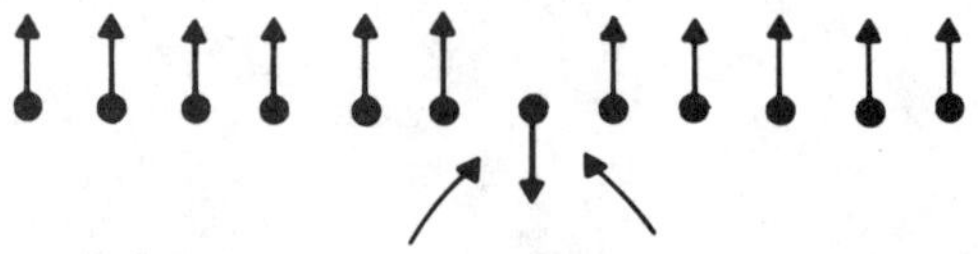

Figure 3. σ_1 flips a spin, creating a kink pair.

The pairs produced in this manner are separated by a minimum distance and if $\lambda>>1$ the number of such pairs is small. That is to say a dilute gas of kink pairs is formed. Obviously if we consider a distant pair of points, say 0 and n it is most likely that an even number of kinks are found on the interval between them. Only in the rare event that 0 or n "splits a pair" will the number be odd. Thus it is most likely that the product $\sigma_3(0)\sigma_3(n) = 1$ and the correlation $\lim_{n\to\infty} <\sigma_3(0)\sigma_3(n)>$ remains near 1. In fact the deviation from unity is just proportional to the probability to find a pair of 0 or n and is proportional to $(1/\lambda)^2$.

As λ decreases two things happen. First the density of such pairs begins to increase. Secondly the separation between kinks in a pair increases. Both effects make it more likely that an odd number of kinks be found on the interval (0,n). Accordingly the long range correlation decreases.

Finally a point occurs at which the distance between pairs becomes comparable to the separation within an average pair. At this point the pairing breaks down. On a large interval it becomes equally likely to find an odd or even kink number. It follows that $<\sigma_3(0)\sigma_3(n)>\to 0$ at large distances.

Beyond this point the addition of a kink to the vacuum does not lead to an unambiguously distinct state. There is so much fluctuation in the kink occupation (kink occupation can be defined as -1^N where N is the number of kinks) that the additional kink is "swallowed" by the condensate. Thus the disorder parameter or expectation value of the kink creator becomes nonvanishing.

III. Z_2 GAUGE THEORY

The Z_2 gauge theory is the simplest prototype of a gauge system. We shall study this system in 3+1 dimensional space time where three dimensional space is replaced by a simple cubic lattice. As usual, the degrees of freedom for a gauge system live on the links of the lattice which are denoted by a site r and a unit lattice vector $\hat{n}$. (The links $(r,\hat{n})$ and $(r+\hat{n},-\hat{n})$ are evidently the same link.) For each link a Pauli spin operator σ $(\sigma_1,\sigma_2,\sigma_3)$ is defined. The Hamiltonian contains two terms analogous to the electric and magnetic energies. The electric energy is a sum over links

$$- \sum_{\text{links}} \sigma_1(r,\hat{n}) \tag{3.1}$$

The magnetic energy is not associated with links but rather with elementary boxes or plaquettes of the lattice. For each such box bounded by four links we define the magnetic energy to be

$$-\lambda\ \sigma_3(1)\sigma_3(2)\sigma_3(3)\sigma_3(4) \tag{3.2}$$

where 1,2,3 and 4 denote the four sides of the box. Thus

$$H = - \sum_{\text{link}} \sigma_1(\ell) - \lambda \sum_{\text{boxes}} \sigma_3(1)\sigma_3(2)\sigma_3(3)\sigma_3(4) \tag{3.3}$$

The symmetry of this Hamiltonian is much bigger than the global Z_2 symmetry described by 2.8. The Hamiltonian 3.2 has a separate invariance for each site of the lattice. Consider the six links $(r,\hat{n}_i)$ originating at site r (i runs from 1-6). The operator

$$G(r) = \prod_{i=1}^{6} \sigma_1(r,\hat{n}_i) \tag{3.4}$$

generates a local gauge transformation which has the following action:

All σ_1 are left unchanged by G(r). However the 6 σ_3 operators belonging to the links $(r,\hat{n}_i)$ are flipped in sign. All other σ_3 remain unchanged. It is easy to see that any such transformation leaves H unchanged. The operators G(r) evidently take on the eigenvalues ±1.

Another important point is that all the G(R) commute with one another and can simultaneously be diagonalized. Since they are all conserved, the specification of an initial state requires us to specify the values of all these G(r) simultaneously.

In what follows we shall see that G(r) always equals 1 in the ground state $|\lambda>$

$$G(r)|\lambda> = |\lambda> \quad (3.5)$$

The freedom to choose G(r)=-1 at some point corresponds to the introduction of an external charge at site r. If we wish to study the dynamics of the pure gauge field without sources then G should be set equal to unity everywhere

$$G(r) = 1 \quad (3.6)$$

This equation is not an operator equation but rather a condition on the states of the system. It states that the meaningful states are those which are invariant under the action of a gauge transformation.

Now consider the behavior of the vacuum for very small λ. For λ=0 the vacuum, $|0>$, is a simultaneous eigenvector of all the σ_1 variables with σ_1=1. We shall say that the electric flux through each link vanishes in such a state.

Excited states are configurations in which some set of $\sigma_1(r,\hat{n})$ are flipped. Those excited links with σ_1=-1 will be said to have Z_2 electric flux through them. The total energy of an excited state relative to the vacuum is 2N where N is the number of links carrying electric flux.

The simplest excitation of the vacuum is a single link carrying electric flux (see Fig. 4). However this configuration

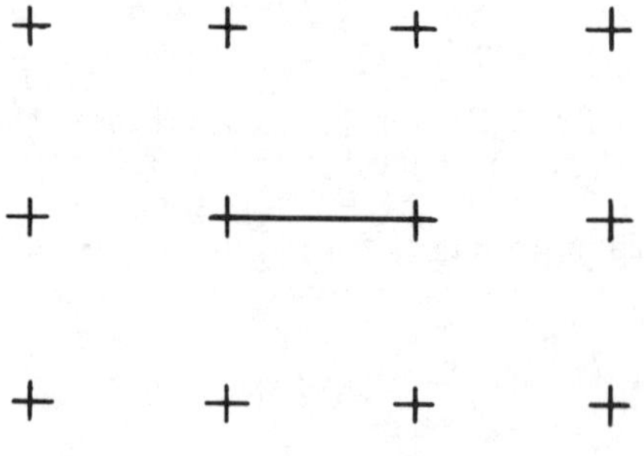

Figure 4. An excitation of the vacuum in which a single link carries electric flux.

clearly violates the condition of gauge invariance (Eq. 3.6) at the ends of the excited links. In general, to satisfy Eq. 3.6 we must superimpose continuous non-ending lines of flux. For example any closed loop of flux (Fig. 5) satisfies 3.6 since every vertex has an even number of excited links radiating from it. The first gauge

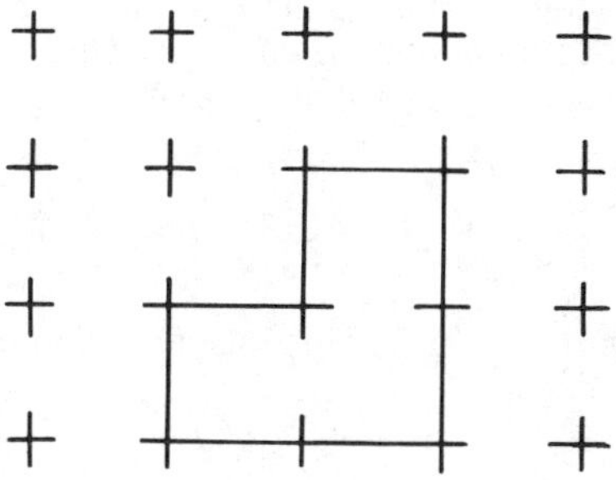

Figure 5. A gauge invariant excitation of the vacuum. An even number of excited links radiate from each vortex.

invariant excited state consists of a single square of excited lines and has excitation energy 8.

The most interesting feature of this model is that it confines. To define this property let us imagine introducing the Z_2 analogue of static sources (electric charges) by modifying Eq. 3.6 to the form

$$G(r) = \eta(r) \tag{3.7}$$

where $\eta(r)=\pm 1$. Wherever $\eta=-1$ we shall say a Z_2 charge is present. Evidently charges are sources of flux which allow a line of flux to terminate on a site. As an example suppose that only sites 0 and r_0 have charges. The original vacuum with $\sigma_1=1$ no longer satisfies 3.7 at $r=0$ and $r=r_0$. A simple class of states satisfying 3.7 can be constructed by allowing a continuous line of flux to run from 0 to r_0 along any path. The state with minimum energy is obtained by finding the path of minimal number of links. In general this path will have a number of links of order the distance (measured in lattice units) from 0 to r_0. The energy will thus be

$$E(0,r_0) \sim 2|r_0| \tag{3.8}$$

Thus a linear confining potential exists between charges.

All of the above conclusions are unchanged for small λ. In particular there continues to exist a linear confining force although the coefficient (string tension) is a decreasing function of λ.

Now consider the behavior for $\lambda>>1$. In this case the magnetic term in H dominates and it is convenient to write

$$\frac{H}{\lambda} = - \sum_{\text{boxes}} \sigma_3(1)\sigma_3(2)\sigma_3(3)\sigma_3(r) - \frac{1}{\lambda} \sum_{\text{links}} \sigma_1(\ell) \tag{3.9}$$

The second term will be treated as a perturbation.

Let us consider the ground state of 3.9 in the case $1/\lambda=0$. Since all the plaquette operators commute they may be simultaneously diagonalized. Therefore the ground state must have each plaquette satisfying

$$\sigma_3(1)\sigma_3(2)\sigma_3(3)\sigma_3(4) = 1 \tag{3.10}$$

We can easily achieve this by choosing the ground state to satisfy

$$1 = \sigma_3(\ell) \tag{3.11}$$

for every link ℓ.

The state described by 3.11 is neither gauge invariant or unique. Assuming no external charges, gauge invariance requires

$$G(r)|\psi> \prod_{i=1}^{6} \sigma_1(r,\hat{n}_i)|\psi> = |\psi> \tag{3.12}$$

But the action of G(r) is to flip the signs of the $\sigma_3(\ell)$ for the 6 links $(r,\hat{n}_i)$. Thus $|\infty>$ can not satisfy 3.11 and be gauge invariant

Evidently the action of a gauge transformation on the state given by 3.11 leads to another state of identical magnetic energy. In fact there are a huge class of degenerate states which minimize the energy, all connected by gauge transformations. Although these states do not satisfy $\sigma_3(\ell)=1$ they do satisfy the gauge invariant equation

$$\sigma_3(1)\sigma_3(2)\sigma_3(3)\sigma_3(4) = 1 \tag{3.12}$$

for every plaquette.

This degeneracy of the ground state is an artifact of the $\lambda=\infty$ limit and is lifted in 6th order in λ^{-1}. To see this we suppose the perturbation $1/\lambda\ \sigma_1(\ell)$ acts 6 times on the 6 links originating at a site. The result is to flip these 6 σ_3's and thus connects the original state to a gauge transform. It is not difficult to prove that diagonalizing the 6th order perturbation in the degenerate subspace leads to a unique ground state given by a gauge invariant superposition of all gauge replicas of the state in 3.11. Defining the state satisfying 3.11 to be $|\psi>$, the true ground state for $\lambda>>1$ is

$$|\infty> = \prod_r \{\frac{1+G(r)}{2}\}\ |\psi> \tag{3.13}$$

Let us now consider the question of confinement for $\lambda>>1$. The introduction of a static source at site r_0 is again accomplished by requiring $G(r_0)=-1$. An eigenvector of the energy satisfying this constraint can easily be constructed by modifying 3.13. Consider the state

$$|r_0\rangle = \prod_{r\neq r_0} \{\frac{1+G(r)}{2}\} \cdot \{\frac{1-G(r_0)}{2}\} \ |\psi\rangle \qquad (3.14)$$

It is easy to see that $|r_0\rangle$ satisfies the following

a) $G(r)|r_0\rangle = |r_0\rangle$ (all $r\neq r_0$)
b) $G(r_0)|r_0\rangle = -|r_0\rangle$
c) $|r_0\rangle$ is in the subspace of degenerate eigenvectors of the unperturbed ($\lambda=\infty$) Hamiltonian (Eq. 3.9). Thus for infinite λ there is no cost in energy for a static source.

A harder thing to prove is that the energy of this state remains finite to all orders in λ^{-1} as the perturbation is turned on. In actual fact the energy difference between the configuration with a source and the vacuum remains zero until 6th order in $1/\lambda$. Thereafter it is finite in each order. We may conclude that for $\lambda >> 1$ the energy of a single isolated source is finite.

The finiteness of the energy of a source can be used to replace the order parameter in defining the existence of two phases of the Z_2 gauge system. Recall that for $\lambda << 1$ the energy of a source is infinite and for $\lambda >> 1$ it is finite. It follows that a point λ_c must exist at which the transition between these behaviors occurs.

IV. ORDER AND DISORDER

The phases of the Ising model in a transverse field are characterized as disordered or ordered according to whether $\langle\sigma_3(0)\sigma_3(r)\rangle \to 0$ (as $r\to\infty$) or not. For the Z_2 gauge system this definition of order is not useful. Consider the correlation function

$$C(\ell_1,\ell_2) \equiv \langle\lambda|\ \sigma_3(\ell_1)\ \sigma_3(\ell_2)\ |\lambda\rangle \qquad (4.1)$$

where ℓ_1 and ℓ_2 are any two distinct links. Also consider a site r_0 such that ℓ_1 ends on r_0 but ℓ_2 does not. According to Eq. 3.5

$$\begin{aligned} C(\ell_1\ell_2) &= \langle\lambda|G^\dagger(r_0)\sigma_3(\ell_1)\sigma_3(\ell_2)G(r_0)|\lambda\rangle \\ &= \langle\lambda|\{G^\dagger(r_0)\sigma_3(\ell_1)G(r_0)\}\{G^\dagger(r_0)\sigma_3(\ell_2)G(r_0)\}|\lambda\rangle \end{aligned} \qquad (4.2)$$

But since ℓ_1 ends on r_0 and ℓ_2 does not

$$\begin{aligned} G^\dagger(r_0)\sigma_3(\ell_1)G(r_0) &= -\sigma_3(\ell_1) \\ G^\dagger(r_0)\sigma_3(\ell_2)G(r_0) &= +\sigma_3(\ell_2) \end{aligned} \qquad (4.3)$$

so that

$$C(\ell_1,\ell_2) = -C(\ell_1,\ell_2) \tag{4.4}$$

Therefore for any distinct pair of links

$$C(\ell_1,\ell_2) = 0 \tag{4.5}$$

Thus there can be no correlation, let alone long range correlation between σ_3's. This is of course a consequence of the local guage invariance of the system.

Let us consider a more general kind of correlation function in order to define a useful concept of long range order in the Z_2 gauge system. Define a region of space Γ of dimension d (d must be less than or equal to the dimensionality of space). The boundary of this region is another region, Γ', of dimension d-1. Now consider a product of spins

$$\prod_{\Gamma'} \sigma_3(\Gamma') \tag{4.6}$$

consisting of all the spins on the boundary Γ'. For example, if Γ is a line Γ' is its two endpoints and 4.6 is the product of two spins. In the Z_2 gauge theory Γ may consist of a surface bounded by a closed loop of links (see Fig. 6). In this case the product

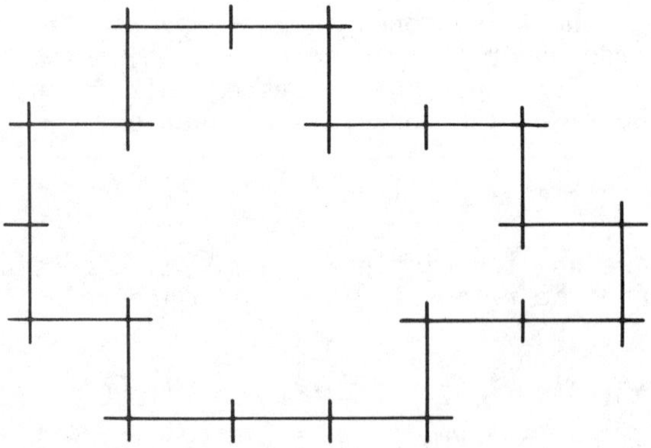

Figure 6. A surface Γ and its boundary Γ' consisting of a closed loop of links.

in 4.6 contains all the σ_3's on the links defining Γ'. Unlike the product of a pair of σ_3's this new object is gauge invariant.

Now define the joint correlation

$$C(\Gamma) \equiv \langle \prod_{\Gamma'} \sigma_3(\Gamma') \rangle \tag{4.7}$$

(in the case where Γ is a line $C(\Gamma)$ is the ordinary correlation function). The behavior of $C(\Gamma)$ as Γ is made larger in size will characterize the global phase of a system. In general this behavior is one of two possibilities:

1) Disordered Behavior: in this case

$$-\log <\prod_{\Gamma'} \sigma_3(\Gamma')> \sim V \tag{4.8}$$

where V is the volume of the region Γ. For example if Γ is a line then 4.8 says

$$-\log <\sigma(0)\sigma(r)> \sim |r| \tag{4.9}$$

Eq. 4.9 expresses the exponential falloff of correlations in the disordered phase of a system like the Ising model. It applies to the analysis of Section II.

For the Z_2 gauge system 4.8 would say that the product of spins on the border of a surface Γ goes to zero exponentially with the area of Γ.

2) The Ordered Phase:

The ordered phase is defined by

$$-\log C(\Gamma) \simeq V(\Gamma') \tag{4.10}$$

where $V(\Gamma')$ is the volume of the boundary of Γ. For example if Γ is a surface $V(\Gamma')$ is the perimeter of its border. For the case Γ=line, the boundary consists of two points and its measure is independent of the volume (length) of the line. Thus in the ordered state of a ferromagnet, Eq. 4.10 says that the correlation $<\sigma_3(0)\sigma_3(r)>$ is constant as $r\to\infty$. Similarly for the Z_2 gauge theory the ordered phase is defined by

$$-\log <\prod_{\Gamma'} \sigma_3> \sim \text{perimeter} \tag{4.11}$$

where Γ' is a closed curve.

In drawing analogies between gauge and spin systems one should always use a rule of correspondence which replaces the 2-point correlation of spin systems with the loop of spins in the gauge system. For example, we can easily prove that for $\lambda >> 1$ the expectation value of the loop satisfies the perimeter law while for $\lambda << 1$ the area law applies. Thus the same language applies to 1+1 dimensional spin systems and 3+1 dimensional gauge systems.

V. LATTICE DUALITY

The dual of a simple cubic lattice is the new lattice of points located at the centers of the cubes of the original lattice (see Fig. 7).

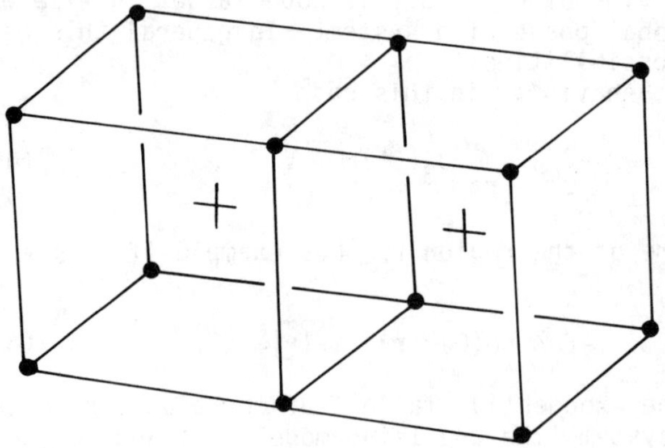

Figure 7. A three-dimensional lattice and two points of the dual lattice. The vertices of the original lattice are denoted by dots. The vertices of the dual lattice, denoted by +'s, are at the centers of the cubes.

Each link of the original (dual) lattice intersects a face of the dual (original) lattice. Thus the replacement of a 3 dimensional lattice by its dual involves

sites ↔ body centers
faces ↔ links

Let us now introduce a set of variables for the dual lattice. For each link of the dual ($\bar{\ell}$ = dual link) define

$$\mu_1(\bar{\ell}) = \sigma_3(1)\sigma_3(2)\sigma_3(3)\sigma_3(4) \tag{5.1}$$

where 1, 2, 3, 4 are the links forming the corresponding plaquette (see Fig. 8). The definition of $\mu_3(\bar{\ell})$ is more complicated. First

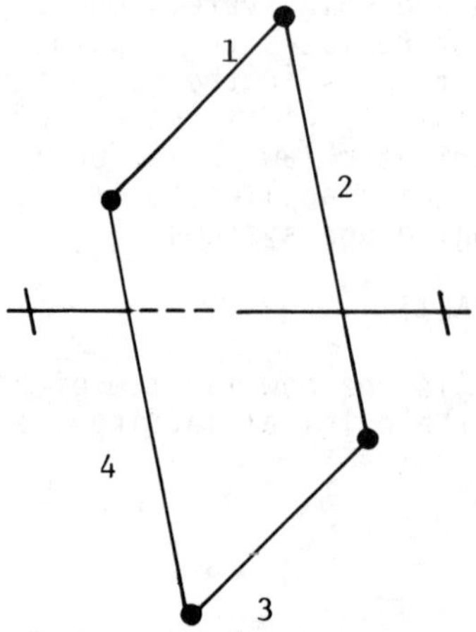

Figure 8. A plaquette of the original lattice and its dual link of the dual lattice.

FIO

define

$$\mu_3(\bar{\ell}) \equiv 1 \qquad (5.2)$$

if $\bar{\ell}$ is a link oriented along the z axis.

For links along the x or y axes $\mu_3(\bar{\ell})$ are formed from infinite products of the σ_1's.

Consider a dual link in the x direction and its related plaquette (Fig. 9).

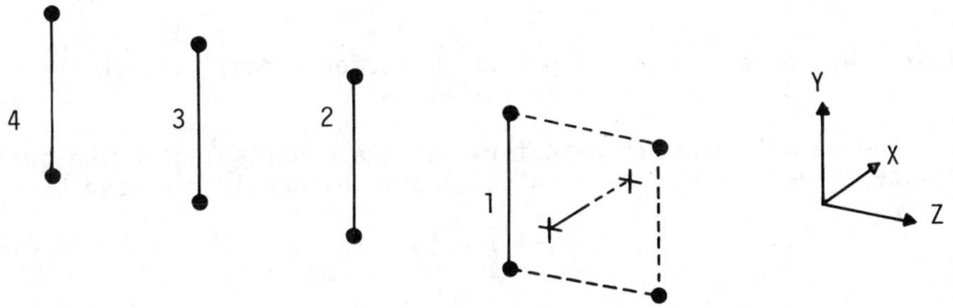

Figure 9. A dual link in the x direction and a portion of the original lattice. μ_3 for the dual link is the product of σ_1's for the links drawn as solid lines.

The plaquette in question is labelled by its lower left hand corner called r. The definition of $\mu_3(\bar{\ell})$ in this case is

$$\begin{aligned}\mu_3(\bar{\ell}) &= \sigma_1(r,\hat{n}_y)\sigma_1(r-\hat{n}_z,\hat{n}_y)\\ &\quad \sigma_1(r-2\hat{n}_z,\hat{n}_y)\sigma_1(r-3\hat{n}_z,\hat{n}_y)\ldots\\ &= \prod_{m=0}^{\infty} \sigma_1(r-m\hat{n}_z,\hat{n}_y)\end{aligned} \qquad (5.4)$$

The infinite product is over the set of links shown as solid lines in Fig. (9).

Similarly for $\bar{\ell}$ oriented along the y axis

$$\mu_3(\bar{\ell}) = \prod_{m=0}^{\infty} \sigma_1(r-m\hat{n}_z,\hat{n}_x) \qquad (5.5)$$

This is shown in Fig. 10.

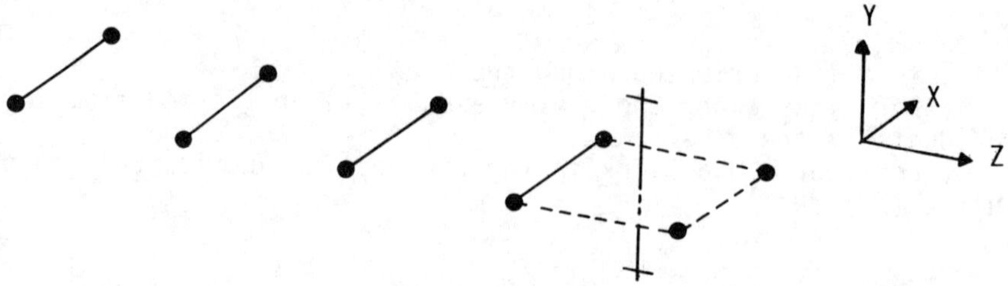

Figure 10. Showing the definition of μ_3 for a dual link in the y direction, similar to Fig. 9.

Now we will rewrite H in terms of the μ variables on the dual lattice. The magnetic term in Eq. 3.3 is evidently replaced by

$$-\lambda \sum_{\ell} \sigma_1(\overline{\ell}) \tag{5.6}$$

The electric term is given by $-\sum_{\overline{b}} \mu_3(1)\mu_3(2)\mu_3(3)\mu_3(4)$. To see this consider a dual box $\overline{b}$ which is identified with a link ℓ in the x direction (see Fig. 11).

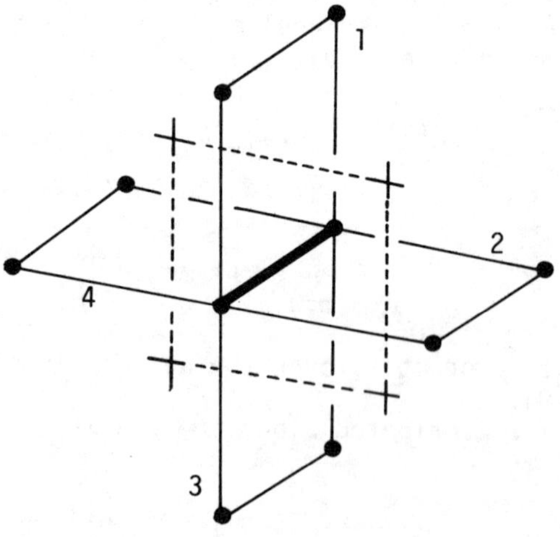

Figure 11. A dual box, the corresponding link of the original lattice, and the four plaquettes including that link.

The object $\mu_3(1)\mu_3(2)\mu_3(3)\mu_3(4)$ has nontrivial factors $\mu_3(2)$ and $\mu_3(4)$ since $\mu_3(1)$ and $\mu_3(3)$ are defined to be 1. Using the definitions of $\mu_3(2)$ and $\mu_3(4)$ we see

$$\mu_3(2)\mu_3(4) = \sigma_1(\ell)$$

Thus for ℓ along the x and y axes the electric term can be identified with the dual magnetic energy $\mu_3(1)\mu_3(2)\mu_3(3)\mu_3(4)$.

For links ℓ in the z direction the argument is less straight forward. Consider the configuration in Fig. 12.

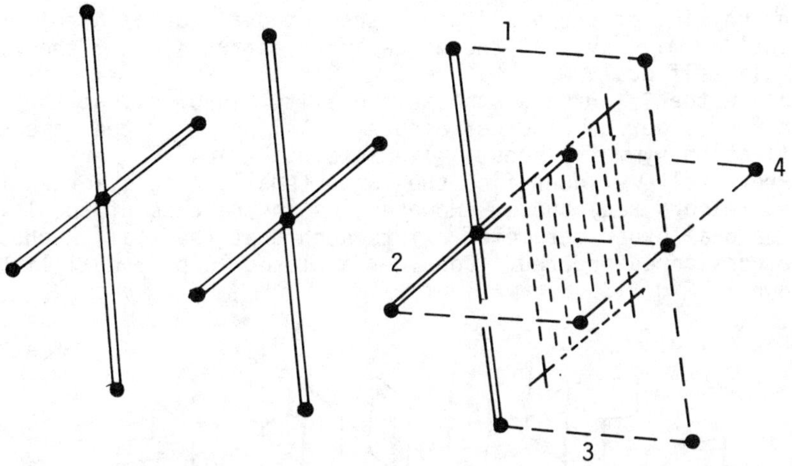

Figure 12. The dual magnetic energy of a dual plaquette parallel to the x-y plane is the product of σ_1's on the links of the original lattice denoted by double lines. Gauge invariance reduces this product to σ_1 on the link through the dual plaquette.

In this case the product $\mu_3(1)\mu_3(2)\mu_3(3)\mu_3(4)$ does not collapse to a finite product of σ_1's. However if we use the gauge condition which says that the product of σ_1's originating at a site is unity then the product collapses to $\sigma_1(\ell)$. Accordingly the total energy is given by

$$\begin{aligned} &-\lambda \sum_{\ell} \mu_1 - \sum_{b} \mu_3\mu_3\mu_3\mu_3 \\ &= \lambda \left[-\sum_{\ell} \mu_1 - \frac{1}{\lambda} \sum_{b} \mu_3\mu_3\mu_3\mu_3\right] \end{aligned}$$

The system therefore appears self dual within the subspace of gauge invariant states. However there are formal differences between the μ system and the σ system. First of all μ_3 is identically 1 for dual links along z. Secondly the condition

$$\prod_{1}^{6} \mu_1 = 1$$

for dual sites is not a subsidiary condition but rather an identity satisfied by μ_1. Nevertheless the physics is identical to that of a Z_2 gauge theory. What happened is that in making the dual transformation we arrived in different gauge, namely the analog of the axial gauge, ($A_z=0$) for the dual variables. Since the gauge invariant physics of the axial gauge and temporal gauge ($A_0=0$) are identical we may say that the gauge invariant physics of the Z_2 theory is self dual.

As in the Ising case, the self duality insures that the transition from order to disorder occurs at $\lambda=1$ and that the same kind of reflection symmetry about $\lambda=1$ exists.

Previously we identified the large (small) λ behavior as the ordered (disordered) phase. However, as in the case of the Ising model, a dual order parameter exists such that the small λ phase can be considered ordered. Consider a closed loop of dual links as shown in Fig. 13.

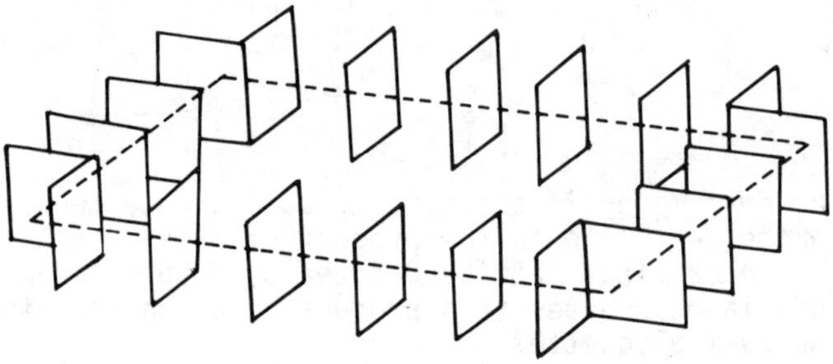

Figure 13. A closed dual loop and the corresponding plaquettes.

If we form the product of μ_3's around that loop the resulting object can be defined as the dual Wilson loop. It consists of a product of σ_1's which can be brought to a form in which the links involved are perpendicualr to the plane of the loop as in Fig. 14

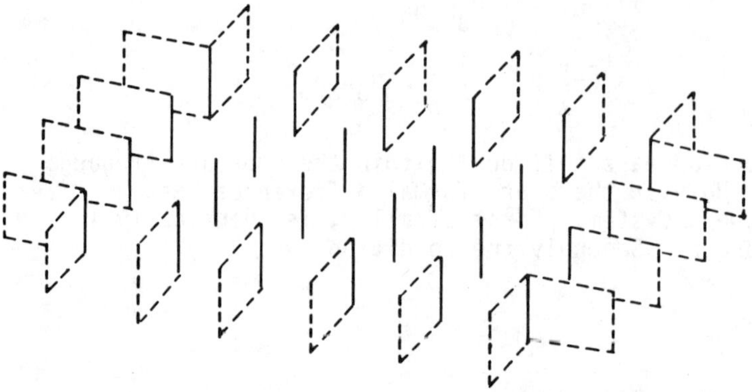

Figure 14. The product of σ_1's on the solid lines of Fig. 13 is equal to the product of σ_1's on the solid lines of this fig., as a result of gauge invariance.

The dual loop evidently behaves as $e^{-\text{Area}}$ for $\lambda>1$ and $e^{-\text{perimeter}}$ for $\lambda<1$.

The dual order variable in the Ising model was identified as a "kink" creation operator in the ordered ($\lambda>1$) phase. In the gauge theory the dual loop operator also creates excitations of the ordered ($\lambda>1$) phase. It is easy to see that the dual loop flips the magnetic fields (products of 4 σ_3's) along the loop and therefore excites a closed loop of magnetic flux. These closed loops of magnetic flux play the same role in the dynamics of the gauge theory as the kinks play in spin systems.

Consider the ground state for $\lambda\gg1$. In this case the magnetic term in H forbids the occurrence of magnetic loops in the vacuum. The Wilson loop is obviously equivalent to a product of magnetic variables over any surface spanning the loop (Fig. 15).

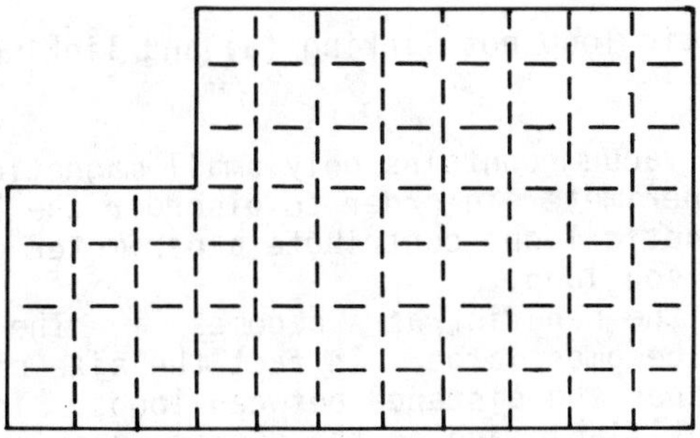

Figure 15. The Wilson loop is equivalent to a product of magnetic variables over any surface spanning the loop.

Thus if no magnetic fields are flipped for $\lambda\gg1$ then the loop has value 1.

Now consider the behavior for somewhat smaller λ. In this case the vacuum is populated with a "dilute gas" of small magnetic loops. For example when the perturbation σ_1 acts on any link it creates a loop of magnetic flux around that link. These magnetic fluctuations begin to disorder the Wilson loop.

Consider the effect of a closed loop of magnetic flux on a large Wilson loop. If the magnetic loop does not link the Wilson loop (Fig. 16a) then the Wilson loop has value 1. Indeed such a magnetic loop pierces the plane of the Wilson loop an even number of times.

If, on the other hand, the magnetic loop links the Wilson loop, the sign is flipped (Fig. 16b).

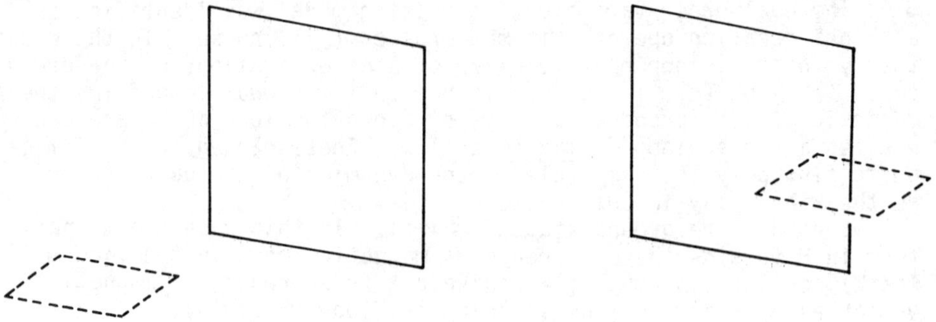

a b

Figure 16. Magnetic loop not linking (a) and linking (b) a Wilson loop.

Obviously, if the vacuum contains only small magnetic loops then they must be near the perimeter in order to disorder the loop. For this reason, small magnetic loops contribute a perimeter dependence to the log of the Wilson loop.

Now consider the behavior as λ becomes ~ 1. The average size of magnetic loops becomes large. In fact the size of such loops eventually approaches the distance between loops. Individual loops are no longer identifiable and on the average there are loops of infinite extent. When this occurs, the linking of the Wilson loop is no longer a perimeter effect and a transition to the disordered phase occurs. We might say that a condensate of magnetic flux has formed.

VI. NON ABELIAN GAUGE THEORY

In a non Abelian gauge theory such as QCD the link variables $\sigma_3(\ell)$ are replaced by SU_N matrices $U(\ell)$. The Z_2 electric fluxes are replaced by chromoelectric fluxes which take on discrete values. The color components of the chromoelectric flux are noncommuting and satisfy the Lie algebra of the color group. For example, if we consider an SU_2 color group, then the 3 components of the chromoelectric flux satisfy

$$[E^{\alpha}(\ell),E^{\beta}(\ell)] = i\,\varepsilon^{\alpha\beta\gamma}E^{\gamma}(\ell) \tag{6.1}$$

The magnetic (plaquette) variables are defined as the trace of the product of 4 U's.

Among the elements of SU_N is a special subgroup called <u>The Center</u>. This is the subgroup which commutes with all the elements of SU_N. For example, in SU_2 it consists of the unit element and the element -1. The element -1 can also be identified with any 2π rotation. For SU_2 the center is Z_2, the group of the Z_2 gauge theory. More generally the center of SU_N is Z_N.

The representations of SU_N fall into 2 classes; those which are invariant and those which are non invariant under the center. For SU_2 the odd (even) representations under Z_2 are just the half integer (integer) angular momenta. For SU_3 the center is trivial for the octet and anything which can be built from it. The 3, $\overline{3}$, 6 and other triality bearing representations transform nontrivially under the center. Thus gluons are center-trivial but quarks are not.

Let us consider the possible long range confining potential between static colored objects. In the strongly coupled lattice QCD (analogous to $\lambda \ll 1$ for the Z_2 case) a linear confining force exists between quark pairs due to the line of electric flux connecting the pair. Translated into the language of Wilson loops, this means an area law for the loop-product.

$$\mathrm{Tr} \prod_{\mathrm{LOOP}} U(\ell) \sim e^{-A} \tag{6.2}$$

Next consider a pair of sources which transform under the adjoint representation. For SU_2 this means color-spin 1. In this case there is no linear growth in the potential energy. Intuitively this is due to the fact that it is energetically favorable to create a gluon pair which combine with and neutralize the sources. If we consider a color spin-1 source at r=0 the minimum energy gauge invariant state is (see Fig. 17)

$$\mathrm{Tr}\, U(1)U(2)U(3)U(4)\, \tau^{\alpha}\, \phi^{\alpha}_{(0)}$$

where $\phi^{\alpha}_{(0)}$ is the creation operator for such a spin 1 source at

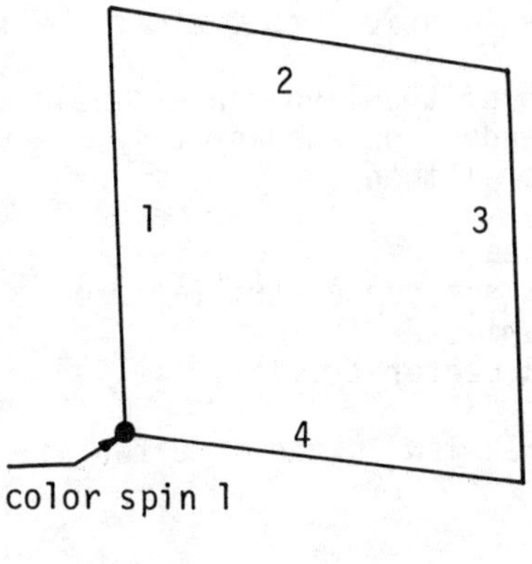

Figure 17. The minimum energy gauge invariant state for a color spin 1 source.

the origin and α is a color-1 index. When a pair of such sources is separated, the lowest energy state consists of a pair of such objects which can loosely be described as bound states of source and a gluon.

The corresponding statement about the Wilson loop involves loop products of matrices in the spin 1 representation. Thus, for each link ℓ the matrix $U(\ell)$ defines an abstract group element. This group element has a representation matrix in each representation. We can call these matrices $\mathcal{U}_R(\ell)$ where R labels a representation. The original U matrices are $\mathcal{U}_F$ where F means fundamental. Indeed all the $\mathcal{U}_R$'s are functions of U. For example in SU_2

$$\mathcal{U}_1^{\alpha\beta} = \frac{1}{2} \operatorname{Tr} \tau^{\alpha} U^{\dagger} \tau^{\beta} U \tag{6.3}$$

the Wilson loop for color R sources is

$$\operatorname{Tr} \prod_{\ell} \mathcal{U}_R(\ell)$$

Evidently the possibility of creating gluons to neutralize a spin 1 source means

$$\operatorname{Tr} \prod_{\ell} \mathcal{U}_1(\ell) \to e^{-\text{Perimeter}} \tag{6.4}$$

Consider now a source with any integer color spin. (In SU_N we replace integer spin representation by center-trivial representations.) Evidently such a source can be neutralized by gluons so that no long range fource exists. On the other hand, any source with 1/2 integer color can only be partly neutralized. For example, a color 3/2 source can be partly neutralized to form a color 1/2 object. The lowest energy state for a pair of distant color 3/2 sources is obtained by forming a pair of such color 1/2 bound states. The resulting long range force is identical to that of the 1/2 color sources.

More generally, the strength of the long range linear confining potential between sources is dependent on whether they are trivial or not under the center. In general then

$$\begin{aligned} -\log \operatorname{Tr} \mathcal{U}_R(\ell) &\to \text{Area} \\ &\quad (\text{R not center-trivial}) \\ -\log \operatorname{Tr} \mathcal{U}_R(\ell) &\to \text{Perimeter} \\ &\quad (\text{R center-trivial}) \end{aligned} \tag{6.5}$$

In fact the coefficient of the area law is the same for all center-non-trivial representations.

A disorder parameter can be defined as in the case of Z_2 theory. It is formed exactly as in the Z_2 case except that the products of σ_1 (see Fig. 14) are replaced by products of factors given by

$$\exp\{2\pi i\ E^{\alpha}\} \qquad (6.6)$$

In Eq. 6.6 E^{α} can be any of the 3 components of E and may refer to either end of the link. In fact the operator 6.6 represents a 2π rotation (center of SU_2) and therefore is independent of the rotation axis. The result of applying a product of such operators over a region (Fig. 14) is to create a quantized line of chromomagnetic flux. The transition to the confining phase of non abelian theories has been described by 't Hooft as a condensation of such quantized magnetic lines.

SEMICLASSICAL METHODS IN QUANTUM CHROMODYNAMICS: TOWARD A THEORY OF HADRON STRUCTURE

Curtis G. Callan, Jr.[#]
Princeton University
Princeton, New Jersey 08540

Roger F. Dashen[†]
The Institute for Advanced Study
Princeton, New Jersey 08540

David J. Gross[#]
Princeton University
Princeton, New Jersey 08540

I. THE SCALE HIERARCHY OF QCD 270

II. SEMICLASSICAL QCD: INSTANTONS 274

III. MASSLESS FERMIONS AND CHIRAL SYMMETRY BREAKING 290

IV. THE DILUTE GAS VACUUM 305

V. INFRARED CUTOFFS AND PHASE TRANSITIONS 319

VI. MODELS OF HADRONS 333

[#]Research sponsored by the National Science Foundation under Grant No. PHY78-01221.

[†]Research sponsored by the Department of Energy under Grant No. EY-76-S-02-2220.

ISSN: 0094-243X/79/550269-71$1.50

I. THE SCALE HIERARCHY OF QCD

These days most physicists are convinced that ordinary hadrons and their interactions are manifestations of an underlying quantum gauge field theory of quarks and gluons based on the gauge group SU_3. This conviction is based mainly on the unique ability of gauge theories to explain, via asymptotic freedom,[1] the observed scaling behavior of certain high-energy inclusive cross-sections. This conviction is definitely not based on our ability to explain in any detailed fashion the old-fashioned aspects of strong interaction physics--spectroscopy, chiral symmetry and the manner of its breaking. Our aim in these lectures is to show that recent advances in our understanding of the dynamics of Yang-Mills theories have changed this situation: We claim that it is now possible to derive from first QCD principles, in at least a semi-quantitatively fashion, many of the salient features of hadron physics. It should be noted that we do not claim to prove "confinement" but rather that the genuinely difficult problem of proving strict confinement can usefully be separated from the problem of computing the properties of lowlying hadrons. Once one understands how to make this separation, a conceptually attractrive picture of hadrons, very reminiscent of the MIT bag model,[2] can be extracted from QCD. In what follows we shall present the technical arguments which underlie these assertions as well as a synopsis of the resulting picture of hadron physics.

To set the stage, we recall the Lagrangian of QCD:

$$\mathcal{L} = \frac{1}{4} \sum_{a=1}^{8} \left(F^a_{\mu\nu}\right)^2 + \sum_{i=1}^{F} \bar{\psi}_i \, (i\not{\partial} + g_o \not{A}^a T_a - m_i)\psi_i$$

$$F^a_{\mu\nu} = \partial_\mu A^a_\nu - \partial_\nu A^a_\mu + g_o f_{abc} \, A^b_\mu \, A^c_\nu \tag{1}$$

where T_a are the generators of SU_3 in the fundamental representation, F equals the number of independent "flavors" of quark (u,d,s,c,b,...), m_i are their "bare" masses and g_o is the basic coupling constant of the theory. Conventional strong interaction physics requires only three flavors of quark (u,d,s) and they are distinguished from the more exotic flavors by having much smaller bare masses ($m_u, m_d \sim 10$ MeV, $m_s \sim 100$ MeV). In fact, the world in which the u, d and s bare masses are precisely zero is not a bad zeroth approximation to the real world[3] (i.e., flavor $SU_3 \times SU_3$ is not a catastrophically badly broken symmetry in the real world). Therefore we shall for the most part discuss the physics of Yang Mills theories with only massless flavors. For pedagogic reasons we shall let the number of massless flavors vary from none to three, but our ultimate aim will be to understand the properties of the "realistic" model with three massless flavors.

The model we will study appears to have only one parameter, g_o. In fact we know that, by virtue of the phenomenon of dimensional transmutation,[4] it actually has in a certain sense no parameters! On the other hand, a mass scale Λ necessarily enters the theory via the renormalization subtraction prescription -- g_o is not meaningful unless we say that it is (for example) the value of the quark gluon vertex at the point where all three external legs have mass Λ (see Fig. 1). But since Λ is an artifact of the renormalization procedure, it must always be possible to compensate for a change in Λ by a change in g_o. This is expressed mathematically by the renormalization group equation for a typical dimensionless physical quantity,[5]

$$\left(\Lambda \frac{\partial}{\partial \Lambda} + \beta(g_o) \frac{\partial}{\partial g_o}\right) F\,(g_o, \Lambda) = 0. \tag{2}$$

Suppose F is the mass ratio between two particles. Then F depends on no other parameters than g_o and Λ and, since it is dimensionless, it can only depend on g_o. But then Equation 2 implies that it does not even depend on g_o -- i.e., the mass ratio of two particles is a pure number, independent of any parameters, and nothing can be learned about it by doing a perturbation expansion in g_o!

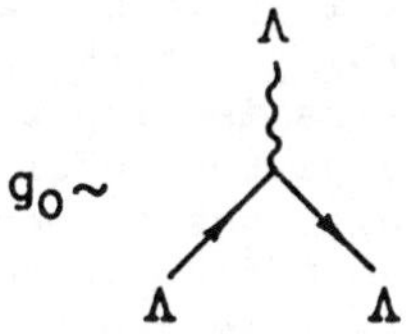

Fig. 1. The definition of gauge coupling constant in terms quark gluon amplitude at external momentum Λ.

If F depends on some other dimensional quantity as well as on Λ (an external momentum variable in a scattering amplitude, for example) the content of Equation 2 is expressed slightly differently. Let the extra dimensional parameter be taken to be a distance scale, ρ. Then Equation 2 implies that

$$\begin{aligned} F(\rho; g_o, \Lambda) &= \bar{F}(\rho\Lambda;\ g_o) \\ &= \bar{F}(1,\ g(\rho\Lambda)) \end{aligned} \tag{3}$$

where $g(\rho\Lambda)$ is a "running" coupling constant normalized so that $g(1) = g_0$ and satisfying the equation

$$\rho \frac{d}{d\rho} g(\rho\Lambda) = \beta(g(\rho\Lambda)). \tag{4}$$

The statement that the theory has no free parameter is replaced by the equivalent statement that different couplings are associated with different distance scales and changing the coupling is equivalent to changing the numerical value of the scale at which a given coupling occurs, a change which can have no physical content.

In the case of QCD, the function $\beta(g)$ is such that $g(\rho\Lambda)$ approaches zero in the limit $\rho \to 0$: for small enough ρ one finds the famous asymptotic freedom result

$$x(\rho) = \frac{8\pi^2}{g^2(\rho)} \simeq (11 - \tfrac{2}{3}F) \ln \frac{1}{\Lambda\rho} . \tag{5}$$

(The combination $8\pi^2/g^2$ is a convenient parameter for our purposes, not least because perturbation theoretic experience shows that $g^2/8\pi^2$ is a good measure of the relative strength of radiative corrections.) This means that the physics of QCD is governed at small scales by small effective coupling and is nothing but the physics of essentially free quarks and gluons. This physics is simple and is what underlies the success of such things as the parton model,[6] but tells us nothing about hadrons themselves. Conversely, one expects the physics at large distance scales to be governed by large effective coupling (the notion of infrared slavery) and to be very remote from the perturbation theory of quarks and gluons. It is of course precisely the large distance scale which must be examined in order to see whether quarks are confined in the strict sense of the word. A wide variety of arguments, based in part on lattice models,[7] suggest that at a large enough distance scale, QCD is a confining theory of "strings" of electric flux. The physics of such strings is relevant to the confinement question and probably also to highly excited, spatially extended states of hadrons, but not to the low-lying hadrons themselves.

Somewhere between these two extremes lies the distance scale at which the basic hadrons form. This distance scale (presumably a modest fraction of a fermi, if one may judge from information on the pion and nucleon charge radii) we shall call the hadron scale and denote by ρ_H. Our main contention is that the physics of the hadron scale, while definitely non-perturbative, is actually weak coupling physics and therefore not beyong qualitative and even quantitative understanding. The reason is that in field theory non-perturbative semi-classical effects can be large enough to change the qualitative physics from that of quarks and gluons to that of low-lying hadrons even for effective coupling so small that the semiclassical approximation makes sense. Put another way, although the problem of confinement is a true strong coupling problem (and therefore genuinely difficult) the problem of hadron structure is an in principle separate question, posed, it turns out, at a distance scale where the effective coupling is weak enough for simple methods to give useful information. By restricting our attention to distance scales equal to or smaller than the hadron scale we will see that it is possible to build a very attractive semiquantitative overall picture of hadron structure. The foundation of this picture is the existence in QCD of a hierarchy of distance scales with quite different physical characteristics: On scales small compared to ρ_H, the effective coupling is small and field correlations, fluctuations, etc. are those of a perturbative theory of quarks and gluons (this is the parton domain); on scales comparable to ρ_H, the effective coupling is still small but non-perturbative semi-classical effects are large enough to bind quarks and gluons into recognizable hadrons with a large mass gap between color singlet and color octet states; finally, on scales large compared to ρ_H, the field correlations are those of a true strong coupling theory and of course very difficult to discuss quantitatively. The physics of this scale must be mastered in order to decide whether or not QCD is truly confining. Our claim will be that since the basic properties of hadrons are established at smaller scales it is possible to understand many basic properties of hadrons without actually understanding the details of the physics of the confinement scale. The rest of these lectures will be devoted to explaining our arguments for these claims.

II. SEMICLASSICAL QCD: INSTANTONS

Before we can attempt to construct hadrons, which are, after all local perturbations on the vacuum, we must learn how to construct the QCD vacuum itself. This problem is most conveniently studied via the Euclidean functional integral technique. This method amounts to writing a path integral for matrix elements, Z, of the imaginary time-evolution operator e^{-HT}. In the limit $T \rightarrow \infty$ one can see, by inserting a complete set of energy eigenstates, that

$$Z_{fi} = \langle\psi_f | e^{-HT} | \psi_i\rangle \underset{t\to\infty}{\longrightarrow} e^{-\varepsilon_0 T} \langle\psi_f|0\rangle \langle 0|\psi_i\rangle \tag{2.1}$$

which is to say that in the large time limit Z automatically picks out the lowest energy, or vacuum, state.

The path integral form for Z in pure Yang Mills field theory is[8]

$$\begin{aligned} Z &= \int [dA]\, e^{-S(A)} \\ S(A) &= \tfrac{1}{4}\int_{-T/2}^{T/2} d^4x \sum_{a\mu\nu} (F^a_{\mu\nu})^2 \end{aligned} \tag{2.2}$$

where: the path integration over gauge field time histories is defined over all space and from time $-T/2$ to time $T/2$ (boundary conditions may be imposed); the integration measure contains the appropriate gauge-fixing and ghost field integrations; the space-time metric is Euclidean (imaginary time) and therefore S(A) is positive.

Since S(A) is positive the dominant contribution to (2.2) should come from fields in the neighborhood of the global minimum of S. With the appropriate choice of gauge-fixing term one can arrange that the global minimum of S be zero and that it occur at $A_\mu = 0$. Since S(A) has an expansion in powers of g,

$$S(A) = S_2(A) + gS_3(A) + g^2 S_4(A), \tag{2.3}$$

where $S_i(A)$ has terms of order $(A)^i$, and since functional integrals over quadratic action functionals are trivial, one usually evaluates Z by expanding e^{-S} in powers of g and doing the resulting Gaussian integrations term by term. This is identical to ordinary perturbation theory and yields an expression for Z which is a power series in g with leading term of order g^o. Were one to construct an explicit wave function corresponding to this construction of the ground state it would be the familiar Gaussian centered at zero field strength.

Polyakov pointed out[9] that local minima of the action function may in fact be just as important, or even more important than the global minimum in computing Z. A local minimum will necessarily have a positive action S_o (the global minimum has $S_o=0$) and will occur at some field history $A^o_\mu(x) \neq 0$. We may expand the functional integral in the neighborhood of A^o_μ,

$$
\begin{aligned}
A_\mu &= A^o_\mu + \delta a_\mu \\
S(A_\mu) &= S_o + \bar{S}_2(\delta a) + g\bar{S}_3(\delta a) + g^2\bar{S}_4(\delta a)
\end{aligned}
\tag{2.4}
$$

(where $\bar{S}_i(\delta a)$ is $O(\delta a^i)$ and $\bar{S}_1$ is not present because A^o is by hypothesis a local minimum) and once again obtain Z as a power series in g. This time, because of the presence of the term S_o, we get

$$Z = e^{-S_o} (a_o + a_1 g^2 + \ldots) \tag{2.5}$$

Since S_o is in fact usually itself $O(g^{-2})$ this contribution would seem to be negligible compared to the perturbative contribution, especially in the weak-coupling limit. The loophole is that there may be many different fields, A^o_μ, corresponding to the same S_o, whose contribution to Z is additive. If for example the field A^o_μ is localized and has a collective coordinate corresponding to space-time location, (2.5) must be replaced by

$$Z = VT\, e^{-S_o} (a_o + a_1 g^2 + \ldots) \tag{2.6}$$

where VT is the total space time volume. Since VT is large, it is not obvious that (2.5) is small compared to the perturbative contribution.

To examine this question more closely, it is best to look at the specific example of a non-trivial local minimum of the QCD action found by Belavin et al.[10] They discovered that the functions

$$A^a_\mu(x) = \frac{2}{g} R_{a\alpha} \frac{\eta_{\alpha\mu\nu}\Delta_\nu}{\Delta^2+\rho^2} \tag{2.7}$$

(where: $\Delta = x-\bar{x}$, $\bar{x}$ arbitrary; $R_{a\alpha}$, with $a = 1,\ldots,8$ $\alpha = 1,2,3$, are arbitrary elements of the adjoint representation of SU_3;[11] $\eta_{\alpha\beta\nu}$ is a fully antisymmetric self dual tensor defined by $\eta_{\alpha ij} = \varepsilon_{\alpha ij}$ and $\eta_{\alpha 0i} = \delta_{\alpha i}$) are solutions of the Euclidean Yang Mills equations in Landau gauge for any values of the position coordinate $\bar{x}$, scale parameter ρ and group orientation matrix R. The field strength tensor derived from (2.7) is

$$F^a_{\mu\nu} = \frac{4}{g_o} \frac{\rho^2}{(\rho^2 + \Delta^2)^2} R_{a\alpha} \eta_{\alpha\mu\nu} \tag{2.8}$$

and shows clearly that the action density of this solution is localized in a space time region of size ρ centered on $x = \bar{x}$. In honor of this space-time localization the solution (2.7) has been dubbed instanton. The field strength tensor (2.8) is self-dual (this is perhaps its most important characteristic). There is in fact another set of solutions, identical to (2.7) and (2.8) except for the replacement $\eta_{\alpha\mu\nu} \to \bar{\eta}_{\alpha\mu\nu}$ ($\bar{\eta}_{\alpha ij} = \varepsilon_{\alpha ij}$ $\bar{\eta}_{\alpha 0i} = -\delta_{\alpha i}$), for which the field strength tensor is anti-self-dual. These solutions are called anti-instantons.

The contribution of an instanton to Z is determined in part by its integrated Euclidean action, which is easily found to be

$$S_o = \tfrac{1}{4}\int d^4x \ \Sigma(F^a_{\mu\nu})^2 = \frac{8\pi^2}{g^2} . \tag{2.9}$$

One must add together the contributions corresponding to independent realizations of the same solution with different values of the collective coordinates. This will give us an expression for the one-instanton contribution to Z of the following form:

$$Z \sim \int d^4\bar{x} \int \frac{d\rho}{\rho^5} \int [dR] \ C(\rho) e^{-8\pi^2/g^2} \ (1 + a_1 g^2 + \ldots) \tag{2.10}$$

where C is a factor including the Jacobian of the transformation to collective coordinates (from which we have removed a factor ρ^{-5} which must be present for dimensional reasons) and the result of doing the Gaussian integration over non-collective degrees of freedom while [dR] denotes invariant integration over the SU_3 group orientation of instanton. The calculation of C was first done by 't Hooft for the case of SU_2[12] and subsequently extended to SU_3 and higher groups by other workers.[13] The result for SU_3 may be expressed as

$$Z \sim \int d^4x \int \frac{d\rho}{\rho^5} C_o \left(\frac{8\pi^2}{g^2(\rho)}\right)^6 e^{-\frac{8\pi^2}{g^2(\rho)}} (1 + a_1 g^2 + \ldots) \tag{2.11}$$

$$= \int d^4x \int \frac{d\rho}{\rho^5} D(\rho)$$

(In passing from (2.10) to (2.11) we have done the [dR] integration - the integrand is independent of the integration variables). The Gaussian functional integration produces divergences and their treatment via a renormalization procedure simply converts the coupling constant to the usual renormalization group running coupling constant. As long as the effective coupling is small we may make the usual asymptotic freedom approximation,

$$\frac{8\pi^2}{g^2(\rho)} = 11 \ln \frac{1}{\Lambda\rho} + \ldots \tag{2.12}$$

where λ is the, for the moment arbitrary, renormalization mass. The factor of g^{-12} comes from the normalization of the 12 collective co-ordinate zero modes appearing in the collective coordinate Jacobian. The dimensionless numerical factor C_o depends on how the coupling constant has been defined--using 't Hooft's Pauli-Villars or dimensional regulation schemes, one finds[14] $C_o = 1.51 \times 10^{-3}$ and $C_o = 1.06 \times 10^2$ respectively. This alarmingly large difference in C_o between the two schemes just amounts to a difference of a factor e between the Λ's appropriate to the two schemes and reflects the fact that the same physics is associated with different values of g in different renormalization schemes. We will see that to a reasonable degree of approximation, physically significant quantities are not affected by this giant change in C_o. Finally, it should be said that while we have no information about the first radiative correction, $a_1 g^2$, in (2.11), we shall assume that it is small for the physically interesting range of effective couplings (what that range is will soon become apparent).

A useful insight into the significance of the instanton is provided by an examination of its appearance in $A_0 = 0$ gauge (this gauge is particularly useful for discussions of the Schrödinger picture quantum mechanics of gauge field theory). To simplify our equations, let us choose the particular solution obtained from (2.7) with $R_{\alpha a} = \delta_{\alpha a}$ and $\bar{x} = 0$. Then it is easy to show that the gauge transformation to $A_0 = 0$ gauge is effected by the matrix

$$U(\bar{x},x_o) = \exp\left\{-2i\int_{-\infty}^{x_o} \frac{dt}{t^2+\bar{x}^2+\rho^2} \sum_{a=1}^{3} T_a x_a\right\} \tag{2.13}$$

$$U(\bar{x},x_o) \begin{cases} \xrightarrow[x_o \to -\infty]{} 1 \\ \xrightarrow[x_o \to +\infty]{} \exp\left\{-2i \dfrac{\sum_{a=1}^{3} T_a x_a}{\sqrt{\bar{x}^2+\rho^2}}\right\} = \bar{U}(\bar{x}) \end{cases} \tag{2.14}$$

The corresponding asymptotic behavior of the spatial components of the gauge field is

$$T_a A_i^a(\bar{x},x_o) \begin{cases} \xrightarrow[x_o \to -\infty]{} 0 \\ \xrightarrow[x_o \to +\infty]{} \frac{1}{i}\bar{U}(\bar{x})\,\bar{\nabla}_i\,\bar{U}^{-1}(x) \end{cases} \tag{2.15}$$

Both the $x_o = +\infty$ and $x_o = -\infty$ fields are pure gauge fields (i.e. correspond to color magnetic field $B_i^a = \frac{1}{2}\varepsilon_{ijk}(\partial_j A_k^a - \partial_k A_j^a + f_{abc} A_j^b A_k^c)$ equal to zero) and are good classical representatives of the Yang-Mills vacuum. Indeed, they are just gauge transforms of one another under the gauge transformation effected by $\bar{U}(\bar{x})$. Therefore the instanton describes a transition from a vacuum state to a gauge-equivalent vacuum state with a transition amplitude of order $e^{-8\pi^2/g^2}$ (and which therefore vanishes in a non-analytic fashion as $g \to 0$. This non-analytic behavior in g (or $\hbar$) is characteristic of barrier penetration amplitudes in ordinary quantum mechanics and suggests that the two configurations of (2.15), while gauge equivalent, are actually separated by a finite energy barrier.

Closer examination of this idea reveals[15,16] that the possible Yang-Mills vacuum states fall into a discrete set of classes characterized by a topological invariant. The general vacuum gauge field is characterized by a matrix $U(\bar{x})$ (as in (2.15)) mapping three-dimensional space into the gauge group. For technical reasons one need only consider mappings which take all points at spatial infinity into the same element of the gauge group: $\lim_{r\to\infty} U(r\hat{n}) = g_o$, where g_o is independent of $\hat{n}$. Therefore the mapping U is, in effect, from three-space with points at infinity identified (or, equivalently, the manifold S_3) into the gauge group. All the maps which can be continuously deformed into one another form a so-called homotopy class and it is a (non-trivial) theorem that there are an infinity of homotopy classes labeled by an integer which counts how many times an SU_2 subgroup of the gauge group is covered as three-space is covered once. In the case of (2.15) the vacuum $A_i=0$ belongs to the n=0 class while the vacuum $A_i \simeq -i\bar{U}\Delta_i \bar{U}^{-1}$ belongs to the n=1 class.

In fact there is a vacuum state for each integer n and in computing vacuum to vacuum amplitudes one must specify the initial and final values of n. What the functional integral approach teaches us is that transitions which are diagonal in n can be seen in ordinary perturbation theory, while transitions which are off-diagonal in n occur via barrier penetration and are therefore non-analytic in g:

$$\langle n \,|e^{-HT}|n\rangle \sim 1$$
$$\langle n+1 \,|e^{-HT}|n\rangle \sim e^{-8\pi^2/g^2} \qquad (2.16)$$

The situation is very similar to that which obtains in ordinary quantum mechanics in the case of a periodic potential (Fig. 2). One may construct the ground state energy by computing the vacuum persistence amplitude $< n \,|e^{-HT}|m >$ with the help of the Euclidean path integral. The diagonal elements $< n \,|e^{-HT}|n >$ are O(1) and dominated by paths which fluctuate near the minimum n. Off-diagonal elements, such as $< n+1 \,|e^{-HT}|n >$, are dominated by whatever path joining q = na and q = (n+1)a minimizes the Euclidean action

$$S = \int dt \,[\tfrac{1}{2}\dot{q}^2 + V(q)] \qquad (2.17)$$

The minimum of the action is non-zero (it can in fact be shown to equal the standard barrier penetration factor, $S_o = \int_{na}^{(n+1)a} dq \,\sqrt{2V(q)}$) and is achieved by a path, $q_o(t-t_o)$, of the type shown in Fig. 3, satisfying the boundary conditions and the Euclidean equations of motion. The corresponding contribution to the vacuum persistence amplitude is $\sim \exp(-S_o/\hbar)$ and is non-analytic in $\hbar$.

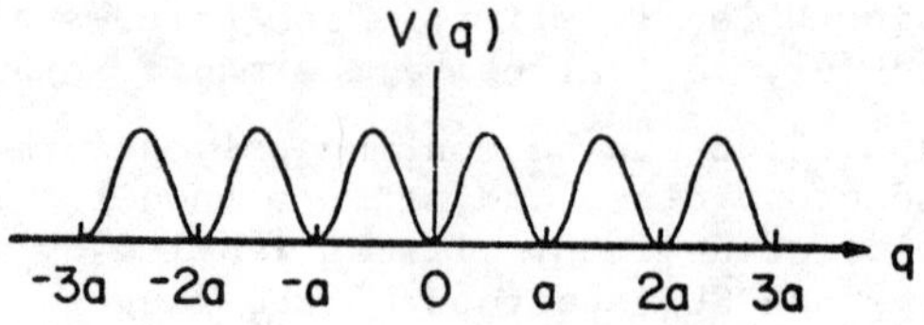

Fig. 2. The periodic potential V(q).

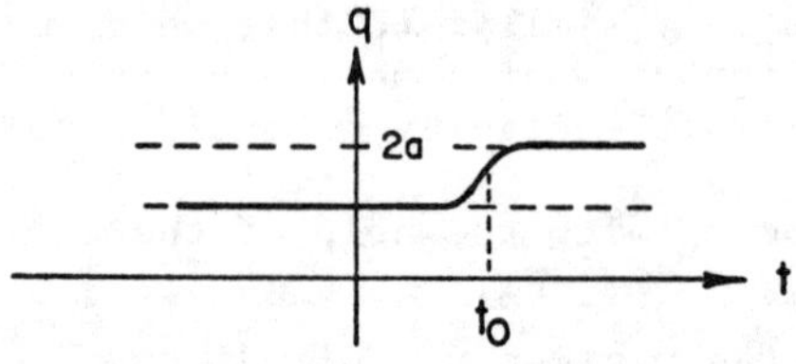

Fig. 3. A typical instanton solution in the potential V(q), connecting the degenerate minima at q=a and q=2a and centered at $t=t_o$. The collective coordinate t_o is arbitrary.

Just as in the instanton case, there is a collective coordinate in the solution $q_o(t-t_o)$, which labels the time at which the tunneling took place. Since it is localized in time, we will call this solution an instanton also. In doing the path integral it is necessary to integrate over this parameter, obtaining, just as in (2.6),

$$< n+1 \,|e^{-HT}|\, n > = cT\ e^{-S_o/\hbar} \tag{2.18}$$

In order to have a complete picture of the effect of vacuum tunneling, we must allow tunneling events to occur not just once, but an arbitrary number of times, in the time interval T. That means that we must consider time histories of the general class shown in Fig. 4. Roughly speaking, the contribution of m_+ tunnelings and m_- antitunnelings (in a tunneling $n \rightarrow n+1$, while in an antitunneling $n \rightarrow n-1$) to a general vacuum persistence amplitude is

$$\langle n' \,|e^{-HT}|\, n \rangle = \sum_{m_+, m_-} \frac{\left(cT\ e^{-S_o/\hbar}\right)^{m_+ + m_-}}{m_+!\ m_-!} \delta_{n'-n,\ m_+ - m_-} \tag{2.19}$$

(the factorials eliminate overcounting of identical configurations, the powers of T come from integrating over the instanton and antiinstanton collective coordinates and the factor $e^{-(m_+ + m_-)S_o/\hbar}$ arises because the action of the instantons and antiinstantons is, to a good approximation, additive). Because of the symmetry of the problem, the true vacuum wave function must be a superposition with equal amplitude of <u>all</u> the states $|n >$ ($|n >$ in effect is a Gaussian centered on the nth minimum of the potential):

$$|\Psi_o \rangle = \sum_{-\infty}^{\infty} |n \rangle . \tag{2.20}$$

The true vacuum persistence amplitude is then

$$\frac{\langle \Psi_o | e^{-HT} | \Psi_o \rangle}{\langle \Psi_o | \Psi_o \rangle} = \sum_{n=-\infty}^{\infty} \langle n | e^{-HT} | o \rangle$$

$$= \sum_{m_+, m_-} \frac{\left(cT\, e^{-S_o/\hbar} \right)^{m_+ + m_-}}{m_+!\; m_-!} \tag{2.21}$$

$$= \exp\left(2c\, e^{-S_o/\hbar}\, T\right)$$

This exponential time dependence means that the net effect of summing over all possible multiple vacuum tunneling configurations is to lower the energy of vacuum by an amount $\Delta E = -2c\, e^{-S_o/\hbar}$ relative to the perturbative energy. Furthermore, the sum over m_+ and m_- in (2.21) is, by the usual statistical mechanics argument, dominated by the value of $m_\pm$ which maximizes the summand. This value turns out to be $m_+ = m_- = e^{-1}\, c\, e^{-S_o/\hbar}\, T$, which means that the sum is dominated by configurations where the density (per unit time) of instantons or anti-stantons is $\sim e^{-S_o/\hbar}$. Therefore the significance of the single instanton contribution is not that it directly determines the vacuum amplitude, but that it determines the density of instantons in the multi-instanton configurations that actually dominate the functional integral. This conclusion concerning the significance of the instanton is applicable in just the same way to the QCD problem.

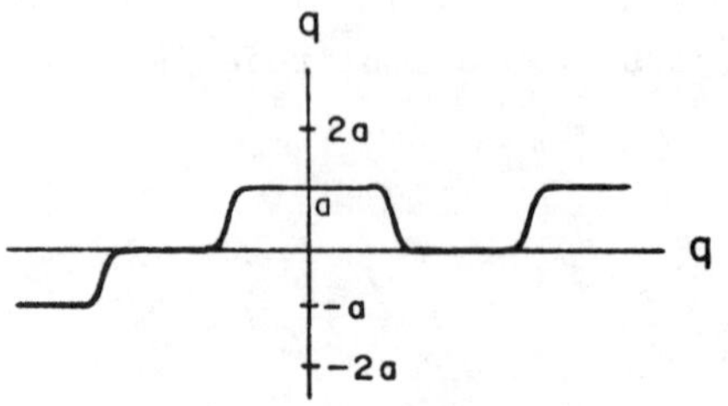

Fig. 4. A typical multi-instanton path history in the potential V(q).

Before returning to the QCD problem we want to make some observations about the nature of the vacuum wave function which is being constructed by the instanton technique. The path integral method just described is in fact equivalent to a certain WKB procedure for solving the Schrodinger equation.[17] On general grounds one expects the wave function to consist of Gaussians of unit amplitude centered on the degenerate potential minima plus exponentially vanishing tails in the forbidden regions between the ninima. The properties of the instanton solution enter into the WKB determination of the amplitude of the wave function in the forbidden region.

The role of the instanton in determining the wave function can be made more striking in the context of a slightly generalized problem in which the quantum mechanical configuration space is two-dimensional and the potential function, $V(q_1, q_2)$, has, for example, two degenerate minima. The situation is shown in Fig. 5, where A and B label the degenerate minima and the contours are lines of equal potential. Here, too, there will be an instanton--a solution of the classical Euclidean equations of motion starting at $q = B$ at large positive time. A possible trajectory is indicated by the dashed line in Fig. 5. The WKB method applied to this problem will produce a ground state wave function consisting of the usual Gaussian of unit amplitude centered at A and B, plus, in the classically forbidden region, a tube of non-zero amplitude centered precisely on the instanton trajectory. The amplitude at the center of the tube decreases exponentially as one moves away from A or B in a fashion determined by the action of the instanton solution and it also decreases a Gaussian fashion as one moves away from any point on the instanton trajectory in a direction orthogonal to the trajectory.[18] So, although the wave function is small everywhere in the classically forbidden region, there is a narrowly defined region about the instanton trajectory where the wave function is least negligible. The instanton therefore identifies the restricted set of points in configuration space where the WKB wave function has significant amplitude. The path integral method is equivalent to the computation of the energy using a trial function with the sort of restricted support in configuration space just described.

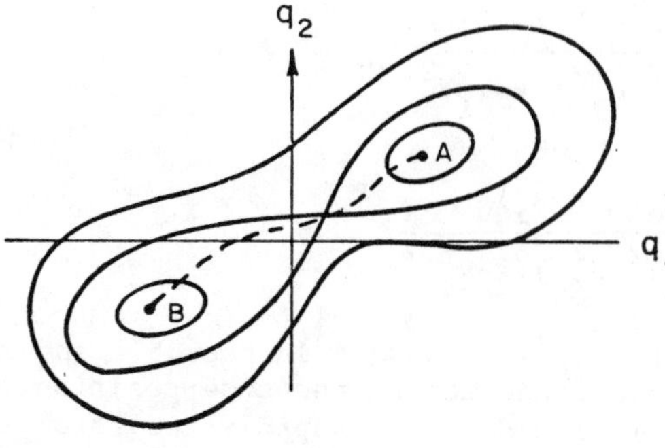

Fig. 5. Equipotential contours in a two-dimensional configuration space with degenerate minima at A and B. The dashed line represents the trajectory of an instanton connecting A and B.

The proper way of treating the QCD instanton should now be clear--in order to construct the vacuum state, we must sum over multiple instanton configurations with all possible values of the instanton collective coordinates (space time location, scale size and gauge orientation). In the quantum mechanical case, one may trivially superpose instantons and antiinstantons because the instanton is a finite range disturbance (the mass scale is provided by the small oscillation frequency about the degenerate minima). In QCD there is no mass scale, and, indeed, the instanton configuration of (2.7) decreases only as x^{-1} at infinity. This slow a rate of decrease would preclude any simple superposition of instantons. Fortunately, the x^{-1} asymptotic behavior of (2.7) is pure gauge, and can be eliminated by a change of gauge. By a simple gauge transformation, closely related to inversion about the origin, we can turn (2.7) and (2.8) into

$$A^a_\mu = \frac{2}{g}\,\frac{2R_{a\alpha}\,\bar{\eta}_{\alpha\mu\nu}\,\Delta_\nu\,\rho^2}{(\Delta^2+\rho^2)\,\Delta^2} \tag{2.22}$$

$$F^a_{\mu\nu} = \frac{4}{g}\,\frac{I_{\mu\mu'}(\Delta)\,I_{\nu\nu'}(\Delta)\,R_{a\alpha}\,\bar{\eta}_{\alpha\mu'\nu'}\,\rho^2}{(\Delta^2+\rho^2)^2}$$

$$I_{\mu\mu'}(\Delta) = g_{\mu\mu'} - \frac{2\Delta_\mu\Delta_{\mu'}}{\Delta^2} \tag{2.23}$$

The expression (2.22) decreases at infinity as x^{-3} and is therefore much better suited to the construction of multiinstanton configurations. The price one pays is that the expression (2.22) has a singularity at $\Delta = 0$. However, since the singularity is pure gauge, it causes no trouble.

The multiple instanton antiinstanton configurations which we shall use to compute the functional integral are simply superpositions of the basic configurations (2.22):

$$A^a_\mu(n_+,n_-) = \sum_{i=1}^{n_+} \frac{2\rho^2_{i+}}{g}\,\frac{R^{i+}_{a\alpha}\,\bar{\eta}_{\alpha\mu\nu}\,(x-x^+_i)_\nu}{((x-x^+_i)^2+\rho^{+2}_i)\,(x-x^+_i)^2}$$

$$+\sum_{i=1}^{n_-} \frac{2\rho^2_{i-}}{g}\,\frac{R^{i-}_{a\alpha}\,\eta_{\alpha\mu\nu}\,(x-x^-_i)_\nu}{((x-x^-_i)^2+\rho^2_{i-})\,(x-x^-_i)^2} \tag{2.24}$$

This is good enough for our present purposes, but it is possible and sometimes necessary to be more careful in constructing the superposition configuration.[19] Since the instanton fields decrease rapidly away from the instanton centers it is also true in this case that the total action of (2.24) is approximately equal to the sum of the individual instanton actions:

$$S(n_+,n_-) = (n_+ + n_-)\, S_o \; . \tag{2.25}$$

Therefore, the entire discussion of the Euclidean vacuum functional given for the periodic potential problem may be carried over unchanged except for the need to integrate over more collective coordinates. The type of gauge field history we will be integrating over is indicated pictorially in Fig. 6, where the blobs indicate the location and scale size of the instantons. Again, because of the symmetry of the problem, the vacuum must be a superposition of the integer-labeled topological classes of vacuum as in (2.20). The vacuum persistence amplitude in this state is the generalization of (2.21) to multiple QCD instanton configurations:

$$Z = \frac{\langle \Psi_o | e^{-HT} | \Psi_o \rangle}{\langle \Psi_o | \Psi_o \rangle} = \sum_{n_+,n_-=0}^{\infty} \left(\int d^4x \int \frac{d\rho}{\rho^5} D(\rho) \right)^{n_+ + n_-} \frac{1}{n_+!\, n_-!} \tag{2.26}$$

$$= \exp\left\{ 2VT \int \frac{d\rho}{\rho^5}\, D(\rho) \right\}$$

($D(\rho)$ is defined in (2.11) and $\int d^4x = VT$ = (space volume) x (time interval)). The effect of the instantons in this case is to lower the energy of the vacuum relative to perturbation theory by an energy per unit volume,

$$\Delta\varepsilon = -2 \int \frac{d\rho}{\rho^5}\, D(\rho) \; . \tag{2.27}$$

The configurations which dominate the sum over numbers of instantons are those in which the space time density of instantons (or antiinstantons) with scale sizes lying between ρ and $\rho + d\rho$ is

$$D(\rho) \frac{d\rho}{\rho^5} = C_o \left(\frac{8\pi^2}{g^2(\rho)} \right)^6 e^{-8\pi^2/g^2(\rho)} \frac{d\rho}{\rho^5} \tag{2.28}$$

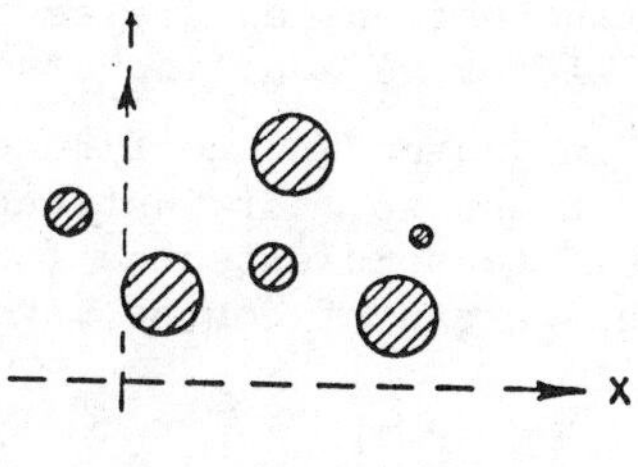

Fig. 6. Schematic representation of a multi-instanton gauge field configuration. The shaded regions represent the cores of instantons of differing scale sizes and locations.

An important new feature of the QCD problem is the need to integrate over instanton scale size. A useful way of quantifying the role played by instantons of various scale sizes is to compute the fraction of space time they occupy. Assigning to each instanton (and antiinstanton) a sphere of radius ρ (and volume $\frac{\pi^2}{2}\rho^4$) and using the density function (2.28) we find for the fraction, $f(\bar{\rho})$, of space time covered by instantons of scale size less than $\bar{\rho}$ the expression

$$f(\bar{\rho}) = 2\int_0^{\bar{\rho}} \frac{d\rho}{\rho^5} D(\rho)\cdot\frac{\pi^2}{2}\rho^4 = \pi^2\int_0^{\bar{\rho}} \frac{d\rho}{\rho} D(\rho) \tag{2.29}$$

As long as $g^2(\bar{\rho})$ is in the asymptotic freedom regime, we may combine (2.11) and (2.12) to reexpress (2.29) as an integral over effective coupling

$$f(\bar{\rho}) = \frac{C_o\pi^2}{11}\int_{x(\bar{\rho})}^{\infty} dx \cdot x^6 \; e^{-x} \tag{2.30}$$

(we have set $x = 8\pi^2/g^2$). We have plotted f as a function of $x(\rho)$ for the dimensional regulation value of C_o in Fig. 7 in order to make two points: The instanton contribution is very sharply cut off at small ρ (large x) by the asymptotic freedom decrease of the coupling. Indeed, if the effective coupling did not decrease at small ρ and $D(\rho)$ had a finite limit as ρ went to zero, then (2.27) and (2.29) would diverge at the limit $\rho = 0$ and arbitrarily small instantons would make arbitrarily large contributions to the vacuum energy. In fact, because of asymptotic freedom, instantons smaller than a certain scale are totally without physical effect and the physics of the vacuum on scales smaller than this limit is purely perturbative. Conversely, at a rather well-defined scale, ρ_c, and effective coupling, $x(\rho_c)\simeq 24$, f passes through unity. Once f is unity, instantons are dense in the vacuum and it is clear that fluctuations on scales larger than ρ_c cannot be usefully described as instantons. This is just another manifestation of the infrared problem of QCD and one suspects (quite rightly) that the physics of confinement is established by the properties of the fluctuations on scales greater than ρ_c. An important point is that the effective coupling associated with ρ_c, the scale of transition between instanton and confinement physics is so small that ordinary radiative corrections to, for instance, instanton effects should be small. This statement is to some extent a function of which coupling constant definition we use--we can of course make the effective coupling at which $f \sim 1$ anything we want by using an arbitrarily eccentric definition of the coupling constant.

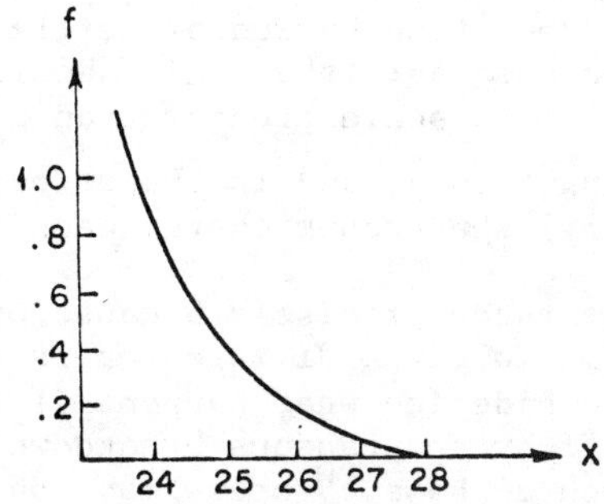

Fig. 7. A plot of $f(\rho)$ versus $x = 8\pi^2/g^2(\rho)$ for the dimensional regulation definition of coupling constant.

Evidently, because of the absence of an intrinsic infrared cutoff, a global description of the properties of the vacuum is going to be very hard to obtain. We will therefore look for situations where an extrinsic cutoff is provided by external conditions and which succeeds in restricting the significant fluctuations to scale sizes which are at worst dominated by instantons of integrated density less than one. In doing this we will see that while the instanton fluctuations on scales smaller than ρ_c are associated with weak effective coupling, they cause very large modifications (compared to perturbation theory) of the properties of the vacuum. In fact, these effects are so large that they provide a mechanism for the formation of hadron bound states of size comparable to ρ_c. Insofar as we are only interested in the properties of such states (in the end we expect them to be the only physical states) the properties of fluctuations on scales much larger than ρ_c cannot be important.

Therefore, in computing the vacuum functional, we will make the approximation of integrating over instanton scale size up to a cutoff, ρ_c, where the integrated instanton density is ~ 1 (we will call this the dilute instanton gas method). This procedure will correctly deal with quantum fluctuations on scales less than ρ_c and will correctly describe both standard perturbative physics (relevant at short distances) and the large non-perturbative effects which are directly responsible for hadron formation. Since short distance physics and hadron scale physics (scales comparable to ρ_c) are treated on the same footing, we will be able to relate the short distance renormalization mass, Λ, to other dimensional parameters characterizing hadrons, such as masses and radii.

Since we will not properly deal with the fluctuations on arbitrarily large scales we will not be able to see the confinement of color. What we will see is the generation of a large, but finite, mass gap between color singlet and color non-singlet states. If the gap is large enough, the properties of the color singlet states computed in this way should not be unduly influenced by the mistreatment of color non-singlet states. We shall see that a simple quantitative criterion for the success of this method concerns the fraction of the vacuum energy density (2.27) contributed by fluctuations on scales large compared to ρ_c -- if the instanton contribution dominates, then our method of computing hadron properties ought to be accurate. It is by no means unreasonable that this should be true, for we shall see that fluctuations on a scale greater than ρ_c are governed by an effective strong coupling theory, and in the strong coupling limit (at least of a lattice theory) the vacuum energy density goes to zero.

The critical scale, ρ_c, makes its appearance precisely because QCD does not have an intrinsic infrared cutoff. To get a different perspective on the above procedure, it is worth considering what happens if the theory acquires an intrinsic infrared cutoff by spontaneous breakdown of the gauge symmetry (QCD itself may have such a phase,[20] or we may add a charged scalar field with the right kind of potential). The gauge field then has a mass, μ_w, and the theory is infrared cutoff at a scale μ_w^{-1}: on scales smaller than μ_w^{-1} the mass of the gluon is ineffective and the vacuum fluctuations are those of the unbroken gauge theory; on scales greater than μ_w^{-1} there are no fluctuations at all because of the screening due to the symmetry breaking. If we set $\mu_w^{-1} \lesssim \rho_c$, then the theory has non-trivial non-perturbative effects due to instantons while the screening causes the instanton density to decrease at large scales in such a way that the integrated density of instantons over all scales is less than one. The computation of the vacuum functional by summation over instanton configurations should be quite accurate.

The vacuum wave function of such a system may be constructed by a WKB procedure very similar to that used to construct the quantum mechanical tunneling wave function discussed earlier.[21] The wave function is defined on the infinite-dimensional configuration space whose points correspond to the various possible functions $A_i^a(\bar{x})$ which define a configuration of the gauge field in $A_0 = 0$ gauge. The points in this configuration space for which $A_i^a(\bar{x})$ is a pure gauge and the associated magnetic field vanishes correspond to classical vacuum configurations. The wave function is, in zeroth approximation, a Gaussian centered about such points. In $A_0=0$ gauge, the instanton solution, $A_i^a(\bar{x},x_0)^I$, is nothing but a sequence of points in configuration space (one for each value of x_0) joining the two vacuum configurations $A_i^a(\bar{x},\pm\infty)^I$. In a WKB approximation, these trajectories define tubes in the non-vacuum region of configuration space where the wave function, while small, is a local maximum (except of course in the direction in function space that points along the instanton trajectory). The smaller the effective coupling associated with the instanton, the sharper the local maximum. Therefore the method of summing over instanton configurations in the functional integral approach is equivalent to taking expectation values in a wave function with restricted support in configuration space of the kind just described. In the true QCD problem, therefore, our method of summing over instanton configurations is also equivalent to using a special vacuum trial function of the above type. The arguments we have given suggest that this trial function should correctly describe quantum effects on scales smaller than ρ_c--the perturbative and the non-perturbative semiclassical scales--but not on scales much larger than ρ_c--the confinement domain.

III. MASSLESS FERMIONS & CHIRAL SYMMETRY BREAKING

As the discussion of the previous section should have made clear, a non-abelian gauge field all by itself is already quite a complicated system. In order to say anything useful about the physics of hadrons, it is necessary to complicate the system even further by adding color triplet fermions (quarks). The complication arises because the qualitative role played by a flavor of quark is quite different according as its bare mass is large or small compared to a critical mass scale (not unrelated to the mass scale ρ_c^{-1} introduced in Sect. 2). For these purposes, the bare mass of a quark is the value of its momentum-dependent renormalization group mass at a large enough momentum that all non-perturbative effects have damped out. This mass is not uniquely defined, but, since it varies only logarithmically with energy,[5] the ambiguity is tolerable. If the bare quark mass is in this sense large, the quark simply plays the role of an external source for color and does not affect the basic dynamics of the gauge field.

On the other hand, if the bare quark mass is small compared to the critical mass, the vacuum dynamics are strikingly modified. The nature of the modification is most simply explained if we take the bare quark mass to be not just small but precisely zero. Because of asymptotic freedom, fluctuations on very short distance scales are those of a perturbative theory of gluons and massless quarks and possess the full chiral symmetry appropriate to the number of massless flavors. At some intermediate scale, ρ_c, weak coupling non-perturbative effects (instantons) first become significant and signal that the gauge field vacuum wave function is developing support away from the perturbation theory vacuum region. The interaction of the massless fermions with the gauge field is such that the fermion vacuum then shifts from the perturbative, explicitly chiral symmetric vacuum, to one in which chiral symmetry is spontaneously broken. The quarks acquire a momentum-dependent spontaneous mass which turns on at a momentum essentially equal to ρ_c^{-1}. The associated Goldstone bosons will have quark-antiquark wave functions with mementum components concentrated at ρ_c^{-1}. The quantities which measure the magnitude of the chiral symmetry breaking, f_π and the spontaneous quark mass, are proportional to ρ_c^{-1}. The quantity ρ_c, just as in the no-fermion case, can be expressed in terms of the asymptotic freedom renormalization mass, Λ.

Again there is the vexing question of how the vacuum wave function behaves on scales large compared to ρ_c, scales where confining effects are significant. Since ρ_c will turn out to be the scale size of physical hadrons we will assume that understanding the nature of fluctuations on scales large compared to ρ_c is not crucial to understanding the properties of ordinary hadrons (though it is of course crucial to understanding confinement!).

The connection between nonperturbative fluctuations and chiral symmetry breaking is absolutely crucial to understanding the phenomenology of light quark hadrons. In particular, the bare quark masses are not zero, but finite, and this connection will allow us to decide a priori when a quark is light enough for there to be a recognizable Goldstone boson associated with it. It will also allow us to compute the not-quite-vanishing mass of the Goldstone bosons in terms of bare quark masses and Λ. Finally, the connection between instantons and spontaneous quark mass generation will play an important role in the dynamics of our eventual bag model of light quark hadrons in general.

Our aim is to see whether the relations between all these phenomenologically interesting quantities which emerge from our picture of the vacuum wave function are roughly consistent with reality. Since the mathematics of dynamical symmetry breaking are notoriously difficult to treat, we will have to make a number of brutal approximations, and guesses, in order to get a useful overview of the situation. This is perhaps tolerable since at the present stage we are looking to evaluate the rough qualitative consistency of our picture rather than its numerical details.

In order to explore these questions let us add N_F flavors of color triplet quark to the pure gauge field system discussed in Sect. 2. Our first task is to compute the ground state energy and we once again elect to do so via the Euclidean functional integral. We have only to modify (2.2) to[17]

$$Z = \int[d(A)]\, e^{-S(A)} \prod_{i=1}^{N_F} \int[d\psi_i d\bar{\psi}_i]\, e^{-S_i(\psi_i,\bar{\psi}_i,A)} \tag{3.1}$$

where

$$S(\psi,\bar{\psi},A) = \int d^4x \;\; \bar{\psi}[+i\gamma_\mu D_\mu(A) - m]\,\psi \tag{3.2}$$

and

$$D_\mu(A) = \partial_\mu - ig_o A_\mu^c T_a$$

$$\{\gamma_\mu,\gamma_\nu\} = -2g_{\mu\nu} \tag{3.3}$$

$$[T_a,T_b] = i \;\; f_{abc} \;\; T_c$$

Since (3.2) is bilinear in the fermi variables, each fermion functional integral in (3.1) can be done explicitly and yields a determinant:

$$Z = \int[dA]\ e^{-S(A)} \prod_{i=1}^{N_F} \text{Det}\ (-i\gamma_\mu D_\mu(A) + m_i)$$

$$= \int[dA]\ e^{-S(A) + \sum_{i=1}^{N_F} \log \text{Det}\ (-i\gamma_\mu D_\mu(A) + m_i)} \tag{3.4}$$

As in Sect. 2, we will imagine computing Z by expanding the A functional integration about the minima of S(A). Integrating about the global minimum, A=0, gives a contribution of O(1) and generates ordinary perturbation theory of quarks and gluons. The instanton local minima of S(A) again make contributions of order $e^{-8\pi^2/g^2}$, but the fermion determinant, even through only O(1), makes a qualitative change in the interpretation to be put on these nonperturbative effects.

The fermion determinant is just the product of the eigenvalues of the operator $-i\gamma_\mu D_\mu(A) + m$. These are known once we have found the complete set of eigenvectors $\{\psi_i\}$ and eigenvalues $\{\lambda_i\}$ of the operator $-i\gamma_\mu D_\mu(A)$:

$$-i\gamma_\mu D_\mu(A)\ \psi_i(x) = i\lambda_i\ \psi_i(x) \tag{3.5}$$

The corresponding determinant for a single flavor of quark is

$$\text{Det}\ (-i\gamma D + m) = \prod_i\ (i\lambda_i + m) \tag{3.6}$$

The eigenvalue set $\{\lambda_i\}$ has a continuum piece, corresponding to non-normalizable eigenfunctions, whose product is divergent and must be renormalized. The renormalized product has been explicitly computed by 't Hooft[22] in the m=0 limit. More important is the fact that when A is the instanton field as specified in (2.7), (3.5) has a single <u>discrete, normalizable</u>, zero eigenvalue solution.[22] The corresponding eigenvector is

$$\psi_0^{\pm}(x) = \frac{\sqrt{2}\,\rho\,\gamma_\mu\,(x-x_I)_\mu\,\chi^{\pm}}{\pi[(x-x_I)^2]^{\frac{1}{2}}\,[(x-x_I)^2+\rho^2]^{3/2}} \tag{3.7}$$

where $\chi^{\pm}$ is a unit normalized Dirac-SU_3 spinor which can be characterized by

$$\chi^{\pm}\bar{\chi}^{\pm} = \left(\frac{1 \pm \gamma_5}{2}\right)[\tfrac{1}{4} + \tfrac{i}{8}\,\eta^{(\pm)}_{\mu\nu\alpha}\,R_{\alpha a}\,T_a\,\gamma_\mu\,\gamma_\nu] \tag{3.8}$$

The $\pm$ sign indicates the form appropriate to instanton and antiinstanton, respectively. The quantities ρ, x_I and R are the instanton collective coordinates. The existence of the discrete zero eigenvalue means that (3.4) has a discrete factor of m and therefore vanishes in the m=0 limit. This is the famous result of 't Hooft that massless fermions suppress instantons. The end result of the properly renormalized calculation of Z in the presence of N_F flavors of fermions with mass m is[12]

$$Z = \int d^4x_I \int \frac{d\rho}{\rho^5} \int [dR]\;\Phi(m\rho)\;C_o\left(\frac{8\pi^2}{g^2(\rho)}\right)^6 e^{-8\pi^2/g^2(\rho)}\;[1+O(g^2)] \tag{3.9}$$

where

$$\frac{8\pi^2}{g^2(\rho)} = x(\rho) \simeq (11 - \frac{2N_F}{3})\;\ell n\,\frac{1}{\Lambda\rho} \tag{3.10}$$

is the asymptotic freedom running coupling appropriate to N_F flavors, and C_o is a normalization constant depending on the coupling constant definition we have chosen. If we define C_o to have the same values as in the no-fermion case, then we find that Φ has the form

$$\begin{aligned} \Phi &= (\kappa m\rho)^{N_F} && \text{for } m\rho \lesssim 1 \\ \Phi &\simeq 1 && \text{for } m\rho > 1 \end{aligned} \tag{3.11}$$

where κ has the value 1.33 (.59) when calculated using Pauli-Villars (dimensional) regularization. As a result we can say that vacuum tunneling on a scale ρ is suppressed if there are fermions in the theory with bare mass less than ρ^{-1} and would be tempted to say that if there are any massless fermions, vacuum tunneling is eliminated altogether.

In fact, even in the zero fermion mass limit, the instanton plays just as important a role in determining the physics of the theory as in the pure gauge theory case. But, since its effect is mainly to redefine the fermion vacuum rather than the gauge field vacuum we must use new techniques to study it. The proper technique depends on the number of massless flavors and introduces an unpleasant, but necessary, degree of complication into the discussion. To keep the discussion manageable, we will concentrate on the cases of one or two massless flavors of fermion interacting with an SU_3 gauge field.

Consider first the one flavor case. Instead of computing the instanton contribution to the vacuum to vacuum amplitude (which we know vanishes in the m=0 limit) let us look at the instanton contribution to the fermion propagator

$$S(x,y) = < \psi(x)\bar{\psi}(y) > = \int[dA]\ r^{-S(A)} \int[d\psi d\bar{\psi}]\ \psi(x)\psi(y)\ e^{-S(\psi,\bar{\psi},A)} \tag{3.12}$$

The integration over the fermi fields is still trivial and can easily be expressed in terms of the complete set of eigenfunctions $\{\psi_i(x)\}$ of the operator $-i\gamma_\mu D_\mu(A)$:

$$S(x,y) = \int[dA]\ e^{-S(A)} \{\sum_i \psi_i(x)\bar{\psi}_i(y) \prod_{j\neq i} \lambda_j\}$$

$$= \int[dA]\ e^{-S(A)}\ \psi_0(x)\bar{\psi}_0(y) \prod_{j\neq 0}\lambda_j \tag{3.13}$$

The sum degenerates to one term precisely because the corresponding eigenvalue is zero. The discussion leading up to (3.9) should make it clear that (3.13) will be identical to the m=0 limit of (3.9) if we remove one factor of m and replace it by $\rho\psi_0^{+}(x)\psi_0^{-+}(y)$:

$$S_+(x,y) = \int d^4x_I \int \frac{d\rho}{\rho^5} \int [dR]\ C_o \left(\frac{8\pi^2}{g^2(\rho)}\right)^6 e^{-8\pi^2/g^2(\rho)}\ \rho\psi_o^+(x)\bar{\psi}_o^+(y)$$

$$= \int d^4x_I \int \frac{d\rho}{\rho^4} \int [dR]\ D(\rho)\ \psi_o^+(x-x_I;\rho,R)\ \bar{\psi}_o^+(y-x_I;\rho,R) \tag{3.14}$$

where we have defined an instanton density function

$$D(\rho) = C_o \left(\frac{8\pi^2}{g^2(\rho)}\right)^6 e^{-8\pi^2/g^2(\rho)} \tag{3.15}$$

We have made explicit the dependence of ψ_o on both ρ and R (group orientation matrix) and for convenience we have normalized the R integration to unity. There is of course a similar expression for the anti-instanton, with ψ_o^+ replaced by ψ_o^-.

By a similar argument one can show that the single instanton contribution to all other multiple fermion Green's functions (<1>, $\langle\psi\bar{\psi}\psi\bar{\psi}\rangle$, ...) vanishes. In short, the instanton or antiinstanton must be regarded as an effective non-local quark-antiquark interaction

$$\mathscr{L}^{\pm}_{EFF} = \int d^4x\ d^4y\ \bar{\psi}(x)\ (i\vec{\partial}_x\cdot\gamma)S_{\pm}(x,y)\ (i\overleftarrow{\partial}_y\cdot\gamma)\ \psi(y) \tag{3.16}$$

Because of the γ_5 structure of $\psi_o^{\pm}$, this has the same chirality properties as a $\bar{\psi}(1\pm\gamma_5)\psi$vertex and we must conclude that an instanton vacuum tunneling of the gauge field is necessarily accompanied by a transition in the fermion sector from the vacuum to a state consisting of a positive helicity quark and negative helicity antiquark (and vice-versa for an antiinstanton). This is why the fermion vacuum to vacuum amplitude (3.2) must vanish.

This sort of event violates the axial U(1) symmetry (associated with the conservation of the current $J_\mu^5 = \bar{\psi}\gamma_\mu\gamma_5\psi$) possessed by the theory at the Lagrangian level, and tells us that if non-perturbative effects play a significant role, chiral invariance will be badly broken.[22] This is fundamentally due to the Adler-Bell-Jackiw[23] anomaly, which says that due to renormalization effects the axial current is in fact <u>not</u> conserved:

$$\partial_\mu J^5_\mu = \frac{g^2}{32\pi^2} \sum_{a=1}^{8} F^a_{\mu\nu} \tilde{F}^a_{\mu\nu} \tag{3.17}$$

The kinematics of nonperturbative chiral symmetry breaking are so clearly explained in the original papers[22,15,16] and later reviews[17] that we feel no need to go into such matters here.

The discussion so far has been carried out by doing perturbation theory about the chirally summetric massless quark vacuum. One sign of this is that we have been using the usual m=0 fermi propagator, which anti-commutes with γ_5. However, every time a vacuum tunneling occurs, the chirality of the fermion system changes, and it is no longer obvious that this is the proper fermion vacuum. The true ground state of the system will presumably have no definite chirality and be characterized by a quark propagator which does not anticommute with γ_5 and therefore has a non-vanishing (and probably momentum-dependent) mass function.

To determine the true quark propagator, $\dot{S}(x-y)$, one inevitably must perform a resummation procedure that treats simultaneously an infinite number of orders of perturbation theory about the naive (chirally symmetric) quark vacuum. We will borrow a technique widely used in studies of dynamical symmetry breakdown in quantum field theory.[24] The fermion action function (3.2) is the sum of a free propagation term and an interaction term:

$$S(\psi,\bar{\psi},A) = \int d^4x d^4y\, \bar{\psi}(x)\, [S_o^{-1}(x-y) + g_o A^a_\mu T_a \gamma_\mu\, \delta(x-y)]\, \psi(y) \tag{3.18}$$

where S_o is the free massless propagator. We rewrite S as a sum of two terms:

$$S(\psi,\bar{\psi},A) = S_1 + S_2$$

$$S_1 = \int d^4x\, d^4y\, \bar{\psi}(x)\, [S^{-1}(x-y) + g_o\, A^a_\mu\, T_a\, \gamma_\mu\, \delta(x-y)]\, \psi(y) \tag{3.19}$$

$$S_2 = \int d^4x\, d^4y\, \bar{\psi}(x)\, [S_o^{-1}(x-y) - S^{-1}(x-y)]\, \psi(y)$$

using the assumed true quark propagator $S(x-y)$. When we do functional integrals such as (3.1) or (3.12) we treat the S_1 piece exactly and, in particular, evaluate the fermion determinant about the instanton as if the fermion had whatever mass function is implied by S. The S_2 piece, which compensates for our having chosen the "wrong" propagator, is expanded as a series of self-mass insertions on the graphs generated by S_1. The propagator, S, is finally determined by the condition that all the graphs (those generated by S_1 and by the expansion of S_2 being taken together) having the topology of a self-mass insertion should cancel against one another. This cancellation has the consequence that the quark propagator, as calculated by the above procedure from (3.12), is exactly the function $S(x-y)$ we started with.

The implied self-consistency condition for S is

$$S^{-1}(x-y) - S_o^{-1}(x-y) = \Sigma\ (S;x,y) \tag{3.20}$$

where Σ is the sum over all amputated two and one-fermion irreducible propagator insertion graphs that can be constructed using the interactions of the theory and the assumed propagator S. Σ of course gets contributions from ordinary quark and gluon exchanges, but we are more interested in the effective interactions arising from integrating over non-perturbative sectors of the gauge field time history.

We have already seen that the lowest-order effect of the single instanton, computed in the chiral vacuum, is to generate an effective fermion self-mass insertion. We can in fact conclude from (3.16) that

$$\Sigma\ (S_o;x,y) = i\vec{\partial}_x\cdot\gamma\ [S_+(x,y) + S_-(x,y)]\ i\overleftarrow{\partial}_y\cdot\gamma \tag{3.21}$$

where we have added together the contributions of instanton and anti-instanton to get the total insertion. We have written $\Sigma(S_o)$ to indicate that the effective interaction has been computed in the chiral fermion vacuum. We could compute $\Sigma(S)$ by redoing the calculation following (3.12) with the replacement of $S(\psi,\bar{\psi},A)$ in (3.12) by the term S_1 defined in (3.19). In the limit of small spontaneous fermion mass it is safe to replace $\Sigma(S)$ in (3.20) by $\Sigma(S_o)$ and we will make this approximation.

By computing (3.14), (3.7) and (3.8) and doing the instanton group orientation integration we obtain the following explicit form of (3.20):

$$S^{-1}(x-y) - S_o^{-1}(x-y) = \frac{1}{4} \int d^4x_I \int \frac{d\rho}{\rho^4} \phi_\rho(x-x_I) \phi_\rho^+(y-y_I) D(\rho)$$

$$\phi_\rho(x) = \gamma\cdot\partial \left\{ \frac{\sqrt{2}}{\pi} \frac{\rho\ \gamma\cdot x}{\sqrt{x^2}\ (x^2+\rho^2)^{3/2}} \right\} \quad (3.22)$$

his is most conveniently reexpressed in momentum space as

$$\tilde{S}^{-1}(p) - \tilde{S}_o^{-1}(p) = 2\pi^2 \int \frac{d\rho}{\rho^2} D(\rho)\ F^2(\rho p) \quad (3.23)$$

where we have defined F by

$$\int d^4x\ e^{ip\cdot x} \phi_\rho(x) = 2\sqrt{2}\ \pi\rho\ F(\rho p) \quad (3.24)$$

The function F has been normalized so that F(0) = 1 and can be shown to decrease for $x \gtrsim 1$ as x^{-3}.

The interpretation is straightforward--the effect of nonperturbative fluctuation is, in first approximation, simply to generate a momentum-dependent fermion mass

$$m(p^2) = 2\pi^2 \int \frac{d\rho}{\rho^2} D(\rho) F^2(\rho p) \quad 3.25$$

This is as direct an expression as one could want of the fact that non-perturbative effects in QCD destroy chiral symmetry. Although this mass is formally $O(e^{-1/g^2})$ (because $D(\rho)$ is of this order) and therefore not visible in perturbation theory, it will turn out not to be particularly small on the scale of hadron masses.

To be quantitative about m(p) we need to know what to do with the upper limit on the integration over ρ (the infrared problem again). Because of asymptotic freedom, the integration converges at small ρ. Because of the presence of $F(\rho p)$, the integration has a strong power law convergence factor which comes in for $\rho > p^{-1}$ and (3.25) should therefore be well-defined for large enough p. In fact, one expects $m(p^2)$ to vanish, because of F, as p^{-6} for large p. This is the expected signal that non-perturbative effects do not influence short distance physics, although we see here that they die away only as a power of momentum.

To evaluate the static fermion mass, m(0), we need some prescription for dealing with instantons of large scale sizes. We shall adopt essentially the same attitude as advocated in Sect. 2 for the pure gauge theory. We have obtained (3.25) by summing over multiple instanton anti-instanton time histories by a method which, as far as the gauge field dependence is concerned, is equivalent to constructing a WKB approximation to the vacuum wave function. For the same reasons as advocated in Sect. 2, we believe that by summing over all such configurations, including instantons up to some maximum scale size, ρ_c, chosen so that the integrated fraction of space time covered by instantons is O(1), one is in effect constructing a trial wave function which accurately treats non-perturbative fluctuations of scale size up to ρ_c. This mistreats the large-scale confining fluctuations, but may be adequate for a discussion of the properties of ordinary hadrons if their physics is determined by fluctuations on scales comparable to ρ_c.

To determine ρ_c we must compute the instanton density. By our self-consistent method of calculation this is to be determined using the self-consistently determined fermion propagator, S(x-y). This means that we must redo the argument following (3.5), attributing to the fermion not a constant mass, m, but a momentum dependent mass function, m(p). In general the answer is complicated, but in the limit of small spontaneous mass, the answer is simple: We replace $\Phi(m\rho)$ in (3.9) by $m_o(\rho)\cdot\rho$, where $m_o(\rho)$ is the expectation value of the spontaneous mass function in the fermion zero-mode wave function (3.7) and is given explicitly by

$$\begin{aligned} m_o(\rho) &= \int \frac{d^4\rho}{(2\pi)^4}\; m(p^2)\cdot\frac{8\pi^2\,{}^2}{p^2}\; F^2(\rho p) \\ &= \int \frac{d^4 p}{(2\pi)^4}\;\frac{8\pi^2\rho^2}{p^2}\; F^2(\rho p)\cdot 2\pi^2\int_0^\rho \frac{d\rho'}{\rho'^2}\, D(\rho')\; F^2(\rho' p) \end{aligned} \tag{3.26}$$

So long as $\rho m_o(\rho)$ is not too big we account for the effect of fermions by appending a factor of $\rho m_o(\rho)$ to all our pure gauge theory formulas for density of instantons of scale size ρ. In particular, the condition determining ρ_c is just

$$1 = f(\rho_c) = \int_0^{\rho_c} \frac{d\rho}{\rho}\, D(\rho)\cdot\rho\; m_o(\rho) \tag{3.27}$$

These equations are not completely transparent, but a certain amount of fiddling with them should convince the reader that with ρ_c chosen as we have advocated, the static spontaneous fermion mass, m(0), is approximately ρ_c^{-1}, to within a factor of a few. It could hardly be otherwise since ρ_c is the only internally defined physical scale and we have only verified that no large dimensionless factors creep in. Eventually ρ_c will turn out to be a typical hadronic scale and ρ_c^{-1} will therefore be a typical hadronic energy of order several hundred MeV or so. This means that the dimensional parameter, m(0), which measures the magnitude of chiral symmetry breaking will be of typical hadronic magnitude and chiral symmetry breaking effects large. At the same time, chiral symmetry breaking effects are scale dependent and turn off at scales small compared to ρ. It is perhaps also worth noting that the spontaneous mass generation phenomena discussed here would also occur if the quark had a non-zero bare mass. If the bare mass is significantly smaller than m(0) then the above picture would be unchanged, while if the bare mass were significantly larger than m(0) the physics would presumably revert to that of pure gauge theory. A key point for future use is that the magnitude of chiral symmetry breaking is directly tied to the density of instantons--if they can be suppressed, chiral symmetry breaking turns off.

This lengthy prelude, dealing with the case of one massless flavor, has provided us with the tools we need to discuss the phenomenologically much more interesting case of two massless flavors. Let us first discuss the effect of a single instanton on the massless quark vacuum. If there is one massless flavor, we know that the instanton acts like the insertion of an effective $\bar{\psi}(1+\gamma_5)\psi$ interaction. Since different flavors are coupled to the gauge field indepently, one expects that with two massless flavors, the instanton must act like the insertion of a four-point vertex-a factor of $\bar{\psi}(1+\gamma_5)\psi$ type for each flavor. This suggests that we study the four point function

$$S(x_1y_1;\ x_2y_2) = \int[dA]\ e^{-S(A)} \prod_{f=1}^{2} \int[d\psi_f\ d\bar{\psi}_f]\ \psi_f(x_f)\ \bar{\psi}_f(y_f)\ e^{-S(\psi_f,\bar{\psi}_f,A)} \tag{3.28}$$

Precisely the line of argument that leads to (3.13) gives the result

$$S(x_1y_1;x_2y_2) = \int[dA]\ e^{-S(A)} \prod_{f=1}^{2} \{\psi_o^+(x_f)\ \bar{\psi}_o^+(y_f)\ \prod_{j\neq 0} \lambda_j\} \tag{3.29}$$

The end result of normalizing with respect to the vacuum and carrying out the renormalization procedure is

$$S(x_1y_1;x_2y_2) = \int d^4x_I \int \frac{d\rho}{\rho^5} \int [dR]\, D(\rho) \prod_{f=1}^{2} \{\rho\, \psi_o^{+}(x_f)\, \bar{\psi}_o^{+}(y_f)\} \tag{3.30}$$

where D(ρ) is given by (3.15). All other multifermion Green's functions can readily be seen to vanish and the instanton must therefore be regarded as a non-local effective four quark interaction

$$\mathcal{L}^{\pm}_{EFF} = \int d^4x_1 d^4x_2 \int d^4y_1 d^4y_2\, \bar{\psi}(x_1)\bar{\psi}(x_2)(i\gamma\cdot\vec{\partial}x_1)(i\gamma\cdot\vec{\partial}x_2) \times \\ S_{\pm}(x_1y_1;x_2y_2)(i\gamma\cdot\overleftarrow{\partial}y_1)(i\gamma\cdot\overleftarrow{\partial}y_2)\, \psi(y_1)\, \psi(y_2) \tag{3.31}$$

In γ_5 and flavor structure this is equivalent to an interaction of the form $[\bar{\psi}_1(1\pm\gamma_5)\psi_1]\,[\bar{\psi}_2(1\pm\gamma_5)\psi_2]$ and means that the basic gauge field vacuum tunneling is accompanied by the production of a positive helicity quark and negative helicity antiquark for each flavor. The consequences we can draw from this are quite different from those appropriate to the single flavor case. At the naive Lagrangian level, the theory possesses an SU(2) ⊗ SU(2) ⊗ $U_5(1)$ chiral symmetry. It is the $U_5(1)$ symmetry, whose current is $J^5_\mu = \sum_{f=1}^{2} \bar{\psi}_f\gamma_\mu\gamma_5\psi_f$, which is destroyed by the effective interaction (3.31). The generators of SU(2) ⊗ SU(2), whose generator matrices are traceless in flavor space, continue to be conserved. There is even a discrete element of the $U_5(1)$ (rotation angle π) which is conserved and which is enough to guarantee that the quark propagator has no mass term if the vacuum is invariant under it.[15] It is in this sense that the instanton solves the "U(1) problem"[22] --it introduces an effective interaction into the theory which violates the unwanted symmetry. The question of what further symmetries are broken spontaneously and associated Goldstone bosons created by the passage from the naive zero chirality vacuum to a fermi vacuum which is stable under tunneling remains to be addressed.

In fact, we expect to find that chiral SU_2 ⊗ SU_2 is broken down to SU_2 by the spontaneous generation of a fermion mass. We explore this possibility by examining the self consistency condition (3.20) for the fermion propagator and asking whether it has solutions with a non-zero mass function.[15,25,26,27] We shall once again make the approximation that the leading contribution to Σ comes from the one-instanton piece of the integration over non-perturbative gauge field configurations. Therefore we must calculate the propagator of a fermion with mass function $m(p^2)$ in the presence of an instanton in the case that there are two flavors of fermion. To do this we must redo the argument following (3.12) for the case of two flavors and a non-vanishing mass function.

For a first orientation we will assume that the spontaneous mass, $m(p^2)$, is small. By including a small mass term in the fermion Lagrangian in (3.12) we readily find, by an extension of the argument following (3.3), that

$$\Sigma(x,y) = \int d^4x_I \int \frac{d\rho}{\rho^4} D(\rho)\cdot[\rho m_o(\rho)]\cdot(i\gamma\cdot\overleftarrow{\partial}_x\, \psi_o(x)\, \bar{\psi}_o(y)\, i\gamma\cdot\overleftarrow{\partial}_y) \quad (3.32)$$

where $m_o(\rho)$ is the expectation value of the mass function in the zero mode of an instanton of scale size ρ and $D(\rho)$ is the instanton density function, as defined in (3.15), appropriate to two flavors. This expression, when substituted into (3.20), leads to the following self-consistency equation:

$$m(p^2) = 2\pi^2 \int \frac{d\rho}{\rho^2} D(\rho)\; F^2(\rho p)\cdot[\rho m_o(\rho)] \quad (3.33)$$

Since $m_o(\rho)$ is a linear functional of $m(p^2)$, this may be reexpressed in the form of a linear integral equation:

$$m(p) = \int \frac{d^4p'}{(2\pi)^4}\; K(p^2,p'^2)\; m(p') \quad (3.34)$$

$$K(p^2,p'^2) = \frac{(2\pi)^4}{p'^2} \int \rho d\rho\; D(\rho)\; F^2(\rho p)\; F^2(\rho p') \quad (3.35)$$

The self-consistency equation is therefore the statement that the mass function is an eigenvector of K with eigenvalue one. If the kernel is too weak, there will be no such eigenvalue and (3.34) will have no solution other than the trivial one with $m(p) = 0$ and no chiral symmetry breaking. The strength of K is essentially determined by the upper limit,

ρ_c, on the ρ integration in (3.35) which in turn is determined by the range of gauge field fluctuations we have elected to integrate over in performing the functional integral. An important question is whether the effective coupling associated with a choice of ρ_c such that K first acquires a unit eigenvalue is small enough for weak-coupling semi-classical arguments to make sense.

The rough indication is that K first has a unit eigenvalue when ρ_c is chosen such that the integrated fraction of space-time covered by instantons, computed using the density function $D(\rho)$, is of order unity.[25] As mentioned in the study of the pure gauge field problem, this happens, for reasonable coupling constant definitions, at small enough scales and effective couplings for weak coupling arguments to appear quite reasonable.

The answer phenomenological questions, one must know not just that there is a non zero spontaneous mass, but also how large it is. We have so far studied the self-consistency equation in the limit of small spontaneous mass and therefore obtained a linear equation, (3.34), which is incapable of fixing the scale of the self-mass. To fix the scale one needs a non-linear equation, which may be obtained by <u>not</u> expanding the self-consistency equation. Unfortunately, no one has yet obtained a useful general form for the finite mass case and no general information about the magnitude of symmetry breaking effects is available.

If we were to indulge in a guess, we would propose the following picture for what happens: The chiral symmetry breaking effects turn on at a scale ρ, identified as described in the last paragraph; as soon as the spontaneous mass turns on, it rapidly becomes as large as it reasonably can, that is to say, approximately equal to ρ_c^{-1}; from that point on, the rough qualitative properties of the gauge field fluctuations are as they would be without the massless fermion degrees of freedom. The system has a single well-defined scale, ρ_c, such that instantons of scale size $\sim\rho_c$ are dense in the vacuum and spontaneous breakdown of chiral symmetry, with generation of a spontaneous quark mass of order ρ_c^{-1} occurs at the scale ρ_c. Associated with the breaking of chiral symmetry are a multiplet of pseudoscalar Goldstone bosons. The Goldstone bosons have quark-antiquark wave functions whose structure in momentum space is related to the spatial scale on which the chiral symmetry breakdown occurs and will therefore be roughly of size ρ_c. The pion decay constant, another measure of the magnitude of chiral symmetry breaking, should scale with ρ_c^{-1}, but it is unclear whether f_π is $\sim\rho_c^{-1}$ or smaller. Finally it is clear that the proper measure of whether a quark bare mass is small enough to permit approximate chiral symmetry relations to be useful is the ratio of the bare mass to the universal scale ρ_c^{-1}.

Whether such a simple picture is really correct is not at all clear, and can only be decided after someone has succeeded in doing a sufficiently meaningful model calculation. When we come to our tentative effort to do phenomenology, we will try to put the various possibilities for chiral symmetry breaking parameters into a useful physical context.

IV. THE DILUTE GAS VACUUM

In Section 2, we proposed an approximation procedure for dealing with the gauge field vacuum functional integral which might be called the "dilute instanton gas" method. We claimed that the properties of the trial vacuum state constructed by this method would differ substantially from those of the perturbation theory vacuum, even though the effective coupling associated with the largest scale we actually encounter is so small that ordinary radiative corrections are negligible. In the last section, we partially substantiated this claim by showing that the chiral symmetry properties of a system of massless fermions are, in both qualitative and quantitative senses, profoundly changed by their interactions with the "dilute gas" gauge field fluctuations. In this section, we will offer further evidence of this kind by showing that a variety of quantities which are more directly representative of the dynamics of the gauge field itself are substantially modified from their perturbative vacuum values.

An interesting physical probe of the properties of the gauge field vacuum is the static potential of heavy quarks. The simplest case to discuss is the interaction energy of a quark antiquark pair in an overall color singlet state. If the quarks are infinitely massive, this energy can be computed from the properties of a simple Wilson loop:[25]

$$e^{-E(R)R} = \lim_{T\to\infty} \left\langle P(\mathrm{tr}\, e^{i\oint dx_\mu A^a_\mu T_a}) \right\rangle \tag{4.1}$$

where the loop integral is taken around a rectangle of dimensions T and R, P is the symbol for path ordering and the expectation is taken over Euclidean Yang Mills fields with the usual Yang Mills action as the weight. If the quarks have finite mass, one cannot do the loop integral over a fixed path, but must also path-integrate over possible quark path histories and include an appropriate quark action in the functional weight. We shall not discuss this development here but it is essential if one is to calculate recoil-order corrections, such as the spin-dependent potential, to the leading static potential.[28,29] How heavy the quark must be in order to be treated as infinitely massive is determined by the properties of the gauge field vacuum in a way we shall see.

If this Wilson loop integral is evaluated by integrating over Gaussian fluctuations about the perturbative vacuum, one obtains the usual perturbation theory answer,

$$E(R)_{\mathrm{pert}} = -\frac{4}{3}\frac{g^2(R)}{4\pi}\frac{1}{R} \tag{4.2}$$

where g(R) is the running coupling constant appropriate to separation R. The nonperturbative, dilute instanton gas fluctuations make an independent contribution of quite different functional form. So long as we neglect correlations between the instanton fluctuations, it is easy to show that the loop functional of (4.1) is an exponential of individual instanton contributions and that E(R) may be expressed as an integral over single instanton contributions for all possible values of the collective coordinates.[25] Explicitly

$$E(R) = 2\int \frac{d\rho}{\rho^2} D(\rho)\, W(\frac{R}{\rho}) \tag{4.3}$$

where

$$W(\frac{R}{\rho}) = -\frac{1}{3\rho^3} \int d^3x_I \, \mathrm{tr}\, [U(\bar{R}-\bar{x}_I)\, U^{-1}(\bar{x}_I) - 1] \tag{4.4}$$

and $U(\bar{x})$ is the matrix characterizing the gauge transformation carried out on the vacuum in $A_0=0$ gauge by the instanton:

$$U(\bar{x}) = \exp \left[2\pi i\, x_a R_{a\alpha} T_\alpha \Big/ \sqrt{\bar{x}^2+\rho^2} \right] \tag{4.5}$$

The function, W, which characterizes the separation dependence of the contribution of an instanton of definite scale size must be evaluated numerically. Such a computation shows that W vanishes at zero separation and approaches a constant at infinity. The first feature is reasonable enough since we are dealing with an overall color singlet system. The second feature is also reasonable since it amounts to a positive mass renormalization of each quark due to its interaction with nonperturbative vacuum fluctuations. To study the mass renormalization effect and the approach to it a bit more carefully, we rearrange the expression for W as follows:

$$\begin{aligned} W(\frac{R}{\rho}) &= \frac{1}{3\rho^3} \int d^3x_I \, \mathrm{tr}\, [(1 + U(\bar{R}-\bar{x}_I)) + (1 + U^{-1}(x_I))] \\ &\quad - \frac{1}{3\rho^3} \int d^3x_I \, \mathrm{tr}\, [(1 + U(\bar{R}-\bar{x}_I))(1 + U^{-1}(\bar{x}_I))] \end{aligned} \tag{4.6}$$

The first term in (4.6) is obviously independent of R and can be identified with the quark mass renormalization term. Since the matrix $1 + U(\bar{x})$ vanishes as x^{-2} for large x, it is easy to show that the second term in (4.6) decreases as R^{-1} for large R. Upon doing the indicated integrals, we find[19]

$$E(R) \xrightarrow[R \to \infty]{} 37\int \frac{d\rho}{\rho^2} D(\rho) - \frac{4\pi^3}{3} \frac{1}{R} \int \frac{d\rho}{\rho} D(\rho) + O\left(\frac{1}{R^2}\right) \tag{4.7}$$

The constant term in (4.7) is, as we have already said, simply twice the mass renormalization, δm_q, of an isolated quark and is a measure of the energy gap between color singlet and color non-singlet states. The $O(R^{-1})$ term has the same form as the perturbative interaction energy, (4.2), and therefore is to be interpreted as a coupling constant renormalization:

$$\frac{g^2}{8\pi^2} \longrightarrow \frac{g^2}{8\pi^2} \left[1 + \frac{\pi^2}{2} \int \frac{d\rho}{\rho} D(\rho) \left(\frac{8\pi^2}{g^2}\right)\right] \tag{4.8}$$

It is as if the effect of the dilute instanton gas is to <u>strengthen</u> the Coulomb attraction of two charges by a multiplicative factor

$$\mu = 1 + \frac{\pi^2}{2} \int \frac{d\rho}{\rho} D(\rho) \frac{8\pi^2}{g^2} \tag{4.9}$$

We will eventually see that μ ($\varepsilon = \mu^{-1}$) plays the role of magnetic permeability (dielectric constant) of the vacuum and that the property $\mu > 1$ is just another manifestation of infrared slavery.

Our immediate concern is whether these nonperturbative effects are big enough to compete with ordinary perturbative effects. The dilute gas trial function is characterized by an effective cutoff, ρ_c, on the instanton scale size integration such that instantons are dense in the vacuum. According to (2.29) this means that

$$\pi^2 \int_0^{\rho_c} \frac{d\rho}{\rho} D(\rho) \simeq 1 . \tag{4.10}$$

In a crude, but not too unreliable, approximation we may say that

$$\delta m_q = 18.5 \int_0^{\rho_c} \frac{d\rho}{\rho^2} \cdot D(\rho) \simeq \frac{1.9}{\rho_c} \left\{ \pi^2 \int_0^{\rho_c} \frac{d\rho}{\rho} D(\rho) \right\} \simeq \frac{1.9}{\rho_c} \tag{4.11}$$

and

$$\mu = 1 + \frac{\pi^2}{2} \int_0^{\rho_c} \frac{d\rho}{\rho} D(\rho) \frac{8\pi^2}{g^2(\rho)}$$

$$\simeq 1 + \frac{1}{2}\left(\frac{8\pi^2}{g^2(\rho_c)}\right) \cdot \left\{ \pi^2 \int_0^{\rho_c} \frac{d\rho}{\rho} D(\rho) \right\} \tag{4.12}$$

$$\simeq 1 + \frac{1}{2} \frac{8\pi^2}{g^2(\rho\ )}$$

Since, as was mentioned in Sect. 2, the value of $8\pi^2/g^2$ associated with the cutoff scale is quite large (it will turn out to lie somewhere between 10 and 20), μ differs by a large factor from its perturbation theory value of 1. We shall see that this suffices to make the qualitative physics of the dilute gas vacuum quite different from that of perturbation theory. It is not possible to say apriori whether the quark mass renormalization is large or small compared to typical hadronic energies since we do not yet know how such energies compare to ρ_c^{-1}! When we solve this problem we will find that ρ_c^{-1} is probably an energy in the neighborhood of 500 MeV and that the mass gap between color singlet and color non-singlet being produced by instanton fluctuations is in the neighborhood of one GeV per quark. When we take the effects of $\mu >> 1$ into account, this energy gap is made quite a bit bigger and we feel reasonably confident that our mistreatment of the confining fluctuations of the theory does not invalidate our picture of the low-lying color singlet hadrons.

We have so far treated the dilute instanton gas as a perfect gas, ignoring any interactions between the "particles" of which the system is composed. This is a potentially dangerous assumption and we should determine what the interactions actually are.

The basic quantity we wish to evaluate is the interaction of an instanton (or antiinstanton) whose collective coordinates have prescribed values with a prescribed external Yang Mills field. Eventually, this external field will be due to other instantons and antiinstantons with prescribed collective coordinates, but we want to be able to discuss other types of external field as well. Since a configuration of this kind is not usually a strict solution of the Yang Mills equations but rather a solution of these equations subject to some constraint, this problem requires a careful examination of the problem of doing constrained functional integrals. Since a careful treatment[19] gives the same result as a naive argument [25] we will content ourselves with the naive argument for the correct answer.

In the naive argument we set $A_\mu = A^o_\mu + \delta A_\mu$ where A^o_μ is the single instanton field, (2.7), and δA_μ approaches the potential of a weak, essentially constant $F^{ext}_{\mu\nu}$ at distances large compared to the instanton scale size, ρ. We divide space into two regions by a sphere of radius $R >> \rho$: Inside, in region I, A^o_μ is much bigger than δA_μ and outside, in region II, the situation is reversed. For $|x| \sim R$ we may choose $\delta A_\mu \sim -\frac{1}{2} F^{ext}_{\mu\nu} x_\nu$, the Landau gauge potential of a weak constant $F_{\mu\nu}$. The contribution of region I to the interaction is

$$S_I\,(A^o+\delta A) - S_I\,(A^o) = \int_{|x|<R} d^4x\,(D_\mu(A^o)\,\delta A^a_\nu)\,F^{ao}_{\mu\nu} + O(\delta A^2) \tag{4.13}$$

which may be integrated by parts, using the equation of motion for A^o_μ, to yield a surface term

$$S_I(A^o+\delta A) - S_I(A^o) = \int_{|x|=R} d\Omega\hat{x}_\mu\;\delta A^\alpha_\nu\;(F^a_{\mu\nu})^o + O(\delta A^2) \tag{4.14}$$

A careful examination of the contribution of region II as well as the $O(\delta A^2)$ terms shows that the interaction is just twice the surface term of (4.14).[25]

This surface term is easily evaluated in terms of the parameters of the instanton and F^{ext}:

$$S_{int} = \frac{2\pi^2}{g}\,\rho^2\,R_{a\alpha}\,\bar{\eta}_{\alpha\mu\nu}\,F^{ext}_{a\mu\nu} \tag{4.15}$$

There is of course a similar formula for the antiinstanton with $\bar{\eta}$ replaced by η. This, it should be noted, implies that an instanton (antiinstanton) interacts only with the anti-self-dual (self-dual) part of the external field. If we substitute for F^{ext} the asymptotic form of the field strength tensor of another instanton or antiinstanton (2.8) we get an expression for the instanton antiinstanton interaction:[25,30]

$$S_{int} = \frac{8\pi^2}{g^2}\,\frac{\rho_+^2\,\rho_-^2}{R^4}\,R^+_{a\alpha}\,R^-_{a\beta}\,\eta_{\alpha\mu\nu}\,\bar{\eta}_{\beta\lambda\sigma}\,T_{\mu\lambda}\,T_{\nu\sigma} \tag{4.16}$$

where R is the separation between instanton and antiinstanton and $T_{\mu\nu} = g_{\mu\nu} - 2\hat{R}_\mu\hat{R}_\nu$. Of course, instantons do not interact with instantons at this level because the instanton field is self dual and, by (4.15), instantons interact only with anti self dual fields. The two results, (4.15) and (4.16), will be the basis for the rest of our discussion.

The long range interactions just found are, it turns out, what one expects of four-dimensional magnetic dipoles. This means that the instanton gas may be thought of as a four-dimensional polar medium and suggests that we try to carry over some of the concepts and methods which are found useful in the study of three dimensional polar media.[14] This will be particularly helpful in developing the notion that the vacuum is characterized by a permeability, μ, and in dealing with the fact that μ is large compared to one.

Let us therefore study the problem of four-dimensional abelian static magnetism in the presence of a prescribed distribution of permanent dipoles. We have a vector potential, A_μ, a corresponding field, $F_{\mu\nu} = \partial_\mu A_\nu - \partial_\nu A_\mu$, and dipole sources $J_\mu(x)$. The energy function is

$$\mathcal{E} = \int d^4x \; \{\tfrac{1}{4} F^2 + JA\} \; . \tag{4.17}$$

The equation for A is $\Box A_\mu = J_\mu$. A general point dipole source function may be written in the form

$$J_\mu = -D_{\mu\nu} \, \partial_\nu \cdot 4\pi^2 \, \delta^{(4)}(x-x_o) \tag{4.18}$$

where $D_{\mu\nu} = -D_{\nu\mu}$ to guarantee conservation. The corresponding fields are

$$\begin{aligned} A_\mu &= D_{\mu\nu} \, \partial_\nu \frac{1}{(x-x_o)^2} = -2D_{\mu\nu} \frac{(x-x_o)_\nu}{(x-x_o)^4} \\ F_{\mu\nu} &= \frac{4}{(x-x_o)^4} \, T_{\mu\alpha}(x-x_o) \, D_{\alpha\beta} \, T_{\beta\nu}(x-x_o) \end{aligned} \tag{4.19}$$

Note that the dipole moment of a four-dimensional source distribution, defined by $D_{\mu\nu} = \int d^4x\, (x_\mu J_\nu - x_\nu J_\mu)$, nessarily comes in two rotationally invariant species defined by the self duality of $D_{\mu\nu}$. The interaction energy of a dipole with an external field is

$$\begin{aligned} \mathcal{E}_{int} &= \int d^4x\, A^{ext}_\mu\, J_\mu \\ &= \int d^4x\, A^{ext}_\mu\, [-4\pi^2\, D_{\mu\nu} \partial_\nu\, \delta^{(4)}(x-x_o)] \\ &= 2\pi^2\, D_{\mu\nu}\, F^{ext}_{\mu\nu}\, (x_o) \end{aligned} \qquad (4.20)$$

If we substitute for $F^{ext}_{\mu\nu}$ the field of another dipole located a distance R away, we get

$$\mathcal{E}^{1,2}_{int} = \frac{8\pi^2}{R^4}\, T_{\mu\alpha}\, (\hat{R})\, D^{(1)}_{\alpha\beta}\, T_{\beta\nu}\, (\hat{R})\, D^{(2)}_{\mu\nu} \qquad (4.21)$$

All of these results are exact transcriptions of what we have already found for instantons if we think of each instanton as possessing a permanent color magnetic dipole moment[14]

$$D^a_{\mu\nu} = -\frac{\rho^2}{g(\rho)}\, R_{a\alpha}\, \bar{\eta}^\alpha_{\mu\nu} \qquad (4.22)$$

the index a indicating an independent moment and independent interaction energy for each component of color.

In a medium with a bulk density of permanent dipoles it is feasible and convenient to discuss the response to external fields and sources via a bulk magnetization density rather than in terms of the individual dipoles. We can define a bulk magnetization tensor by summing over the moment densities of all the dipoles in the system.

$$M_{\mu\nu}(x) = \sum_i D^{(i)}_{\mu\nu}\, \delta(x-x^{(i)}), \qquad (4.23)$$

and use M to define a field

$$G_{\mu\nu} = F_{\mu\nu} - 4\pi^2\, M_{\mu\nu} \qquad (4.24)$$

which only has the truly external sources as its source. That is, if $\partial_\mu F_{\mu\nu} = J_\nu^{ext} + J_\nu^{dipole}$, because of (4.23) we have that $\partial_\mu (4\pi^2 M_{\mu\nu}) = J_\nu^{dipole}$ and therefore $\partial_\mu G_{\mu\nu} = J_\nu^{ext}$. The distinction between F and G is the familiar one between magnetic induction and magnetic field.

In the absence of external sources G vanishes and so, presumably, should the net magnetization. This suggests the usual weak field approximation

$$M_{\mu\nu} = \chi G_{\mu\nu} \tag{4.25}$$

which in turn implies that

$$\begin{aligned} F_{\mu\nu} &= \mu G_{\mu\nu} \\ \mu &= 1 + 4\pi^2 \chi \end{aligned} \tag{4.26}$$

The quantity χ is what is usually called the susceptibility and μ is the permeability of the medium. One can also show in the usual way that the total energy of the system, including the orientation energy of the dipoles, in the presence of some external source J_μ^{ext}, is

$$\mathcal{E} = \int d^4x \, \{ \tfrac{1}{4} F_{\mu\nu} G_{\mu\nu} + J_\mu^{ext} A_\mu \} \tag{4.27}$$

If one has a particular spatial distribution of μ, and a particular external source, J^{ext}, then the equations to be solved are the familiar material medium set:

$$\partial_\mu G_{\mu\nu} = J_\nu \, , \qquad F_{\mu\nu} = \partial_\mu A_\nu - \partial_\nu A_\mu \, , \qquad F_{\mu\nu} = \mu G_{\mu\nu} \tag{4.28}$$

A particular example which we will be interested in later on is the problem of determining F at the center of a spherical vacuum cavity (F^c) in a medium of permeability μ in which F has some prescribed uniform asymptotic value (F^E). Textbook manipulations give the result

$$F^c_{\mu\nu} = \frac{2}{\mu+1} F^E_{\mu\nu} \, . \tag{4.29}$$

Thus, the field in such a cavity is reduced relative to the asymptotic field if $\mu>1$. Another example of interest concerns placing a dipole moment $D_{\mu\nu}$ at the center of such a cavity (taken to be of radius R) under the condition that there be no asymptotic field at infinity. Then one finds inside the cavity, over and above the dipole field itself, a uniform reaction field, F^R:

$$F^R_{\mu\nu} = - \frac{\mu-1}{\mu+1} \frac{4D_{\mu\nu}}{R^4} . \tag{4.30}$$

This represents the reaction of the medium to the presence of the dipole in the cavity.

Before embarking on a calculation of μ, we should make a few qualitative remarks. Since the susceptibility is arising from permanent dipoles we expect it to be positive and μ to be greater than one. This causes external sources to be antiscreened: $F_{\mu\nu}$ is proportional to μ and the energy of a current configuration is proportional to μ as well. If μ is big enough, i.e., if the system is strongly paramagnetic, the physical effect would be essentially the same as confinement. It is also worth noting that it is rather easy to see that quarks and scalar fields tend to screen external sources and to make $\mu < 1$: Quark time histories are represented in Euclidean Abelian gauge field theory by current loops which contribute to the partition function a factor tr $(\exp i \int dx_\mu A_\mu)$. If the loop is small enough we may write $\int dx_\mu A_\mu = \Sigma_{\mu\nu} F_{\mu\nu}$ where $\Sigma_{\mu\nu}$ is the area element of the loop. Comparing with (4.20), we see that it is as if the loop were a dipole with imaginary moment proportional to $\Sigma_{\mu\nu}$. Since, as we shall shortly see, the susceptibility is proportional to the square of the dipole moment, such current loops have negative susceptibility and lead to screening behavior.

We are now ready to calculate the permeability of the dilute instanton gas vacuum. With the help of techniques suggested by the dipole gas analogy, we will be able to make a reasonable calculation of μ even when it is very different from its perturbation theory value and strong coupling effects quite significant. The first step is to calculate the response of a single dipole to a weak external field. It has a dipole moment given by (4.22) and an interaction with an applied field given by (4.20). Then the orientation-dependent part of the Boltzmann factor for the instanton is

$$P(\rho,R) = \frac{\exp\{\frac{2\pi^2}{g(\rho)} \rho^2 R_{a\alpha} \bar{\eta}_{\alpha\mu\nu} F_{a\mu\nu}\}}{\int [dR] \exp\{\frac{2\pi^2}{g(\rho)} \rho^2 R_{a\alpha} \bar{\eta}_{\alpha\mu\nu} F_{a\mu\nu}\}} \tag{4.31}$$

where we have normalized the invariant group integration measure by $\int[dR] = 1$. The mean dipole moment of the instanton is

$$\langle D^a_{\mu\nu} \rangle = \int [dR]\ P(\rho,R)\left(-\frac{\rho^2}{g(\rho)} R_{a\alpha}\ \bar{\eta}_{\alpha\mu\nu}\right) \tag{4.32}$$

In the weak field limit we may evaluate this by expanding P in powers of F. Since $\int[dR]\ R_a = 0$, the first term which survives is O(F) and easily computed:

$$\langle D^a_{\mu\nu} \rangle = \frac{4\pi^2}{g^2(\rho)}\ \frac{\rho^4}{8}\ (F^a_{\mu\nu} - \tilde{F}^a_{\mu\nu}) \tag{4.33}$$

There is a similar result for the antiinstanton, with $F-\tilde{F}$ replaced by $F+\tilde{F}$. To obtain the dipole moment density tensor, M, we must multiply by the appropriate instanton and antiinstanton density function, add together the instanton and antiinstanton pieces and integrate over ρ:

$$M^a_{\mu\nu} = \int\frac{d\rho}{\rho^5}\ \bar{D}(\rho)\cdot\frac{4\pi^2}{g^2(\rho)}\cdot\frac{\rho^4}{4}\ F^a_{\mu\nu} \tag{4.34}$$

$\bar{D}(\rho)$ is the true instanton density function, not necessarily equal because of effects of the medium, to the "single particle" density function $D(\rho)$ (2.28).

This result is exact so long as F is small enough for higher order terms in the expansion of (4.31) in powers of F to be negligible and so long as F is interpreted as the local field at the position of the dipole-not necessarily the same as the average field in the medium. If the densi is small, this distinction is not important and we may convert (4.34) directly into a "dilute gas" expression for the susceptibility and permeab ity[25]

$$\mu_{DG} = 1 + 4\pi^2\ \chi_{DG} = 1 + \frac{\pi^2}{2}\int\frac{d\rho}{\rho}\ \bar{D}(\rho)\left(\frac{8\pi^2}{g^2(\rho)}\right) \tag{4.35}$$

This is identical to (4.9) and we may apply here all the remarks concerning the expected magnitude of μ in the dilute gas trial function which were made earlier in this section. The conclusion is that the instanton gas is such a strongly polar medium that even when the dipole density is not too large, the permeability is much too large for a simple linear approximation to work.

The problem is to take into account the collective effect of the medium on the determination of the response of a single instanton to an applied field. Since μ is really the zero-momentum limit of the gluon wave function renormalization, the most straightforward procedure would be to set up and solve an integral equation for the gluon propagator. Technically, this would be quite similar to the method used in Sec. 3 to study the fermion propagator in the presence of instantons. It is easier, and physically more transparent, to use instead a method invented by Onsager[31] to study polar dielectrics of modest density and large dielectric constant. The method is alleged to be exact if the only interparticle forces are dipolar.

The basic problem is to compute the mean dipole moment of a single dipole, given that a uniform field, $F^{ext}_{\mu\nu}$, is impressed on the medium. We imagine that space is divided into two regions--a small sphere containing only the dipole of interest and the rest of space. In the outer region the dipoles are regarded as a continuous medium with dipole moment μ. The inner region is a vacuum cavity with a single dipole in it. The field strength, F^{local}, which actually polarizes the dipole is not F^{ext}, but the local field strength in the vacuum cavity. According to (4.29)

$$F^{local} = \frac{2}{1+\mu} F^{ext} = \frac{2\mu}{1+\mu} G^{ext} \tag{4.36}$$

and therefore (4.34) can be written

$$M^a_{\mu\nu} = \frac{2\mu}{1+\mu} \quad \frac{1}{8} \int \frac{d\rho}{\rho} \bar{D}(\rho) \cdot \frac{8\pi^2}{g^2(\rho)} G^{a\ ext}_{\mu\nu} \tag{4.37}$$

What multiplies G^{ext} in (4.36) is the susceptibility, χ, and if we use the general definition of μ, $\mu = 1+4\pi^2\chi$, we obtain an equation for μ:

$$\mu = 1 + 4\pi^2 \cdot \frac{2\mu}{1+\mu} \int \frac{d\rho}{\rho} \bar{D}(\rho) \cdot \frac{8\pi^2}{g^2(\rho)} \frac{1}{8} . \tag{4.38}$$

This equation is easily solved for μ:

$$\mu = \eta + \sqrt{\eta^2 + 1}$$
$$\eta = \frac{\pi^2}{2} \int \frac{d\rho}{\rho} \bar{D}(\rho) \frac{8\pi^2}{g(\rho)^2} = \mu_{DG}{}^{-1} \tag{4.39}$$

This expression should be a reasonable approximation to μ, even when η is large, so long as the instanton density is small enough that higher order interactions than dipolar are not too important. We will use it withoug apology even when the integrated fraction of space time covered by instantons is of order one. This is not unreasonable since a numerical calculation shows that the interaction between an instanton and an antiinstanton follows the dipole law almost to the point where the instantons overlap (R=2 for instanton and antiinstanton of same scale size, ρ). For large η, $\mu \approx 2\eta$, and a comparison of (4.39) with (2.29) tells us that in the dilute gas trial function, $\mu \approx 8\pi^2/g^2(\rho_c)$. As we have already pointed out, the value of $8\pi^2/g^2$ associated with the cutoff scale, ρ_c, is typically rather large (somewhere between 10 and 20 for the Pauli-Villars coupling definition).

Evidently the instanton fluctuations manage to render the vacuum strongly paramagnetic. We will argue that paramagnetism this strong is already enough to produce structures more like hadrons than those of perturbative QCD. Strict confinement, and, presumably, perfect paramagnetism could only be obtained by integrating over the strong coupling fluctuations on scales larger than the cutoff scale, ρ_c. An interesting aspect of this is that the instantons appear to interpolate between the weak coupling, perturbative fluctuations and the strong-coupling confining fluctuations. The passage between the two regimes occurs in a very narrow range of scales: an examination of the integrals entering into the determination of μ shows that all the significant contributions come in a range of scale sizes differing by a factor two at most. On the lattice at any rate, the strong coupling limit is trivial and one might hope to really solve the theory by somehow matching the dilute instanton gas, which already appears to be in the strong coupling domain, and a lattice strong coupling expansion. For the moment no one knows how to do this!

The previous discussion has focussed on the effect of the medium (i.e., other instantons) on the response of a particular instanton to an external field. An equally interesting effect of the medium is on the vacuum density of instantons in the absence of any external field. The point is that while the density function is proportional to e^{-S_I}, where S_I is the action of the instanton, S_I is no longer equal to the "free-space" value, $8\pi^2/g^2$. The instanton disturbs the surrounding medium, which reacts back on the instanton and lowers (as it turns out) the action of the instanton. This has the effect of increasing the density associated with a given value of the coupling and therefore decreasing the value of the coupling associated with ρ_c.

To compute the change in the instanton "chemical potential" caused by the medium, we will use a model similar to the one used in the Onsager calculation of μ: The medium is replaced by a continuum with dielectric constant μ and the dipole under consideration is placed in a vacuum cavity of radius R in the medium. The field in the cavity will consist of the dipole field itself, plus a uniform reaction field given by (4.30) and proportional to the dipole moment, $D_{\mu\nu}$. The change in chemical potential of the instanton is essentially the energy of interaction of the dipole, $D_{\mu\nu}$, with the corresponding reaction field. A detailed examination[14] shows that

$$\delta S_I = -6 \left(\frac{8\pi^2}{g^2(\rho)}\right) \frac{\mu-1}{\mu+1} \left(\frac{\rho}{R}\right)^4 \tag{4.40}$$

where ρ is the scale size of the dipole and g is the associated effective coupling. This obviously depends only on the strength of the dipole, not on its orientation and is, as advertised, negative.

This expression depends on μ, but is used to determine the density function, $\bar{D}$, which in turn determines μ by (4.39). For the moment, we are interested in situations where we know that μ is going to be large and may simply replace the factor $(\mu-1) / (\mu+1)$ in (4.40) by 1.

Somewhat more troublesome is the question of how to choose R. This we determine as follows: In computing interactions between instantons, we have treated them as point dipoles. In doing the functional integral one should not however integrate over arbitrarily small separations between dipoles: At some separation, the finite size of the instanton comes into play and causes the dipole approximation to the interaction to cease being accurate. Numerical calculations show that the dipole approximation to the interaction of an instanton and antiinstanton of size ρ is quite accurate down to a separation between centers of about 2ρ! For smaller separations the interaction goes smoothly to zero as the instanton and antiinstanton merge. It therefore seems reasonable to cut off the singular dipole interactions by not counting configurations with dipoles separated by less than twice their scale size. Equivalently, in the argument leading to (4.40) we should take the cavity around an instanton of scale size ρ to have radius $R \sim 2\rho$.

We have somewhat arbitrarily chosen $R = 2.2\rho$, so that $6(\rho/R)^4 \simeq \frac{1}{4}$. Therefore, since μ is large, we may set $\delta S_I = -\frac{1}{4}\,(8\pi^2/g^2(\rho))$ and the instanton density function, corrected for the effects of the medium, is

$$\bar{D}(\rho) = C_o \left(\frac{8\pi^2}{g^2(\rho)} \right)^6 e^{-\frac{3}{4}(8\pi^2/g^2(\rho))} \tag{4.41}$$

The constant C_o depends on the coupling constant definition and has the values given in Sect. 2. The effect of this renormalization of the chemical potential is to decrease the effective coupling, and the scale, at which the integrated instanton density reaches unity. For the Pauli-Villars case, one finds that $8\pi^2/g^2(\rho_c)$ is about 14 while for the dimensional regulation definition one finds that $8\pi^2/g^2(\rho_c)$ is about 36!

Obviously, it is impossible to make this kind of discussion very precise. Nonetheless it does make clear the sense in which certain vacuum properties, most notably the vacuum permeability, are radically changed by the instanton fluctuations. In the next sections we will try to develop this rather rough picture of the properties of the vacuum into a semi-quantitative picture of the properties of hadrons.

V. INFRARED CUTOFFS AND PHASE TRANSITIONS

We have seen in the preceding sections that the naturally occurring density of instantons in Euclidean space-time makes the vacuum behave like a medium of large permeability. It is implicit in our whole approach that the permeability, μ, at a point depends on the instanton density, n, at that point. In the undisturbed vacuum it is obvious that this density, and, therefore, the permeability are uniform in space. In the presence of external sources, however, it is almost certain that the most probable configuration is one in which n and, therefore μ, vary from point to point. The thermodynamic reason for this is that while any spatial variation of n raises the free energy, the attendant variation of μ may lower the energy of the color field attached to the sources.[14,32] Conversely, the density function of individual instantons should depend on the field produced by the sources. A systematic exploration of this question will lead us to some very useful insights into the behavior of the QCD vacuum which will be essential to our understanding of hadrons.

As a warm-up problem, we will look at the density of instantons in the presence of a uniform background color scalar field. This problem has been considered by 't Hooft[12], and we will borrow his results (at the price of temporarily descending from an SU_3 gauge group to SU_2). To the action of the Yang-Mills field we add a piece describing an I=1 color scalar

$$\mathcal{L}^{s} = -D_{\mu}\,\phi^{*}\,D_{\mu}\,\phi - V(\phi), \tag{5.1}$$

where D_μ is the covariant derivative and $V(\phi)$ has a minimum at $\phi^2 = F^2$. We can still find instanton solutions of appropriately constrained field equations: the gauge field will look much the same as before, the new element being a scalar field configuration approaching $\phi^2 = F^2$ far from the instanton center. It is rather easy to show that, at least in the limit of small F, the action of the instanton plus scalar field system is

$$S_{TOT} = \frac{8\pi^2}{g^2} + 4\pi^2 F^2 \rho^2 \tag{5.2}$$

which is to say that the interaction with the scalar field <u>increases</u> the action of an instanton. This <u>decreases</u> the instanton density function from its zero-field value, $D(\rho)$:

$$D_F(\rho) = D(\rho) \quad e^{-4\pi^2 F^2 \rho^2} \quad . \tag{5.3}$$

An examination of the limits of validity of this expression shows that, although it was obtained for small F, it continues to be reliable even when the overall exponential suppression factor is quite tiny.

If, therefore, we had some external sources producing a slowly varying background scalar field, the density of instantons would indeed vary from point to point in the specific fashion indicated in (5.3). Not only is the density decreased, but all integrations over instanton density (to compute, for instance, the integrated fraction of space time covered by instantons) are sharply cutoff in the infrared. The density function, $D_F(\rho)$, now has a sharp maximum at some value, $\rho(F)$, which can be made as small as one likes by increasing F. The corresponding coupling, $gF = g\ (\rho(F))$, can be made small enough that the dilute gas picture of the role of non-perturbative vacuum fluctuations is as accurate as one likes.

Therefore, a background scalar not only introduces the possibility of spatial variation of the instanton density, it also provides an infrared cutoff which, if the scalar is strong enough, makes the dilute gas picture reliable. That a scalar field provides an infrared cutoff should be no surprise: By the Higgs mechanism, a scalar field gives the Yang Mills field a mass M and eliminates vacuum fluctuations on scales greater than M^{-1}.

Now let us return to pure Yang-Mills theory and consider the case of a weak background color field, $F^{ext}_{\mu\nu}$, such as might be produced by heavy quark sources. The interaction of an instanton of moment, $D_{\mu\nu}$, with $F^{ext}_{\mu\nu}$ has already been calculated in (4.20) and the modified density function is evidently

$$D_{Fext} \quad (\rho) = D(\rho) \quad e^{2\pi^2 \, D^a_{\mu\nu} \, F^{a\,ext}_{\mu\nu}} \tag{5.4.}$$

Since we are generally interested in quantities which have been averaged over group orientation we shall average (5.4) over group orientation of the instanton. For arbitrary F^{ext} we get something quite complicated, but for small F^{ext} we obtain

$$\left\langle\left(\exp 2\pi^2 D^a_{\mu\nu} F^{a\,ext}_{\mu\nu}\right)\right\rangle \simeq \left\langle 1 + 2\pi^2 DF + \tfrac{1}{2}(2\pi^2 DF)^2 + \ldots\right\rangle$$
$$= 1 + \frac{2\pi^4\rho^4}{g^2(\rho)} (F^-_{ext})^2 + O(F^4_{ext}) \tag{5.5}$$

where $F^{\pm}_{\mu\nu} = \frac{1}{2}F_{\mu\nu} \pm \frac{1}{4}\varepsilon_{\mu\nu\lambda\sigma} F_{\lambda\sigma}$, ρ is the instanton scale size and $g(\rho)$ is the associated effective coupling. To leading order, we may re-exponentiate (5.5), obtaining

$$D^{\pm}_F(\rho) = D(\rho) \exp\left\{\frac{2\pi^4\rho^2}{g^2(\rho)} (F^{\mp}_{ext})^2\right\} \tag{5.6}$$

where the upper (lower signs refer to instanton (antiinstanton). The exponential form is certainly accurate when the argument of the exponential is small, but a more elaborate calculation shows that the exponential form is a reasonable approximation to the exact answer over a wide range of values of the argument.

While the instanton density is modified by an external color field, the net effect appears to be the reverse of that of a scalar field: The density is increased, with larger instantons being more affected than smaller ones. This is consistent with our conclusion in Sect. 4 that the effect of the medium is always to increase the instanton density. In reaching this conclusion we assumed that F^2_{ext} was positive. Now $(F_{ext})^2$ is the Euclidean version of the Minkowski invariant, $F^2 = \vec{B}^2 - \vec{E}^2$, which is potitive or negative according as the field is magnetic or electric. By choosing F^2_{ext} positive (as we must with real Euclidean fields) we are actually studying the effect of external magnetic fields on the instanton density. If the external source produces a color electric field E^a_i (in Minkowski space) we must set $F^2_{ext} = -E^2$. This amounts, in the Euclidean treatment, to choosing E imaginary. The necessary factor of $\sqrt{-1}$ is in fact present in the Euclidean path integral (4.1) for external color charges.

For a background color electric field we must write

$$D_E(\rho) = D(\rho) \exp\left(-\frac{\pi^4\rho^4}{g^2(\rho)} E^2\right) \tag{5.7}$$

(the distinction between instanton and antiinstanton is no longer necessary-- a purely electric field affects both in the same way). This result is qualitatively the same as for the background scalar field: The instanton density is reduced and the reduction factor is larger, the larger the instanton scale size. Consequently, an electric field provides an infrared cutoff, and, if it is strong enough, makes the dilute gas approximation an accurate representation of the effects of non-perturbative fluctuations. Since strong color electric fields do exist inside hadrons, we can expect this effect to play a significant role in the physics of hadrons.

There are a variety of other external influences which provide an infrared cutoff on the non-perturbative fluctuations. One, which we will not be able to explore here, but does play a role, is the chemical potential of quarks: As we have seen in Sects. 3 and 4, instantons change the mass of a single quark. Conversely, if there is a density of quarks present, this effect can be interpreted as changing the action and therefore the density function of an instanton. Under the right circumstances, this effect provides an infrared cutoff. Since the inside of a hadron is characterized not only by color electric fields, but by a non-zero density of quarks, this effect should also play a significant role in hadron physics.

We can reproduce (5.7) by a different line of argument which treats the instanton as part of a medium and is capable of dealing with the collective effects which become significant at some finite value of the instanton density. In the dilute gas approximation (whose validity will be justified after the fact) the Euclidean vacuum functional is identical to the partition function of a two-component ideal gas:

$$Z(\xi) = \sum_{n_+ n_-} \frac{1}{n_+! n_-!} \left[V \int \frac{d\rho}{\rho^5} D(\rho) \right]^{n_+ + n_-} \tag{5.8}$$

where $D(\rho)$ is the instanton density function and V is the space-time volume. Since we will consider variations of the density, it is convenient to introduce an artificial activity, $\xi(\rho)$, which we must set equal to one at the end of the calculation:

$$Z(\xi) = \sum_{n_+ n_-} \frac{1}{n_+! n_-!} \left[V \int \frac{d\rho}{\rho^5} D(\rho)\, \xi(\rho) \right]^{n_+ + n_-} \tag{5.9}$$

In thermodynamic parlance, Z defines a pressure, $P(\xi)$, by $Z = e^{PV}$. The combined density of instantons and antiinstantons is obtained by varying the pressure with respect to ξ:

$$2n(\rho) = n_+(\rho) + n_-(\rho) = \frac{1}{V}\,\xi(\rho)\,\frac{\delta}{\delta\xi(\rho)}\,\ell nZ \qquad (5.10)$$

$n_\pm(\rho)$ are defined by

$$< n_\pm > = \int\frac{d\rho}{\rho}\,n_\pm(\rho).$$

In order to obtain the physically relevant instanton density, we must set $\xi(\rho) = 1$. In order to regard n rather than ξ as the independent variable we introduce the free energy per unit volume, which is the Legendre transform of the pressure:

$$F(n) = \int\frac{d\rho}{\rho}\ \ell n\ \xi(\rho)\cdot 2n(\rho) - P(\xi). \qquad (5.11)$$

Then

$$\frac{\delta F}{\delta n(\rho)} = 2\ell n\ \xi(\rho) \qquad (5.12)$$

and setting $\xi = 1$ is equivalent to finding the density function $n(\rho)$ which stationarizes the free energy. In the dilute gas case we easily find that

$$F(n) = F_o(n) = \int\frac{d\rho}{\rho}\cdot 2n(\rho)\ [\ell n\ \frac{n(\rho)}{n_o(\rho)} - 1] \qquad (5.13)$$

where $n_o(\rho) = D(\rho)\ \rho^{-4}$. As expected, the variations of F_o with respect to $n(\rho)$ vanish when $n(\rho) = n_o(\rho)$, the ideal gas instanton density.

In the presence of an external color field, the free energy of this system is changed by the addition of an electrostatic energy term. From the discussion of Sect. 4, we know that this energy is $\frac{1}{2}ED = E^2/2\mu = \mu D^2/2$, where D is the displacement field associated with E and μ is the permeability of the dilute gas medium. This is of course a linear approximation and is valid only for small enough E. After the fact we can verify that all our interesting effects occur at small enough E for the linear approximation to be valid. Nonlinear effects of a crucial nature are of course included in the physics of μ. There is a subtlety associated with the question whether the free energy is to be regarded as a function of E or D. The same issue arises in the study of the electrostatics of ordinary material media and the textbook analysis applicable to that case[14] applies here. We choose to take E as the independent variable and use the energy functional

$$F(n,E) = F_o(n) - \tfrac{1}{2}ED \tag{5.14}$$

whose first variations satisfy

$$\delta F = 2\int \frac{d\rho}{\rho} \ln \xi(\rho)\, \delta n(\rho) - D\delta E \,. \tag{5.15}$$

Since physics corresponds to $\xi(\rho) = 1$, the equation which determines the density appropriate to a given E is

$$\left(\frac{\delta F}{\delta n(\rho)}\right)\Big|_E = 0 \,. \tag{5.16}$$

Again the physical density makes the free energy stationary (and presumably minimizes it).

To calculate the instanton density in an external field we therefore must minimize $F = F_o(n) - E^2/2\mu$ with respect to n. We will use (4.39) to describe the dependence of μ on $n(\rho)$, although in the end it will turn out that our interesting results come from a region where the dilute gas formula for μ,(4.34), would be just as accurate. The basic equation for n is

$$0 = \frac{\delta F}{\delta n}\Big|_E = \frac{\delta F_o}{\delta n(\rho)} + \frac{E^2}{2\mu^2}\,\frac{\delta\mu}{\delta n(\rho)} \tag{5.17}$$

From (5.13) we have

$$\frac{\delta F_o}{\delta n(\rho)} = 2 \ln \frac{n(\rho)}{n_o(\rho)} \tag{5.18}$$

and from (4.39) we have

$$\frac{\delta\mu}{\delta n(\rho)} = \frac{2\mu^2}{1+\mu^2}\,\frac{\delta\eta}{\delta n(\rho)} = \frac{2\mu^2}{1+\mu^2}\,\frac{\pi^2}{2}\left(\frac{8\pi^2}{g^2(\rho)}\right)\rho^4 \tag{5.19}$$

the dilute gas picture makes real sense. To interpret the instability we have found, it is necessary to make an educated guess about what the D(E) curve will do as we extend it to small values of E and D, where the instanton density is not cutoff.

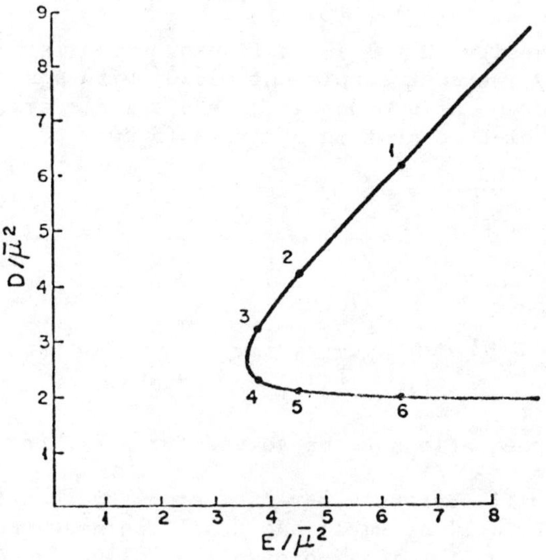

Fig. 8. D versus E for the Pauli-Villars coupling definition.

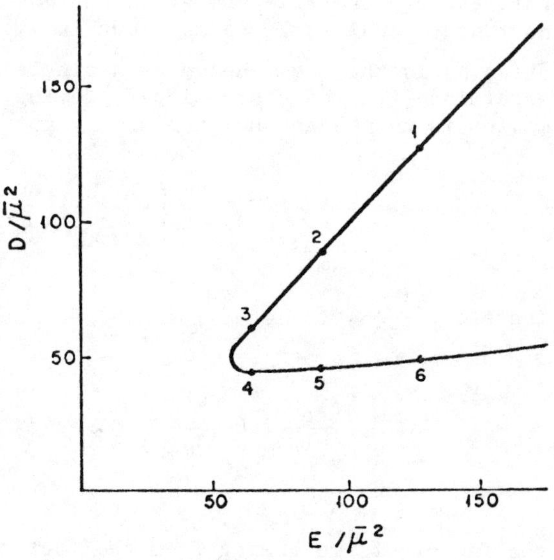

Fig. 9. D versus E for the dimensional regulation coupling definition.

Therefore the equilibrium density function is

$$n(\rho) = n_o(\rho)\ \exp\{-\frac{2}{1+\mu^2}\frac{\pi^2E^2}{8}\left(\frac{8\pi^2}{g^2(\rho)}\right)\rho^4\} \tag{5.20}$$

In the limit of small density, $\mu \simeq 1$ and this expression is identical to (5.7). More generally, we must supplement (5.20) with equations which show how μ is determined by n in order to have a soluble system. To determine μ as a function of E we must incorporate (5.20) into (4.39):

$$\mu = \eta + \sqrt{1 + \eta^2}$$

$$\eta = \frac{\pi^2}{2}\int\frac{d\rho}{\rho}\left(\frac{8\pi^2}{g^2(\rho)}\right) D(\rho)\ \exp\left[-\frac{2}{1+\mu^2}\frac{\pi^2E^2}{8}\ \frac{8\pi^2}{g^2(\rho)}\ \rho^4\right] \tag{5.21}$$

This is a system of nonlinear equations to be solved for $\mu(E)$, from which $n(\rho)$ may be determined.

The solution of (5.21) will obviously have the property that for large E the integrated density will be small and μ will be near one. As we decrease E, the instanton density will increase and collective effects due to the medium will become important. The most important of these is the modification of the instanton density due to its interaction with the medium, and we would like to include it in our calculation. Because of this effect, the free energy in the absence of an external color field is no longer equal to F_o (5.13), but is given by some new function $\tilde{F}_o$. $\delta F/\delta n(\rho)$ is the free energy of a single instanton and we have already determined (in the discussion following (4.40) that the effect of the medium is to change the free energy per instanton by

$$-6\left(\frac{8\pi^2}{g^2(\rho)}\right)\left(\frac{\mu-1}{\mu+1}\right)\left(\frac{\rho}{R}\right)^4 . \tag{5.22}$$

Therefore $\tilde{F}_o(n)$ is determined by

$$\frac{\delta\tilde{F}_o}{\delta n(\rho)} = \frac{\delta F_o}{\delta n(\rho)} - 12\left(\frac{8\pi^2}{g^2(\rho)}\right)\left(\frac{\mu-1}{\mu+1}\right)\left(\frac{\rho}{R}\right)^4 . \tag{5.23}$$

the factor of two coming from the fact that varying $n(\rho)$ varies both the instanton and the antiinstanton density simultaneously.

The complete free energy function in the presence of an external electric field is now $F(n,E) = \tilde{F}_0(n) - E^2/2\mu$ and the equation (5.20) for the equilibrium density function is replaced by

$$n_E(\rho) = n_0(\rho)\ \exp\left\{-\frac{2}{1+\mu^2}\ \frac{\pi^2 E^2 \rho^4}{8}\left(\frac{8\pi^2}{g^2(\rho)}\right) + 6\left(\frac{\mu-1}{\mu+1}\right)\left(\frac{\rho}{R}\right)^4\left(\frac{8\pi^2}{g^2(\rho)}\right)\right\} \tag{5.24}$$

As our previous discussion should lead us to expect, the interaction with the medium increases the instanton density. The considerations discussed after (4.40) lead us to take the cavity radius, R, equal to 2.2 times the scale size of the instanton in the cavity. Therefore the ratio ρ/R in (5.24) can be replaced by the value 1/2.2 and we may use (5.24) together with

$$\mu = \eta + \sqrt{1+\eta^2}$$

$$\mu = \frac{\pi^2}{2}\int \frac{d\rho}{\rho^5}\left(\frac{8\pi^2}{g^2(\rho)}\right) n_E(\rho) \tag{5.25}$$

to determine μ as a function of E and from that to determine $n_E(\rho)$.

The combined equations (5.24) and (5.25) cannot be solved analytically, but it is not hard to obtain a useful numerical solution.[14] We have done this for the cases of Pauli-Villars and dimensional regulation coupling constant definitions in order to verify that the physics is invariant to changes in coupling constant definition. Since there is a factor 10^5 difference between the two schemes in the normalization constant in $D(\rho)$, one is entitled to worry about this point. A suggestive way of plotting the result is as an equation of state, or D versus E curve. Since $D = E/\mu$, this contains the same information as a plot of μ versus E. Furthermore, according to (5.15), $D = -(\partial F/\partial E)_{\xi=1}$ and we may use the D(E) plot to learn something about how the free energy is varying from point to point.

The results of the calculation are plotted in Fig. 8 for the Pauli-Villars case and in Fig. 9 for the dimensional regulation case. The key qualitative feature of both curves is the existence of a point of inflection (let us denote it by (E_i, D_i)) where $(\partial E/\partial D)$ changes sign. Standard thermodynamic arguments show that the region where $(\partial E/\partial D)$ is positive is thermodynamically stable, while the region where $(\partial E/\partial D)$ is negative is thermodynamically unstable. Therefore the portion of the curve for $D < D_i$ is not physically meaningful, at best representing a metastable regime. We have computed and plotted these curves only for values of E large enough that the net instanton density is so small that

The generic behavior indicated in Fig. 10 seems to us almost inevitable: For E and D near zero we are simply exploring the properties of the vacuum. As we argued in Sect. 4, the permeability of the vacuum is necessarily very large, perhaps even infinite. Therefore the D(E) curve must emerge from the origin with very small slope, eventually curving up and joining on to the negative slope portion of the dilute gas curve we have calculated. There are two positive-slope, thermodynamically stable, regions, labelled I and III. These are clearly two possible phases of the vacuum: Phase I exists for $E > E_i$ and is characterized by $\mu = E/D \simeq 1$ and therefore by low instanton density; Phase II exists for $E < E'_i$ (E'_i being the lower inflection point) and is characterized by $\mu \gg 1$ and high instanton density. This is the phase we tried to characterize in Sect. 2 and 4. Although we can't be precise about its properties, we have developed some semiquantitative notions. The region labelled II is metastable and doesn't have any standing as a thermodynamic phase.

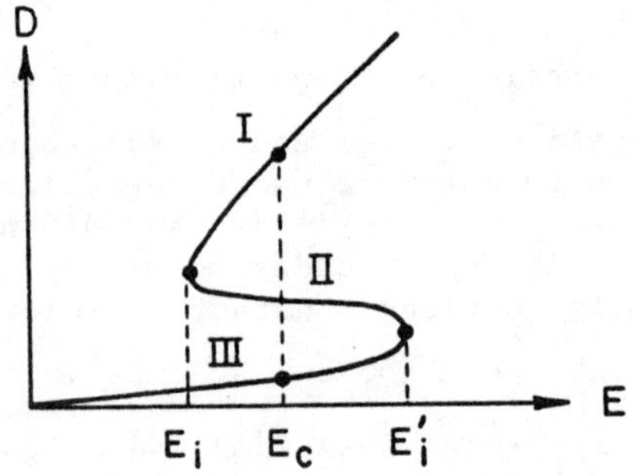

Fig. 10. Generic form of the D versus E curve, including its extension into the dense phase.

The regions of E in which phases I and III can exist appear to overlap. In the overlap region, one or the other of the two phases will have the lower free energy and be the stable phase. There should be one critical value of E at which the free energies are identical and the two phases can coexist. Then phase I is truly stable only for $E > E_c$ and phase III is stable only for $E < E_c$, as indicated schematically in Fig. 10.

The picture so constructed is that of a system which undergoes a first-order phase transition and may coexist in a "dilute" and a "dense" phase only at a critical value, E_c, of the external field. The significance of the point of inflection that we find in the dilute gas treatment is that it is a definite signal for a two-phase structure and puts a lower bound on the value of E_c.

To give some idea of the quantitative properties of the dilute phase we have listed in Tables I and II the values of some relevant quantities at the points labelled 1 through 6 in Figs. 8 and 9 respectively. For each point we have listed the electric field, the value of μ, the fraction of space time covered by instantons (f), and X_p, the value of $8\pi^2/g^2(\rho)$ at the peak of the instanton scale size distribution. In Fig. 11 we have plotted the density profile appropriate to point 2 in Fig. 8 just to give an impression of how sharply peaked the cutoff size distribution is.

TABLE I

	μ	E/Λ^2	x_p	f
1	1.01	6.4	17	.002
2	1.06	4.5	13	.008
3	1.16	3.8	13	.02
4	1.62	3.8	12	.08
5	2.12	4.5	12	.13
6	3.16	6.4	13	.22

The values of selected quantities at points 1-6 on the D versus E curve for the Pauli-Villars scheme plotted in Fig. 8. x_p is the value of $8\pi^2/g^2$ at the peak of the instanton scale size distribution and f is the integrated fraction of space time covered by instantons.

There are several points to make about these tables. The first is that the curves for Pauli-Villars and dimensional regulation are quite similar in shape and that corresponding points on the curves are characterized by roughly comparable values of the dimensionless physical quantities f and μ. Presumably the correspondence would be even better if we could take radiative corrections to instanton effects into account. The values of E do not correspond, but are related to one another roughly by a rescaling of a factor of 3 or 4 of the renormalization mass scale, λ. So, in a crude first approximation, the physics emerging from out treatment is invariant to a substantial change in the definition of the coupling constant. The second important point is that the dilute phase really is dilute! In the Pauli-Villars scheme, at the inflection point, the integrated fraction of space time covered by instantons is only .02 and is even less at larger values of E. It would appear that in the dilute phase, ordinary perturbation theory notions are valid to an excellent degree of approximation. It is also important not only that the density is small throughout the dilute phase region, but also that the relevant value of the effective coupling is small enough for our neglect of radiative corrections not to be too unreasonable. All these things taken together imply that the dilute gas arguments which establish the existence of the instability (and therefore of the first-order phase transition) in particular and the properties of the dilute phase in general are quite reliable.

The possibility of phase coexistence at some critical field E_c will be very important to our picture of hadron structure, and we want to be more explicit about "boundary conditions". The condition for determining E_c in the first place is that the free energy function $F(n,E) = \tilde{F}_o(n) - E^2/2\mu$ should be the same in both phases, or

$$\tilde{F}_o\Big|_I - \frac{E_c^2}{2\mu_I} = \tilde{F}_o\Big|_{III} - \frac{E_c^2}{2\mu_{III}} \tag{5.26}$$

Now in phase I, the instanton density is very low so that $\mu_I \sim 1$ and $\tilde{F}_{oI} \sim 0$. In phase III, on the other hand, the density is large and μ_{III} is very much greater than one. Therefore, to a good approximation,

$$\frac{E_c^2}{2} = -\tilde{F}_o\Big|_{III} \tag{5.27}$$

This condition amounts to the statement that at the phase boundary the pressure due to the magnetic fluctuations of the normal vacuum is balanced by the pressure of the electric field which establishes the dilute phase. This pressure will eventually be identified with the bag constant of the MIT bag model.

Equation (5.11) implies that, at $\xi = 1$, $\tilde{F}_o$ is equal to the vacuum energy density as determined by the basic Euclidean vacuum functional. $\tilde{F}_o$ is indeed negative (the Euclidean vacuum fluctuations lower than the vacuum energy) as required for (5.26) to make sense. To evaluate $\tilde{F}_o$ exactly, we would have to know how to deal with non-perturbative fluctuations on arbitrarily large scales. In Section 2 we proposed to evaluate quantities such as $\tilde{F}_o$ by integrating over instanton configurations up to a cutoff scale size chosen so that the integrated space-time density of instantons is order one. This at least provides a lower bound on $-\tilde{F}_o$ and has a good chance of being semiquantitatively accurate. We therefore propose to identify E_c by

$$\frac{E_c^2}{2} = 2\int_0^{\rho_c} \frac{d\rho}{\rho^5}\, \bar{D}(\rho) \tag{5.28}$$

where $\bar{D}$ is the instanton density function corrected for the effect of interaction with the medium (4.41) and ρ_c is set by the condition,

$$1 = \pi^2\int_0^{\rho_c} \frac{d\rho}{\rho}\, \bar{D}(\rho). \tag{5.29}$$

In the Pauli-Villars case, and with our model for the effect of the medium on $\bar{D}$, this gives $E_c \sim 11\lambda^2$. This is the value we shall use in our numerical discussions of the next section. We can at least be pleased that E_c so defined is larger than E_i!

In order to have mechanical equilibrium at the phase boundary not only must the field have a critical value but certain vector conditions must be met. Consider the case of a plane spatial boundary between an inner region with $\mu = 1$ and an outer region with $\mu >> 1$. According to the magnetostatic equations of Section 4 the normal component of D and the tangential component of E must be continuous:

$$\begin{aligned} D_\perp^{out} &= \frac{1}{\mu} E_\perp^{out} = E_\perp^{in} \\ D_\parallel^{out} &= E_\parallel^{in} \end{aligned} \tag{5.30}$$

As a result, the energy density outside may be expressed in terms of the inside fields:

$$\mathcal{E} = \frac{1}{2\mu}\left[(E_{\perp}^{out})^2 + (E_{\parallel}^{out})^2\right] = \frac{\mu}{2}(E_{\perp}^{in})^2 + \frac{1}{2\mu}(E_{\parallel}^{in})^2 \qquad (5.31)$$

In the limit $\mu \to \infty$, the energy density outside can only be finite if $E_{\perp}^{in} = 0$. A similar analysis of the magnetic field shows that one must have $B_{\parallel}^{in} = 0$. Both conditions are contained in the covariant condition $F_{\mu\nu}^{in} n_{\nu} = 0$, where n_{μ} is the spacelike normal to the phase boundary. These are precisely the boundary conditions of the MIT bag model, and insofar as $\mu = \infty$ (perfect paramagnetism) is a good approximation to the permeability of the normal vacuum, we are entitled to use the same condition at the boundary between dense and dilute phases. These phase equilibrium conditions are summarized graphically in Fig. 12. For comparison, we also show the more familiar conditions at the boundary of a perfect diamagnet (ordinary superconductor).

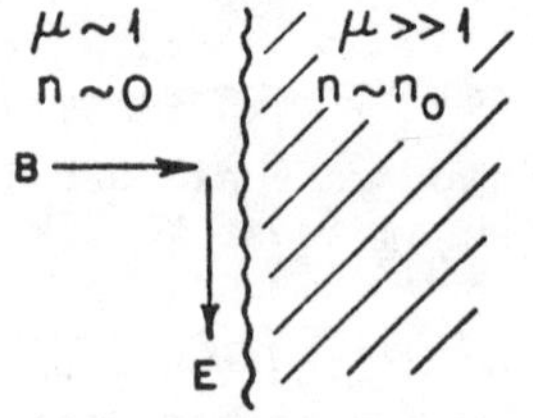

Fig. 12 (a) Boundary conditions on color electric and magnetic fields at a boundary between the dilute and dense phases.

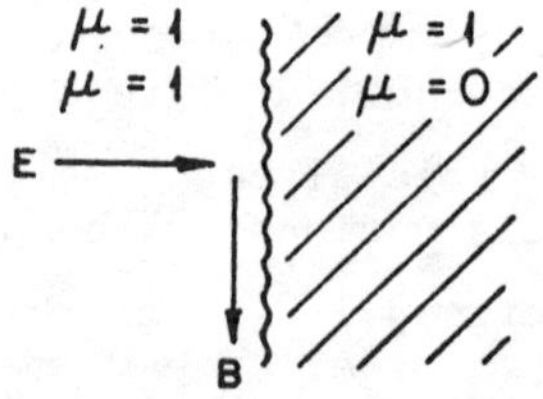

Fig. 12 (b) Boundary conditions on ordinary electric and magnetic fields at a boundary between normal (unhatched) and superconducting (batched) phases.

V1 MODELS OF HADRONS

We would now like to try to convert these results and conjectures into some sort of overall picture of the physics of hadrons. We believe that the results of the last section provide strong evidence that the basic mechanism of the MIT bag model can be understood as a consequence of the non-perturbative phenomena which inevitably play an important role in the QCD vacuum. Our basic framework for discussing hadron physics will therefore be that of the bag model. Since we "derive" the bag mechanism from first principles, we can hope to relate the various bag model parameters to one another and to the short-distance physics of QCD as well as to see what are the limits of validity of the bag picture. Furthermore, since we have kept intact the underlying chiral properties of light quarks, we should be able to see how Goldstone bosons, and chiral physics in general, fit into a bag picture of hadrons. We are of course trying to extract a lot of information from a rather limited understanding of the complicated physics of QCD and we cannot expect our numbers to be in universal agreement with the facts! We can at least hope to sharpen our notions of what are the aspects of hadrons physics that are difficult to understand in the context of QCD.

The vacuum phase structure we found in the last section tells us most directly about the properties of heavy quark "hadrons" in a pure gauge theory. Let us therefore consider a color singlet state constructed out of a static quark-antiquark pair separated by a distance R. This will provide a potential which could be used in a Schrödinger equation to predict the spectrum of heavy quark bound states. The quarks produce a background color field which we expect to be small enough compared to the field fluctuations in the vacuum for our quasi-abelian treatment to make sense. Following the line taken by the bag model we replace the quarks by static abelian sources of charge Qg, where

$$Q^2 = -\left(\sum_{a=1}^{8} T_a^{(1)}\, T_a^{(2)}\right)_{\text{singlet}} = \frac{4}{3} \qquad (6.1)$$

and g is an appropriate value of the running coupling, and solve the linearized "electrostatic" equations

$$\begin{aligned} \overline{\nabla}\cdot\overline{D} &= gQ\,[\delta(\overline{x}-\overline{R}/2) - \delta(\overline{x}+\overline{R}/2)] \\ \overline{\nabla}\times\overline{E} &= 0 \\ \overline{D} &= \overline{E}/\mu \end{aligned} \qquad (6.2)$$

Because μ is large in the normal vacuum and the electrostatic energy is $\mu D^2/2$, the electrostatic energy will be of order μ if the sources are actually embedded in the normal vacuum. Therefore it will be energetically favorable to lower the electrostatic energy by forming a bubble of dilute phase around the sources inside which $\mu \sim 1$. At the boundary between the dilute and normal phases the boundary conditions discussed in the last section (vanishing normal component of E, transverse component equal to E_c) must be met and will keep D from penetrating into the normal vacuum. The size and shape of the bubble, as well as the distribution of E inside the bubble, will be determined by the requirement that the electrostatic equations and the boundary conditions be simultaneously satisfied.

A particularly simple case arises when R becomes large. In that limit the bubble must become a cylinder of diameter d (except near the quarks where its shape is complicated) in which E has a constant magnitude (since $\nabla\times E=0$) and is parallel to the axis of the tube. Evidently, the constant value of E must be equal to E_c. The diameter of the tube is fixed by demanding that the flux of D be equal to gQ. Since the dilute phase permeability is nearly unity, the flux of D and E are nearly the same and we have the condition $gQ=\pi d^2E_c/4$. Our rather crude procedure doesn't tell us very clearly what value to take for g. Rather arbitrarily, we have decided to let g be the value, g(d/2), associated with instantons whose scale size, or radius, is one-half the diameter of the flux tube since that seems a reasonable measure of the scale on which the coupling is defined. The resulting equation for d is

$$d = \left(\frac{4}{\pi}\frac{Q}{E_c}\, g(d/2)\right)^{1/2} \tag{6.3}$$

The property of the flux tube which is of direct physical significance is its energy per unit length, ε. This may be related by classic string model arguments to the universal Regge slope, $\alpha'=.9\ \text{GeV}^{-2}$, the relation being $\varepsilon=1/2\pi\alpha'$. The free energy per unit volume of the dilute phase (exclusive of electrostatic energy) is greater than that of the dense phase by $1/2E_c^2$ (it costs energy to exclude instantons--this energy cost is essentially the bag constant). On top of this there is an electrostatic energy density of $1/2E_c^2$. Therefore the total energy per unit length is

$$\varepsilon = 2\left(\frac{E_c^2}{2}\right)\left(\pi\,\frac{d^2}{4}\right) . \tag{6.4}$$

In order to make specific quantitative comparisons we must adopt a particular coupling constant definition. In what follows we shall use the Pauli-Villars definition. According to the arguments of Sect. 4, the value of E_c is $11\Lambda^2$ and the critical scale in the dense phase, defined as the scale size at which the free energy integral peaks, is $.24\Lambda^{-1}$ (the actual cutoff in the scale size integration--defined by setting the integrated fraction of space time covered by instantons euqal to one--is $.3\Lambda^{-1}$). These numbers are uncertain, to say the least, because our understanding of the dense phase is so poor. We intend all our numbers to be taken with a very large grain of salt and as at best general ballpark estimates.

With these inputs we may solve (6.3) for d, finding $d=.55\Lambda^{-1}$. Then we may use (6.4) and the relation between ε and α' to conclude that Λ=78 MeV. These numbers begin to give us some feeling for the scale of the significant fluctuations in the normal vacuum and how that scale compares with typical hadron dimensions.

Evidently the inverse critical scale in the normal vacuum is approximately 300 MeV. Because of the rapid decrease in importance of non-perturbative fluctuations with decreasing scale size, we can conclude that non-perturbative contributions to vacuum physics must become insignificant at scales two to three times smaller than this critical scale. This is precocious scaling with a vengeance.

The diameter of the flux tube is somewhat more than twice the critical scale in the normal vacuum which is about as small as it could possible be for the phase boundary notion to make any sense. It is quite possible that we have overestimated the critical field, in which case the ratio of flux tube diameter to the vacuum critical scale would be larger: If $E_c \simeq 6\Lambda^2$, the same sort of analysis leads to the conclusion that this ratio is nearer four than two. In any event some significant effects of finite thickness of the phase boundary are to be expected, although there appears to be a good chance that they do not qualitatively change the picture.

If the separation between the sources is comparable to the quantity d, we must solve the more general problem of the heavy quark bag. The quantity which plays the role of the bag constant is $1/2E_c^2=(2.8\Lambda)^4$. With our identification of Λ, we have that $\sqrt[4]{B}$=215 MeV, which is not far from the value that standard bag model considerations favor[2]. If we had chosen $E_c \sim 6\Lambda^2$ rather than $11\Lambda^2$ we would have found $\sqrt[4]{B}$=190 MeV, an interesting indication that the basic dimensional parameter of the bag picture is not unduly sensitive to our lack of understanding of the dense phase.

When the sources are near enough together that we have a bag rather than a flux tube, we expect that the bag radius, rather than its diameter, will be roughly equal to the quantity d. The bag boundary would then appear to be reasonably well-defined, although effects of finite "surface thickness" must be significant. What is certainly true is that if we consider sources so close together that the resulting bag size is less than the critical scale of the noraml vacuum, the bag picture must break down and we must regard the heavy quarks as being directly exposed to the fluctuating fields of the normal vacuum (as in the calculation of the heavy quark potential in Sect. 4). Since the inverse critical scale is about 300 MeV, the bag picture must break down when the quark separation is of order .5 fermi or so. An important general remark is that, as we mentioned in Sect. 1, the basic scale size of hadronic bound states is of the same order of magnitude as the scale size of the important quasiclassical non-perturbative fluctuations in the vacuum. It is for this reason that we can hope to say something useful about hadrons without understanding the large scale vacuum fluctuations responsible for confinement.

Another important point is that the effective coupling appropriate to gluon interactions within the bag is determined since we know the bag size in terms of the parameter Λ. If a typical bag scale is $.5\Lambda^{-1}$, the associated effective coupling is somewhere in the neighborhood of $g^2/4\pi=1$. This is not out of line with standard bag phenomenology and lets us say that we understand from a fundamental point of view why the effective coupling that governs "fine-structure" effects in hadron physics is so surprisingly small. In this connection it must be pointed out that, although our calculation of properties of the dilute phase showed that the instanton density is essentially zero, that calculation is for a homogeneous dilute phase. If it has a finite size, interaction with the outside medium will increase the density of instantons and might produce a density large enough to cause significant non-perturbative effects on top of the perturbative "fine-structure" effects. The pure bag model is subject to this caution as well.

Finally, we would like to turn to the question of light quark bags. Here we have only the most qualitative notions of what is going on, although we think the situation can be improved with a finite amount of work. In the discussion of strings and heavy quark bags in the pure gauge theory we made fairly heavy use of our crude estimate of the free energy of the dense phase. This quantity determines the various lengths and energies of interest in terms of the renormalization mass Λ. This is what allows us to compute the ratio of the bag radius to the typical instanton size in the vacuum and to determine the effective coupling governing fine structure effects inside the bag. As we explained in Sect. 3, when there are massless fermions present, the dynamics of the normal vacuum is tied up with chiral symmetry breaking problems and the establishment of a reasonable guesstimate for the vacuum free energy is a much harder problem than for the no fermion case. If there is only one massless fermion flavor, fermion mass generation is a direct effect of the non-perturbative effective interacitons, and we should be able to construct the vacuum free energy associated with our standard choice of instanton cutoff scale size. Unfortunately this has not yet been carried out. For the physically interesting case of two (or possibly three) massless flavors real dynamical chiral symmetry breaking is at work and we are much less certain of our ability to do anything concrete. Efforts in this direction are under way.

With massless fermion flavors, a key element in the bag picture is the mass generation effect. Form the arguments of Sect. 3, one expects that an isolated massless fermion acquires a mass of order ρ_c^{-1} from the instanton fluctuations in the vacuum (where ρ_c is the typical instanton scale size in the vacuum). On the other hand, if instantons are excluded from a region of size R, the quarks within that region will be massless. Furthermore, if the spontaneously generated quark mass outside (ρ_c^{-1}) is larger than the kinetic energy cost ($\sim R^{-1}$) of confining the quarks inside, then the position-dependent quark mass term in the Dirac equation will effectively confine the quark wave function to the inside region (the non-perturbative fluctuations produce a mass bag). In the limit that the spontaneous mass is infinite, the quark wave function is strictly confined to the bag and the MIT boundary condition $i\not{n}\psi=\psi$ is applicable. The suppression of instantons cost an energy per unit volume, B, equal to the difference between the normal and perturbative vacuum free energy--the quantity which we must learn how to express in terms of Λ in order to make useful quantitative comparisons. At the same time the confinement of the massless quarks inside a spherical bag of radius R produces a kinetic energy of approximately 2/R per quark. Hence, in zeroth approximation, the energy of a nucleon bag of radius R is

$$E(R) = \frac{6}{R} + \frac{4\pi}{3} R^3 \qquad (6.5)$$

$$R = (6/4\pi B)^{1/4} \qquad (6.6)$$

These are the familiar MIT bag equations[2], and we cannot add much to them until we determine B.

If we blindly take the value of B appropriate to the pure gauge theory, we get $R \sim .3\Lambda^{-1}$ which is uncomfortably small compared to the scale size of fluctuations in the normal vacuum. If R really were this small in units of Λ^{-1}, the efficiency of the mass bag would be somewhat doubtful since the confined quark kinetic energy would be somewhat greater than the spontaneous mass. At the same time the bag boundary would be extremely ill-defined and the effective coupling determining fine structure effects would be quite small ($\alpha \sim 1/2$). We can't reach any serious conclusions, however, until we know B.

Finally, a word about the pion is in order. If we have two massless flavors, and there is spontaneous chiral symmetry breakdown, then there will be a Goldstone pion. It will be a chiral fluctuation of the normal vacuum and the dimensional parameters determining its properties will be determined by the dominant fluctuation scale of the normal vacuum. In the case of the pure gauge theory we found that this critical scale satisfied $p_c^{-1} \sim 300$ MeV. If that number is roughly appropriate to the massless fermion case, then a pion charge radius of .5 fermi and $f_\pi = 100$ MeV do not seem to be totally unreasonable numbers. That same mass is also a measure of how heavy a massive quark has to be to behave like a truly massive quark with no vistiges of chiral properties. Consequently, a strange quark with a mass of 200 MeV is light enough that we should expect some remnants of $SU_3 \times SU_3$ chiral symmetry to remain visible.

These matters obvioulsy require and deserve vast amounts of elaboration. We hope that we have convinced the reader that there exists a framework in which to discuss questions of hadron phenomenology without totally losing quantitative touch with the underlying dynamics of QCD and without having to solve the genuinely hard problems of confinement and the true large distance behavior of QCD.

REFERENCES

1. D. J. Gross and F. Wilczek, Phys. Rev. Lett. 30 (1973), 1343. H. D. Politzer, Phys. Rev. Lett. 30 (1973), 1346.
2. An excellent recent review of the status of the MIT bag model is given by P. Hasenfratz and J. Kuti, Phys. Reports C (1978), 75.
3. For an up-to-date view of the status of the quark mass problem, see S. Weinberg, "The Problem of Mass", to appear in a festschrift in honor of I. I. Rabi, to be published by the New York Academy of Sciences.

4. S. Coleman and E. Weinberg, Phys. Rev. D7 (1972), 1888.
5. See the contribution of D. J. Gross in "Methods in Field Theory", R. Balian and J. Zinn-Justin eds., North Holland Publishing Co., New York/Amsterdam, 1976.
6. For a modern discussion of the parton model, see the contribution of R. Field to this volume.
7. K. Wilson, Phys. Rev. D10 (1975), 2445.
8. See the contribution of B. W. Lee in "Methods in Field Theory", R. Balian and J. Zinn-Justin eds., North Holland Publishing Co., New York/Amsterdam, 1976.
9. A. Polyakov, Phys. Lett. 59B (1975), 82.
10. A. Belavin, A. Polyakov, A. Schwartz and Y. Tyupkin, Phys. Lett. 59B (1975), 85.
11. The basic instanton solution of Ref. 10 is an SU_2 object. Our SU_3 instanton is just the SU_2 instanton confined to a particular SU_2 subgroup of SU_3.
12. G. 't Hooft, Phys. Rev. D14 (1976), 3432.
13. An explicit calculation of the determinant for SU_N may be found in Y. Basilov and S. Pokrovsky, Nucl. Phys. B143 (1978), 431.
14. C. G. Callan, Jr., R. Dashen and D. J. Gross, "A theory of Hadronic Structure", to be published in Phys. Rev. D20 (1979).
15. C. G. Callan, Jr., R. DAshen and D. J. Gross, Phys. Lett. 63B (1976), 334.
16. R. Jackiw and C. Rebbi, Phys. Rev. Lett. 37 (1976), 172.
17. S. Coleman, "The Uses of Instantons", Harvard Preprint to be published in the Proceedings of the 1977 International School of Subnuclear Physics, Ettore Majorana.
18. T. Banks, C. Bender and T. T. Wu, Phys. Rev. D8 (1973), 3346.
19. H. Levine and L. Yaffe, Phys. Rev. D19 (1979), 1225.
20. A persuasive exposition of the idea that spontaneous breakdown of the guage symmetry and confinement are mutually exclusive possibilities for QCD is given in G. 't Hooft, Nucl. Phys. B138 (1978), 1.
21. H. J. de Vega, J. L. Gervais and B. Sakita, "Wave Functions and Energy for Vacuum and Heavy Quarks from WKB Schrödinger Equation for Massive Gauge Theories with Instantons", Preprint PAR-LPTHE 78/09, July 1978.
22. G. 't Hooft, Phys. Rev. Lett. 37 (1976), 8.
23. S. L. Adler, Phys. Rev. 177 (1969), 2426. J. S. Bell and R. Jackiw, Nuovo Cemento 60A (1969), 47.
24. J. Cornwall, R. Jackiw and E. Tomboulis, Phys. Rev. D10, 2428 (1974).
25. C. G. Callan, Jr., R. Dashen and D. J. Gross, Phys. Rev. D17 (1978), 2717.
26. D. Caldi, Phys. Rev. Lett. 39 (1977), 121.
27. R. Carlitz, Phys. Rev. D17 (1978), 3225.
28. F. Wilczek and A. Zee, Phys. Rev. Lett. 40 (1978), 83.
29. C. G. Callan, Jr., R. Dashen, D. J. Gross, F. Wilczek and A. Zee, Phys. Rev. D18 (1978), 4684.
30. D. Förster, Phys. Lett. 66B (1977), 279.

31. For a general discussion of this method, see H. Fröhlich, "Theory of Dielectrics", Oxford University Press (London), 1949.
32. The mechanism we are suggesting has also been proposed in similar form by E. V. Shuryak, Phys. Lett. 78B (1978), 150.

AIP Conference Proceedings

		L.C. Number	ISBN
No.1	Feedback and Dynamic Control of Plasmas (Princeton) 1970	70-141596	0-88318-100-2
No.2	Particles and Fields - 1971 (Rochester)	71-184662	0-88318-101-0
No.3	Thermal Expansion - 1971 (Corning)	72-76970	0-88318-102-9
No.4	Superconductivity in d- and f-Band Metals (Rochester, 1971)	74-18879	0-88318-103-7
No.5	Magnetism and Magnetic Materials - 1971 (2 parts) (Chicago)	59-2468	0-88318-104-5
No.6	Particle Physics (Irvine, 1971)	72-81239	0-88318-105-3
No.7	Exploring the History of Nuclear Physics (Brookline, 1967, 1969)	72-81883	0-88318-106-1
No.8	Experimental Meson Spectroscopy - 1972 (Philadelphia)	72-88226	0-88318-107-X
No.9	Cyclotrons - 1972 (Vancouver)	72-92798	0-88318-108-8
No.10	Magnetism and Magnetic Materials - 1972 (2 parts) (Denver)	72-623469	0-88318-109-6
No.11	Transport Phenomena - 1973 (Brown University Conference)	73-80682	0-88318-110-X
No.12	Experiments on High Energy Particle Collisions - 1973 (Vanderbilt Conference)	73-81705	0-88318-111-8
No.13	π-π Scattering - 1973 (Tallahassee Conference)	73-81704	0-88318-112-6
No.14	Particles and Fields - 1973 (APS/DPF Berkeley)	73-91923	0-88318-113-4
No.15	High Energy Collisions - 1973 (Stony Brook)	73-92324	0-88318-114-2
No.16	Causality and Physical Theories (Wayne State University, 1973)	73-93420	0-88318-115-0
No.17	Thermal Expansion - 1973 (Lake of the Ozarks)	73-94415	0-88318-116-9
No.18	Magnetism and Magnetic Materials - 1973 (2 parts) (Boston)	59-2468	0-88318-117-7
No.19	Physics and the Energy Problem - 1974 (APS Chicago)	73-94416	0-88318-118-5
No.20	Tetrahedrally Bonded Amorphous Semiconductors (Yorktown Heights, 1974)	74-80145	0-88318-119-3
No.21	Experimental Meson Spectroscopy - 1974 (Boston)	74-82628	0-88318-120-7
No.22	Neutrinos - 1974 (Philadelphia)	74-82413	0-88318-121-5
No.23	Particles and Fields - 1974 (APS/DPF Williamsburg)	74-27575	0-88318-122-3
No.24	Magnetism and Magnetic Materials - 1974 (20th Annual Conference, San Francisco)	75-2647	0-88318-123-1
No.25	Efficient Use of Energy (The APS Studies on the Technical Aspects of the More Efficient Use of Energy)	75-18227	0-88318-124-X
No.26	High-Energy Physics and Nuclear Structure - 1975 (Santa Fe and Los Alamos)	75-26411	0-88318-125-8

No.27	Topics in Statistical Mechanics and Biophysics: A Memorial to Julius L. Jackson (Wayne State University, 1975)	75-36309	0-88318-126-6
No.28	Physics and Our World: A Symposium in Honor of Victor F. Weisskopf (M.I.T., 1974)	76-7207	0-88318-127-4
No.29	Magnetism and Magnetic Materials - 1975 (21st Annual Conference, Philadelphia)	76-10931	0-88318-128-2
No.30	Particle Searches and Discoveries - 1976 (Vanderbilt Conference)	76-19949	0-88318-129-0
No.31	Structure and Excitations of Amorphous Solids (Williamsburg, Va., 1976)	76-22279	0-88318-130-4
No.32	Materials Technology - 1975 (APS New York Meeting)	76-27967	0-88318-131-2
No.33	Meson-Nuclear Physics - 1976 (Carnegie-Mellon Conference)	76-26811	0-88318-132-0
No.34	Magnetism and Magnetic Materials - 1976 (Joint MMM-Intermag Conference, Pittsburgh)	76-47106	0-88318-133-9
No.35	High Energy Physics with Polarized Beams and Targets (Argonne, 1976)	76-50181	0-88318-134-7
No.36	Momentum Wave Functions - 1976 (Indiana University)	77-82145	0-88318-135-5
No.37	Weak Interaction Physics - 1977 (Indiana University)	77-83344	0-88318-136-3
No.38	Workshop on New Directions in Mössbauer Spectroscopy (Argonne, 1977)	77-90635	0-88318-137-1
No.39	Physics Careers, Employment and Education (Penn State, 1977)	77-94053	0-88318-138-X
No.40	Electrical Transport and Optical Properties of Inhomogeneous Media (Ohio State University, 1977)	78-54319	0-88318-139-8
No.41	Nucleon-Nucleon Interactions - 1977 (Vancouver)	78-54249	0-88318-140-1
No.42	Higher Energy Polarized Proton Beams (Ann Arbor, 1977)	78-55682	0-88318-141-X
No.43	Particles and Fields - 1977 (APS/DPF, Argonne)	78-55683	0-88318-142-8
No.44	Future Trends in Superconductive Electronics (Charlottesville, 1978)	77-9240	0-88318-143-6
No.45	New Results in High Energy Physics - 1978 (Vanderbilt Conference)	78-67196	0-88318-144-4
No.46	Topics in Nonlinear Dynamics (La Jolla Institute)	78-057870	0-88318-145-2
No.47	Clustering Aspects of Nuclear Structure and Nuclear Reactions (Winnepeg, 1978)	78-64942	0-88318-146-0
No.48	Current Trends in the Theory of Fields (Tallahassee, 1978)	78-72948	0-88318-147-9
No.49	Cosmic Rays and Particle Physics - 1978 (Bartol Conference)	79-50489	0-88318-148-7
No.50	Laser-Solid Interactions and Laser Processing - 1978 (Boston)	79-51564	0-88318-149-5
No.51	High Energy Physics with Polarized Beams and Polarized Targets (Argonne, 1978)	79-64565	0-88318-150-9
No. 52	Long-Distance Neutrino Detection - 1978 (C.L. Cowan Memorial Symposium)	79-52078	0-88318-151-7
No. 53	Modulated Structures - 1979 (Kailua Kona, Hawaii)	79-53846	0-88318-152-5
No. 54	Meson-Nuclear Physics - 1979 (Houston)	79-53978	0-88318-153-3
No. 55	Quantum Chromodynamics (La Jolla, 1978)	79-54969	0-88318-154-1